AF894028

Hark to the Bell Buoy's Song

Letters from Monhegan Island 1976 to 1985

by Alta Ashley

Hark to the song of the bell buoy
Which swings and sings close to Monhegan's shore,
Responding to the varied moods of sea and sky
With dulcet tones or din of clashing steel,
Those moods which shape the fortunes of
The hardy folks who choose to live,
Caressed by summer breeze or buffeted by ocean storm,
On this rugged island far at sea.

Grey Gull Publications, Damariscotta, Maine

This book is dedicated to the "Dear Friends"
who have shared the joys and woes
of my life on Monhegan

Acknowledgements: Thanks to the Lincoln County Publishing Company and Frances Brown who helped in the production of this book.

Other Books by The Author:

Time Before The Boat
© 1985
IBSN 0-9614592-1-2

Under The Grey Gull's Wing
© 1984
IBSN 0-9614592-0-4

© 1990 — Grey Gull Publications
1st Printing: June 1990 — 1,500 Copies
Price: $9.95

ISBN 0-9614592-4-7 HARK TO THE BELL BUOY'S SONG

About the Author

Alta Ashley, now 85 years old, has been acquainted and in love with Monhegan Island since the age of two when she was a summer visitor to the island. "Hark to the Bell Buoy's Song" is her third book about Monhegan and is a sequel to "Time before the Boat". Both books are collections of her letters to friends telling of island and family life, especially off season. This book covers from 1976 through 1985 during which time Monhegan celebrated the 200th anniversary of our country and waged a three-year fight with New England Telephone Company, as well as pioneering in solar (photo-voltaic) energy.

In 1985 Dr. Ashley, a retired physician, sold her home on Monhegan and built her solar powered dream house in Damariscotta, Maine where she now works in her Grey Gull Publications on a word processor she is still struggling to master since bringing out her first book "Under the Grey Gull's Wing" in 1985. The makings of at least two more books are still in the files.

FOREWORD

Perhaps I should have titled this book "Life with Phoebe" because from the time she came to live with me on Monhegan Island in the spring of 1976 until she died in the spring of 1985 she was my constant and most loyal companion, announcing visitors, scaring deer from the garden or going everywhere, on foot, by boat or car, Phoebe was there. But because this book covers not only this companionship but the ups and downs of life on the island brought on by the ever changing moods of the sea, so exemplified by the ring of a bell buoy, "Hark to the Bell Buoy's Song" was chosen for the title. During the ten years covered by these letters, any changes in island life, as well as life at High Mooring, occured, notably the loss of telephone service and the struggle which ensured to obtain "reasonable, adequate and reliable" service for the island; the gradual change from oil lamps to home generators, and to reliance on solar energy for households; and the coming of age of the museum at the Lighthouse as a well-organized repository of island history independent of its parent Monhegan Associates, Inc. The letters end with my giving up residency on the island and moving into my solar "dream house" in Damariscotta, Maine with the new dog, Agatha, and a tawny kitten, Christie.

Monhegan, Me. 04852
Jan. 6, 1976

Dear Friends:

So far I have been able to put the right year down. Happy New Year!

A lot has happened since my last general letter, so I'm typing this in single space. Now that trap setting is about over there is time to do other than essential things, such as writing these letters.

First of all, the Monhegan House is **not** for sale. Jeanie and Victor Lord have decided they will run it for the coming year, at least. Elva turned it over to Jeanie in the fall.

Most of you probably know by now that Dwight Paddock Stanley arrived Dec. 17th, weighing in at 8# plus. He missed being born on his great grandfather Dwight's birthday by only a few days. So far I haven't seen him, but those who have say he is a **beautiful** baby.

Thanksgiving was lots of fun in Baldwinsville. The weather was clear and warm for the entire five days, making travelling easy. David Gardner and I drove out in my car and whom should we encounter at a filling station outside of Albany but the Odoms on their way to Cleveland. I recognized their camper as we drove by them and we both stopped at the next station. Doug was riding in the camper and didn't believe Ben and Harry, riding in front, when they said we had passed them - not until he saw me at the station. Elizabeth and Bruce drove out later in the day, the Sutherlands - Scotty, Cindy, Christopher and young Ian came out from Connecticut in their station wagon. That meant that all of the family except Tommy was there; also Aunt Helen and Franklin decided not to try to come. It was the first time that Priscilla had ever had a family get-together. Susan Jackson was missing, too, having a big weekend in Annapolis and taking in the Army-Navy game. We searched for her on TV but didn't find her. All told, there were fourteen of us gathered around a twenty-five pound turkey which was done to a T. Six month old Ian, of course, was satisfied with milk and baby food.

Baldwinsville is just outside Syracuse and we were quite surprised not to be snowed in sometime in the five days were were there. David and I started back early on Sunday so I could get all the way to Thomaston and not have to risk missing the Laura B or the Monhegan

(Doug said I could come back to the island with them). There was a high wind that day and I could hardly control the car crossing the Hudson River bridge, David took over from there, I dropped him in Concord and proceeded to get lost trying to find I-495 but managed to reach Thomaston about nine, just as it began to snow. The next day traffic was at a standstill. The Odoms were marooned in Cape Elizabeth so we didn't get back to the Island until the next day - all that driving Sunday for naught! The ocean, which had been whipped into 10 foot waves the day before, was a flat calm. I just don't see how it calms down so fast. It was good to come back on the Monhegan. The Laura B was still off the run and Ivan's boat was crowded with returning islanders.

At Christmas time I had planned to go down and stay with Aunt Helen in Allston and spend Christmas Day at the Gardners but weather reports were not good and Elizabeth called to say for me not to try. It would be a hard day for her, the first Christmas without Dave. I wanted to be there, but took her advice and well I did. We had another huge storm before and after Christmas. The one before started out as snow and ended as freezing rain, putting a thick smooth glaze of ice on everything. I had dinner at the Boehmers' and thought I'd never make it to their house down at Lobster Cove, even though armed with a cane. The Craigs came by and gave me a hand, much appreciated. I really felt my age that day. Peter and Rickie came back with me so they could make a phone call, so I got back with only one fall.

The day after Christmas the temp. rose to 48 degrees and the wind to 48 knots. The Laura B,now on the run, made the trip - the worst Vernon Burton says he ever made - but Doug and Harry were marooned again in Cape Elizabeth by freezing rain and wind and didn't get back until Saturday. That night they and Rita came up here for exchange of presents from under the tree Harry cut for me before they left two weeks before. They had brought a new stove for the cottage which was much needed during the cold weather of the week before.

We had our community holiday dinner and visit from Santa the Thursday before Christmas. It was lots of fun. The following Tuesday Dougie and Debbie held a combination Christmas-House Warming at their new home which is very nice. They are heating with a wood stove and the sun and hope to have a wind mill operational for their power. Right now they are getting power from storage batteries they charge up with a 1.5 K generator. This runs lights and their water pump. They have a huge cistern which filled to the brim during the huge rains we have had all fall. Several people held open house during Christmas Day. Holiday time is fun here on the island and I'm glad I didn't try to go ashore.

In my November letter I told about how most of my official hats had fallen off my head. It was quite a surprise to find that on Saturday, December 6th a crown fell on it. I had gone to Portland that day to attend some medical meetings. Because I was late getting started I hadn't picked up a paper and was taken quite with surprise when several doctors began to congratulate me. There was one paper left at the hotel desk where we were meeting and there on the front page was an article by Charles Seib of the Washington Post in which he named me ''Queen of County Correspondents''. The article continued on the back page of the first section and had a picture of me. I about died when I saw that. Later that day I went shopping at Jordan Marsh's and several people stopped and stared, so I did my errands and escaped to the anonimity of my car - then I found I still was wearing my identification tag from the meeting!

As Trap Day 1976 approached it seemed as though it would go about as smoothly as any had for years. Everyone was in good health, though Harry and Doug had ''little colds''. All boats and trucks were functioning. The only problem was the weather which on Tuesday didn't look too bad. Even the tide was cooperating.

New Year's Eve was a quiet one because everyone went to bed early (except Rita, my two renters in the cottage and me) to be ready for setting the following morning. Marty Moses was here to help the Odoms. Bob Burton was going with Vernon for the first few days of fishing, Bob Boody with Don Cundy, Jeff Dolan with Robert Rothstein, Victor Lord and Dave Wallace with Steve Rollins, + Ron Hubbard and Paul Horn with one of the other men. There are thirteen boats in the fleet, although Alfred Stanley will go later with Sherm and Shermie will be only part time when he is on leave from his job on the tug of the Boston Fuel Company. Things were all ready for everyone to start at dawn on January 1st. Traps had been moved to the dock late Tuesday after the bait brought out by the Bajuba had been trucked to fish houses. The weather began not to look too good by evening of the 31st and fingers began to cross.

By morning of the first the wind was blowing hard from the north, it was bitter cold and there was a big surge at the dock. Don Cundy and some of the others had put rocks aboard their boats and were in danger of being swamped if they loaded on traps should the wind turn southerly. There was no idea of setting that day but all the Stanleys, Don and Steve did load their boats. Doug and Marty put a load of traps on their truck and loaded their jeep with rocks (those needed to weight down the dry traps). The next morning it was still cold and the winds were blowing around 30 knots. Harry Odom had laryngitis and a fever and so was out of the running. Daniel Boxer was enlisted to work with Doug and Marty Moses. Then the jeep wouldn't run. Soon after nine o'clock it was decided that setting would begin as soon as all the boats were loaded, for the wind was beginning to let go.

Seven of the boats loaded from the dock, one at the head of the dock and one on the south side. The wind was too strong to allow loading from the north side. Dougie Boynton loaded from Swim Beach under the supervision of Donald Goose, Ray Phillips' pet goose, his mate the grey goose, a Canada goose who has joined them, and five mallards. It was quite a sight. Everyone was ready at ten o'clock, then Sherm found that his overboard water discharge hose was frozen solid. It took nearly an hour to free it of ice, meanwhile the boats strained at their moorings like horses at the gate. The Laura B came in on the dot of eleven with the usual large order for the Store, so Doug had to row Marty in to the dock to take care of the order for they found that Daniel couldn't drive the truck! Before Doug could get back to his boat with Daniel (who also knew nothing about handling lines or setting traps!) Don Cundy took off. Sherm had gotten his engine going by that time. Victor Lord slipped the first trap off from Steve's boat within a minute, right by the dock, and the season had begun at 11:05. By the time Doug had dropped his lines every other boat was well on its way to the setting grounds.

In the early afternoon Robert Rothstein had trouble with his engine and returned to his mooring to fix it, inadvertantly dropping a valve from his carburetor overboard and so was finished for the day.

Early Saturday morning, despite rough weather, Shermie towed Robert to Port Clyde and on the way back struck something which broke his drive shaft so Sherm had to go after him in the Phalarope. The wind had shifted into the south, blowing with gale force and sending huge seas through the harbor making reloading of the boats almost impossible. About two o'clock, when Don Cundy came in for more traps, his boat took a big sea, jamming his right arm between it and a wharf piling, possibly breaking his elbow. That accident discouraged the rest of the men and all boats were back on their moorings soon after. It was so rough most of them brought their skiffs to the dock instead of to Fish Beach. Only about half the traps were under water, even though the men had made an informal agreement to set only 600 traps as a maximum.

Sunday, with the wind still blowing a gale, Marty took Don in to have his arm looked at. It was found he had ruptured a muscle but no bones had been broken. So, Don went back to setting on Monday.

Today there has been almost no wind and the mercury is climbing towards 30 and by night fall most of the traps will have slipped under the water. The Lords had their first lobster feed on Saturday. We will have ours probably tomorrow. Doug never hauls until all his traps are set. Harry has recovered and is out with Doug today while Marty is working on the dock.

One more mishap - Bryan Hitchcock blew out his hauler on the first trap he tried to haul on Monday and had to go in shore to have it fixed.

Now for a little general news. We have had some horrendous rains, no snow since that before Christmas. The roads have washed out badly. This road washed down to ledge in the three inches of rain we had a week ago when most of you got snow. I don't know which is worse. One good thing, though, is that the Meadow is full of water and has made a good skating rink. The Ice Pond has a good surface, too, and the kids are making use of both places.

Tom and Irene Martin came out after Christmas - no,the week before - and spent two weeks here in Doug and Harry's fish house. We thought they would freeze to death,but they managed to keep it warm with two space heaters. They saw many sea birds, including an osprey and a small flock of Canada geese. Tom said I could have seen a sage thrasher out back. I hadn't dared believe my eyes. They also saw dicksissels and other odd birds I now can't remember.

Mort and Joan Bernstein came back after Thanksgiving but plan on leaving for the rest of the winter tomorrow, if the storm predicted isn't too bad. We have had a lot of heavy weather. The mail boat missed the day after Thanksgiving, December 8th (both times I was trying to get back here) and the Monday before Christmas. Other days I think Earl wished he hadn't made the run, especially the day after Christmas. I was going in yesterday but 40 knot winds at 5 degrees changed my mind. Now today there is a flat calm. Any windmill would not be generating power today, but would have stored up plenty over the past week. I've grown tired of turning my indicator back each day from 40 knots or over, once from 54. Joan isn't doing well. I think she had a mild shock at the time of their 40th anniversary and doesn't go out at all. I've heard nothing more about Wrenna Rothstein.

The floors in the cottage have been sanded and finished with urethane. It was too cold when I applied the last coats so there are many bubbles, but they will probably keep the rugs in place. The kitchen floor is western pine - very pretty - the others are knotty pine and look great. The natural finish makes the room much lighter than it was with those homely brown floors. I got some plastic sheets (rigid) to put up over the livingroom windows and also for the north window in this apartment and in my bedroom. They are so clear it doesn't look as though there was anything covering the windows. I hope they work out and cut down on the heating bill which is ENORMOUS. I have new space heaters in this part of the house and in the cottage. They have fans which work when the generator is on, distributing the heat quite well. Yesterday morning, however after the fan had been off all night, it was 56 degrees in this livingroom with a north wind at 5 degrees. But then I didn't have my plastic up. Today I've had windows open!

In March I hope to take off a few weeks and go to Mexico by way of Florida to see my cousins the Weinhagen in Ft. Lauderdale and the Johnstones in Port Charlotte. I'll have to leave Laura behind and am in hopes someone here can care for her. She is doing very well now on her cortisone, even runs away on her own and has decided she can climb the stairs once more. The Riddles will be in Cuenavaca then so I can visit them briefly. It all will depend on how and when Mother's estate is settled, I'm the executrix.

Enough for now. In writing this I typed two pages with the liner in place and had to retype them! The best for 1976 for all of you. Do write when you can.

As ever,
Alta

Monhegan, Maine
Jan. 6, 1976

Dear Cousins, nieces, nephews et al:

Aunt Helen will be celebrating her 90th birthday on January 18th. Franklin will be having a birthday - his 70 what? on the 25th so some of us, chiefly Elizabeth Gardner, thought we who can should get together, probably at the Colonial Inn in Concord, Ma. on Sat. the 17th to start them off. We know you all can't make it, but we want as many as can to try to come. Aunt Helen and Franklin love to go out to dinner together, usually once a week. Since neither of them are interested in having more THINGS we thought that a money tree and possibly dinner checks (I can't for the moment think of the proper term) at their favorite restaurant would be the thing to give them.

We also thought that for those of you who won't be able to join us it might please Aunt Helen to have a picture, your house, or even your pet rock!

Let us hope that the series of storms and arctic gales which have marched so frequently across the country for the past several weeks will let up for a while and give us a long spell of good weather around the 17th and that a lot of us will be able to reach Concord for the reunion. The family parties we have had when Elizabeth John and the Weinhagens have come to Boston have been such fun. I've been able to get to only one, but that one was great. It, too, was a birthday party for Aunt Helen three years ago.

Perhaps the best plan is for you to send your gift and pictures (even if you plan to come) to Elizabeth Gardner.

Mrs. David W. Gardner
85 Peter Spring Road
Concord, Mass. 01742

Elizabeth's phone is 617-369-3398. Mine is 207-372-9610

Enclosed is a copy of a letter such as I have been sending friends for the past fourteen years. I thought you might like to have a look into island life in the winter. There is scarcely ever a dull moment on this little piece of rock which raises its heads so proudly above the Atlantic Ocean.

Happy New Year! Let's hope many of you will gather at the Concord Inn on the 17th.

Love to all,
Alta

Monhegan, Maine 04852
January 24, 1976

Dear Cousins who couldn't make it to Concord:

The birthday party for Aunt Helen last Saturday was so much fun and so many of you who couldn't come were so generous I am writing to thank you and to fill you in on the activities.

It was a bitter cold night but fortunately no snow was falling, although there was plenty of ice, making going on foot quite hazardous. But weather didn't deter the nineteen assorted nieces and nephews who gathered at Concord Inn to celebrate Aunt Helen Woodbury's ninetieth birthday (a day early), as well as Franklin's 75th (eight days early).

As usual Aunt Helen and Franklin got there a bit early and the rest of us rather late so Aunt Helen was in a dither, thinking she and Franklin had come on the wrong day. But that soon passed as people began to gather. We had a private dining room with cocktails and hors d'ouvres before dinner. Helen and Robert Woodbury footed the bar bill, which was HORRENDOUS. By the time we sat down to dinner everyone was in a gay mood and alcoholic deafness was well in control! Even the children, Cindy Gardner Sutherland's Christopher (5) and Ian (7 mos) and Peter Woodbury's Allyson (3) were in good spirits and remained so until the party came to a close after ten o'clock! Elizabeth, Priscilla and Tommy paid for the dinner which was a la carte. The food was great. My roast beef could have been cut with a fork; those that had the scallops said they were **divine**.

Aunt Helen and Franklin sat at the bottom (or top) of a squared horseshoe and we all spread out on either side. The children spent most of their time chasing each other under the tables or rolling on the floor with great delight. No tears!

There was much picture taking before and during the meal at the end of which Aunt Helen was presented with an envelope containing $150, which she carefully stowed in her wallet one bill at a time, to help defray the costs of the meals she and Franklin share each Sunday.

Elizabeth Gardner had assembled the pictures, you all so thoughtfully sent, on a bulletin board as much as possible by generation and family. She and Franklin received the cards you sent, too. There were even some old pictures like the five generation picture of (my) Great-grandfather Woodbury, Grampa (I.F.) Woodbury, Aunt Gertrude Dowling with Franklin on her knee and my mother (Alice) with Tommy and Gertrude taken about 1900, judging from Franklin's age, when the picture was taken.

Franklin suggested that we all go to the Allston Congregational Church the following day, which suggestion was well received. Then, after a cake with one candle was brought in, cut and served, and devoured the party broke up. It was quite a night and Aunt Helen seemed to weather it better than any of us, as usual.

Those present at the party were: five generations

1st. Aunt Helen

2nd. Franklin, Alta, Robert and Helen

3rd. Tommy Ashley, Eliz. A. Gardner, Priscilla A. Jackson, Carol, Peter & Karen Woodbury;

4th. Cynthia, Scott, Sutherland Bruce and David Gardner; Allyson Woodbury

5th. Christopher and Ian Sutherland

Guest: Arthur Vash, a friend of Carol Woodbury.

Those attending church were Aunt Helen, Franklin, Robert Helen, Peter, Tommy, Elizabeth, Priscilla, Cindy and Scott - Christopher and Ian were in the crib room. We made up more than half the congregation! The minister was quite impressed and announced that the family had gathered to honor Aunt Helen's ninetieth birthday in the church her father had built (and from whose tower Robert had played his trumpet before service and from whose choir loft I had started my choir career. Emma sang in the choir, too). The same baptismal font was there, the same radiator hissed throughout the service, and Grampa's picture still hung beside that of Mr. Quint in the big Sunday school room. About the only change was the size of the congregation which certainly had dwindled since my day.

Priscilla, Tommy and I nearly didn't make the service. Elizabeth sent me off from Concord to pick up Priscilla and Tommy and I got lost trying to find Belmont. When I did I circled around the streets near 4 Douglas Road, passed the same man with a dog twice and nearly gave up when I found the house, nearly a half hour late. Then I drove down Western Ave. instead of No. Beacon, but got back on the right track at Everett St., then couldn't find a place to park, finally settling on a snowbank under a "No parking at any time" sign. No ticket.

We all, especially Aunt Helen, wished you all could have been at the dinner and at the church service. You were not forgotten in our conversation, anyway. The pictures helped a lot.

Most of the credit for the success of the evening must go to Elizabeth Gardner who made all the arrangements and did such a good job at assembling the pictures with notations. I think she intends to return those which you requested. Aunt Helen didn't want to keep any of them. She is trying to clear out things rather than accumulate more.

Thank you one and all again for helping to make the party so enjoyable for us all and particularly for Aunt Helen.

Now a little local news. Our winter weather didn't really get going until the day before it officially began, but it hasn't let up ever since. We have had one storm after another. Yesterday took the prize. It was -10 and didn't get up to 0 above all day and the wind blew so hard no one could get his or her house warm. Most stayed in bed unusually late or thought of things to bake to help with heating. My livingroom was 49 degrees with a fire in the Franklin as well as my heater. I even had a new electric heater on and got it up to 58 degrees by noon. Although it is nearly as cold today there is no wind, so heating is no problem. Alternate snow and rain, then deep freezes has covered the island with ice. Ashley Glacier out on the road has formed for the first time in three years making oil delivery dubious and I'm down to 1/4 of a tank. We may have to bring up oil in five gallon cans and fill the big barrel with them! Well, this, too, will pass.

Please forgive the quality of this letter. I used a new ribbon which just wouldn't behave. Then I got the generations all mixed up. I hope you can make something of it.

Love to all,
Alta

Monhegan, Me. 04852
February 15, 1976

Dear Friends, Family, and Foes:

It doesn't seem possible that there are two no-boat days in a row and I have no dealines to meet. We would have had three had not the boat missed for the seventh time since Thanksgiving because of Friday's storm. Earl was down with the flu and Ivan felt it was too rough to even take the boat off her mooring in Port Clyde! You can imagine how rough it was here.

In glancing through my weekly columns it seems as though the temperature has swung from the forties and fifties to single numbers or below at forty-eight hour intervals since the first of the year. The roads are a complete **mess** and look more like stream beds than anything. Ashley Glacier has formed, melted, then reformed periodically. This morning's weather report said that gale warnings were up but a look at the ocean revealed not a ripple and my wind indicator needle was scarcely quivering. But now, at ten o'clock there are whitecaps and the boats out fishing are bobbing about.

The storm on Groundhog day was the worst in anyone's memory. All the night before the house groaned and shook like an old barn. I had noted that the day before there was no wind. The temperature fell from a high of 50 degrees at dawn to 10 degrees by nightfall and the wind peaked between 57 and 60 all day and hit 63 a couple of hours before high water at midnight. That was when green water went right over Bob and Mary Burton's house down by the dock. The wind was just west of south all day, causing havoc in the harbor. Lobster cars belonging to Rusty Court, Dougie Boynton and Don Cundy landed up in Calf Cove. Rusty's was tangled in so much warp it didn't make shore but the other two were smashed. Another of Rusty's, empty, went out the southern end of the harbor with the going tide and hasn't been found. Bryan Hitchcock's did the same thing but came back in and stopped unharmed in front of Carl's fishhouse.

Several houses lost shingles, a screen blew off the porch of the Mooring and a window blew out in the other part of Rita White's house. The barometer fell to 28.35, the lowest I have ever seen it. By Friday it had climbed to 31.00, another record. The Burton house took an awful beating. The sundeck came loose, shingles were torn off the two sides facing the harbor, and water poured in through the walls. Mary took Lance up to Winnie and Vernon's for the night, leaving Bob with a CB set on which to call for help if needed. When the tide was the highest the wind had shifted just enough toward the west that the seas slackened somewhat. At high water on Monday noon I tried to go to the dock to take pictures and couldn't make it against the wind over the crest of Wharf Hill. The dock is a complete mess with huge holes as well as many snarls of buoys and traps, seaweed, traps debris and other flotsam. One old trap turned up and in it were two empty lobster claws at least six inches long a piece. The lobster must have been caught there for years.

Don Cundy's boat dragged her mooring and she came down on Doug's mooring buoy (stump). It was far too rough in the harbor to do anything but watch and pray through the night. At dawn the boats were both still there and had not come together. Don went out, despite the rough water, and moved his boat away. Later Doug found the stump had been chafed and his painter weakened, so he is on a borrowed mooring until it is calm enough to replace them. Meanwhile Craig Sproul's mooring balls and painter landed on Smutty Nose on Wednesday after another windy night. Fortunately he had taken his boat in to Boothbay Harbor when he and Judy went in a day or two before, otherwise he certainly would have lost his boat. The men figure they have lost 25% of their gear and many more hours spent just hunting for submerged buoys and disentangling the many snarls. Victor Lord has a bait box for a lobster car, since he has only a few traps out. When the smack came in on Thursday he looked in it and found one lobster, and it was dead. According to Victor the poor thing died of loneliness! the fishing this year is worse than ever. The storm added to the decline in lobsters by littering Middle (Swim) Beach with dead lobsters measuring from two inches up. Kittywakes, Bonapart gulls and many ducks came in to feast on them.

Danny Bates had on his wet suit on Thursday with the temperature below freezing. He had to go down to take warp out of the propellars (wheels) on Steve Rollins' and Dougie Boynton's boats. There is so much warp around after the storm it is hard to avoid it. He'll have to go down to repair Craig's mooring, he thinks, and wishes the water would warm up a bit. Brrrrr.

I've been spending money like the good old days when I earned per week what I get per month and three different sets of expected renters have either cancelled or not shown up so I'm having to count my pennies. I bought a double mattress and box spring for the cottage, an electric heater for this part - it doesn't affect the current, luckily - and a CB set. We don't need telephones with this type of communication available except one has to speak in code, since the whole island is listening. Now we can reach people from the Boehmers at Lobster Cove to the Mersfelders at Green Point.

Other expenditures have been for plexiglass storm window installed on the inside. Paul Horn is busy today making frames for them. What a difference they make. Now I wish I knew how to keep the north wind from howling through the front door of the cottage. I guess I'll have to put up a closed in porch - mud room between the two buildings. I've been keeping Bob Boody's family in steak and cookies with all my mechanical and electrical problems. He just reattached the cable which came down in a wind-ice storm. While he was on the generator shed roof he poked the weathervane which has been pointing optimistically due west for two years. By the way, the golden fish on June Day's house went flying around in circles the day of the storm.

At the time Bob Ives was ordained in the church I was asked to find out whether or not any other ordinations had taken place there and found almost no records, not even a list of ministers! Then I began asking about the school and found another dearth of records. Last fall, during a bicentennial meeting I brought up the idea of getting some material together to record Monhegan history from where Jenney left off in his "Fortunate Island". The idea was well received but I didn't get around to doing anything until after the first of the year. When I came back from Aunt Helen's birthday party I stopped to see Ivy Dodd at the Courier-Gazette and learned about a special edition of that paper to come out in mid-March, but the deadline for materials was yesterday, which meant Wednesday here because of our mail schedule.

Well, for the past two weeks I've sat at this typewriter and pounded out 37 pages, double spaced and three copies, on six different subjects and submitted scads of pictures for possible use. This place looked worse than usual as I searched through albums, folders, old negatives for suitable illustrations. Ivy let me have Saturday as a deadline and then the boat didn't run and the stuff won't now arrive in her office until Tuesday! Luckily I'd sent in all but my last story. I wrote about the store fire, paraphrased my two tugs story and the Museum brochure, wrote a story on the USS Arizona and the Flying Cloud replica when both oversized ships nearly entered the harbor when they were lost in peasoup fogs, Alice Stoddard's show covering sixty years, and a story of our family coming to Monhegan in 1908 or 9 (Edwardian Style).

Elise Nelson wrote about the Memorial Library and I think sent it in because it hasn't come here. I've asked Ted Edison to write about Monhegan Associates, Inc., and Isabel Farrell to send in materials on the church and school. There are about eight other facets that should be covered. When the material is all put together we will have copies available, probably at the library and possibly at the museum, for our gala weekend this coming July 4th, then trust someone will print it. (Under Grey Gulls Wing). Isabel sent reprints of some old newspaper articles which have some wonderful things in them. Did you know that Henry Trefethern thought people were stealing his sheep, so to get even he imported two red foxes which multiplied so fast they had to have a fox hunt after a few years? Such tales shouldn't be lost in the morgues of newspapers but should be in books accessible to interested readers.

There's not much island news except for the storms. The alternate rains and freezes have washed out the roads to where they look more like brook beds than places to walk or drive a truck. The Meadow has water in it way up to this end of it, but not like it was when they flooded it winters before the town water system went in. Middle Beach is **gone**.

Last night Doug and Harry joined Rita and me here for dinner.

I tried to make some cookies in the shape of pink hearts but had no cutter. They came out perfectly round and a very, very pale pink. It didn't deter Harry, he wolfed down nearly all the two dozen I cooked. Then the two men fell asleep in their chairs and slept all through Doc, Mary Tyler Moore, Bob Newhart and Carol Burnett even though Rita and I laughed uproariously, for the shows were particularly funny. Doug said he wasn't sleeping, just resting his eyes. Does he need to snore to do that? Harry twisted his back so he had the seat of honor - the platform rocker Mother used to have. He slept even better than Doug did.

One bicentennial project for the island is getting out a new cookbook. The ones in charge kept asking D. and H. for their recipes. Harry found an 1845 cookbook up in the Underhill cottage so I copied out several recipes from it, including one for a wedding cake using 15 pounds of eggs! Doug put in my father's recipes for Welsh Rabbit and Blushing Bunny. Anne Court, Mary Burton and Rickie Boehmer are doing this. Rickie and Peter left, after the storm, for Ecuador. Timmy and Louisa have been there since November living with cousins. Rickie's mother is Ecuadorian, you know. They will all return in the spring, with imports from Ecuador which they will sell in their shop along with the food such as they had in their Manna shop last summer.

The mail contract is up for renewal this July and people are wondering if Earl will keep it. There are others interested who say they will give the passengers better accommodations. The postal service is talking of delivery to Matinicus, Vinalhaven and North Haven by helicopter so such service might come here. Only problem is lack of blind flight guides on any of the islands.

The freight car on Manana took an unscheduled trip to the boathouse from the winch house in December. Thursday a helicopter brought a new car over. Now the Coast Guard boys don't have to tote everything up the hill to their quarters. Must close and go out for forsythia and pussy willows to force.

As ever,
Alta

February 18th '76

Dear Alta,

You truly deserve your title of "Queen of the Country Correspondents" - the articles this winter have been great and really give a graphic account of this spectacular winter. Have mixed feelings about wanting to be there - it would be great to see the majestic sea but hard going on land.

Glad you found little Laurie - you must have been very concerned. We had a surprising experience with George - he seemed to be having a heart attack - breathing hard and moaning for about an hour. We learned later he had spent the evening cavorting with and chasing a little feisty 6 months old kitten named "Sugar" who lives across the street. It happened twice under similar conditions - guess he doesn't know he's an older "man"-!

Can't believe we'll be thinking about going to Monhegan in about two months. We're not ready - there always seems to be a long winter ahead and there you find yourself rushing to get paintings framed and plans made for the summer.

I'm hoping we can get the Kent Show hung at the museum in early July. Sally Kent Gorton is sending me a photograph and some biographical material. It will be on as the Special Exhibit all summer.

Thank you again for all the news -

Our best
Anne

P.S. Have been taking figure drawing and Chinese brush work - really enjoying it.

Letter written for the Pine Tree CP and published in the March-April 1976 edition.

Monhegan, Maine 04852
February 20, 1976

Dear Sid:

You have set an almost unsurmountable task before me. How can I ever write something without a deadline? In college my Monday themes were always scratched out in the class preceeding English 101, or whatever number it was. I started to write about Monhegan's unique animal life, but having tipped over the vase containing forsythia sprigs being forced for the second time in two days, I guess I'll tell you of what a day or so on Monhegan might be.

We are a tight little island with limited resources so most of us husband our wherewith all. First of all I'll tell you what we do have.

Practically every house has a wonderful view out of more than one window, though not all can see the ocean. But it takes only a few steps to see it, and one can always hear it when the seas are running. Fresh, salty air and beautiful sunsets and sunrises are taken for granted. Although there may be some personality clashes there is a common bond between the seventy or so permanent residents, much like that running through a large family (about half the people are related by blood or marriage, anyway, though there have been no close intermarriages in the nearly two hundred years of the present settlement's history). When disaster strikes, large or small, neighbor helps neighbor, ungrudgingly and the fishermen go to the aid of one in trouble and watch the harbor until every boat is back on its mooring and all skiffs safely on Fish Beach. They even tell one another if something looks awry with the moorings, boats or lobster cars. Whenever there is a storm and fishing gear becomes tangled whoever hauls the tangle and involved traps returns lobsters caught to the men from whose traps they are removed. If any man breaks the lobstering regulations the men handle the situation themselves without calling the warden service and also without visciousness. A man shapes up or ships out.

When you want electric power **you** turn a switch, **we** start a generator. Most year round houses have a generator even if only to operate a few lamps on a 32 volt system or large ones capable of operating anything except large heating units. Many are diesels, some gasoline and others are run by propane gas. Gas has to be delivered in 100 pound cyclinders and is quite expensive with the added costs of boat freight and delivery charge for each cylinder. Oil costs are high, too, because of delivery costs. No one uses his generator twenty-four hours a day, not even the Store with its many freezers, compressors etc. Most of us turn on our generators at sundown and turn them off at bedtime. Mine comes on to pump water from my deep well so I have a sign:

> If you want water for frolic and fun
> Don't flush the toilet for just number one!

or the generator would be going on and off most of the day. We can get Channel 6 on our battery run FM radios, so one doesn't need TV for the Today Show or the evening news.

There is no municipal power and the town water supply is seasonal since the pipes cannot be buried deeply enough to keep them from freezing. Most houses have dug or driven wells (all of limited capacity especially in summer), or cisterns filled from water coming off the roof. The Store has two 5000 gals. cisterns which they fill from the roof or from two driven wells and so their laundramat goes all year round, when the generator is on.

One new house is installing a windmill. Several of us, including me, are waiting to see how this one works before we invest in one ourselves. But windpower would appear to be the answer to our energy problems. I've been keeping track of the wind velocities for the past month during which time there were no windless days and only twelve days it didn't reach 35 knots sometime in the day. The highest was 63!

When the men fish their traps and find crabs and fish instead of or along with lobsters they bring them home. Seldom do the islanders eat lobsters, especially now they are almost as scarce as hens' teeth. We women spend hours, at least I am slow enough to do so, picking out crabmeat which is used for lunchboxes.

Several families have large and productive gardens in which they raise enough vegetables to last them well into the winter. I still have a bit of rutabaga in my freezer and had tomatoes until Christmas. One gal has a cherry tomato plant which is still producing abundantly in her front hall.

Two things we don't have because of lack of 24-hour power are direct dial telephones and gun-type furnaces. There are ten telephones, five of which are in summer homes, and all are pay stations - the only kind of system that can be maintained on batteries. It costs a **fortune** for long distance calls, especially intrastate where the rate is the same every day and every hour, even for the overtime. All our heaters are pot burners, since gun-type burners require electric power to operate. Auxilliary heat comes from Franklin or parlor stove, fireplaces and Aladdin heaters, all of which were used to little avail the day the wind-chill factor was -75.

During the six off-season months our boat brings mail three days a week. Seven times since December first the trip has been delayed a day because of high winds and freezing spray. When it does come it brings the newspapers, but they never go back to Thomaston to pick up mail or papers which could be brought out on the odd day. Passenger accommodations leave much to be desired. Thank goodness for snowmobiling apparel. The big moment of the day is the arrival of the boat and distribution of the mail in the post office at the top of wharf hill. This hill is so steep that mail sometimes has to be carried up and down it on a sled when there is a lot of ice and snow. During the off season we are allowed in the P.O. during the sorting of mail, but not during the summer. If it is going to rain it always seems to take mail and boat time to do so, drenching everyone.

Our roads were built for horse and oxen traffic, not for the twenty trucks which now ply them from nearly one end to the other. This year's heavy rains, alternating freezing and thawing have reduced them to rocky stream beds and horrendous ruts. I can't use my golf cart all winter because of the rough going.

The island is kidney-shaped with the village occupying the pelvis. The rest of the island is wild, kept so by the ever vigilance of the Monhegan Associates (Inc.), an organization formed to preserve this wild beauty. There are eighteen miles of trails which wander through virgin woodlands and up over the steep cliffs which rise right out of the water on the "back" side of the island. There is an ice pond from which ice was harvested until 1974 - no need for ice now the store and hotels have ice machines. There are two meadows, one in the center of town and covered with buck bean (frost flower to islanders) in May and one at Lobster Cove which is covered with cranberries. Deer and pheasants are the only wildlife plus the ever present wharf rats which gobble up quantities of Decon each year.

This didn't turn out to be what I intended. Someone came in to have her blood pressure checked and I got off the track. Hope you can use some of this. Let me know your deadline.

Sincerely,
Alta Ashley

3699 Doyle Road
Baldwinsville
New York 13027

February 23, 1976

Dear Aunt Alta,

I do have more information on windmills.

I would enjoy acting for you during the process of designing a system, preparation of specifications for obtaining quotations and reviewing the quotations.

Of course, I would need a fair amount of information on your present system, your needs and your location. It would be about 6 months before I could have a bid package ready.

There would be no charge for this work and I would not give it the highest priority.

If you are interested, let me know and I will start by preparing a list of information that I would need.

Very truly yours,
Jim
James M. Jackson

(This is in answer to Jims letter of Feb 23)

February 28, 1976

Dear Jim:

Your letter was waiting for me when I returned to the island yesterday and, needless to say, I am delighted with your offer to do the designing, preparation of specifications and reviewing quotations for a windmill. I've been trying to find someone locally but don't know where to turn. Your offer is better.

Since the first of the year I have been keeping a record of wind forces as registered on my anomometer, which is supposed to register only about 75 percent of the gust force. There has been no day when the wind hasn't blown, registered that is, 25k or better, usually better.

If the needle is still it is at the change in wind and it soon picks up and it blows steadily 10k or more most of the time. Up on the hill where the windmill would be set it never stops blowing, even in summer, because it is in a funnel which leads from the SW end of the harbor across a low wetland and up the hill - a veritable wind tunnel.

At present I have two LP gas generators on demand power. One is a 1 cylinder air-cooled 3.5 KW, the other is a 4 cyclinder water cooler 6.5 KW, used as a stand-by. Both are Kohlers. Except for a large frigidaire clothes washer in the cottage, everything in both houses can be run off the 3.5 KW. The motors are powered by car batteries.

At present I use LP gas for refrigeration (2), hot water (2), and kerosene for the stoves in each house. I'm sure if there is electricity available for little or no cost (once installation is paid for) there will be a gradual conversion to electric power for everything. I forgot, both kitchen stoves are LP gas. Both houses are seven room affairs. My refrigerator has a small freezer top.

The generators are housed in a separate house which could be used for the necessary batteries. There is a cement floor on the shed-like addition which measures approx. 5x7. The small generator sits on that at present. I would probably keep one generator as stand-by. Of course, there are two TV's. All radios are battery powered. There is a vacuum cleaner in both houses and a washing machine (clothes) in both - no toasters because the generator isn't on in the morning. It costs nearly a dollar per hour for generator power, so it is run as little as possible.

I forgot the 3/4 horse power deep well pump needed for my water supply. It dims the lights when it comes on. It is activated by a 75 watt bulb and delayed action switch.

The prevailing wind is SW, even in winter. That can be the heaviest and the lightest wind. Winds from the east are always strong so that the mill would get plenty even though the hill is behind it to the east. Most of the trees around the foot of the lot are now dead or dying. The generator house is higher than the road and land to the north, although it doesn't look so in the postcard I'm enclosing. The lighthouse is the highest point on the island. I suppose the tower would extend to about the level of the Wiley house roof, just behind my property. The people in the inn to the right are interested, too, and might go in with me, if that is wise. We are all waiting to see how the Boyntons come out with their apparatus which they have obtained from California.

Bob Boody who is an ex-Coast Guard officer is a mechanic and has a welding apparatus. I hinted that he should think of building a windmill from scratch.

By the way, my property line extends almost to the Wiley house but not much farther south than the generator house, in case guy wires have to be installed.

Any more information? I expect the cost will be $5,000 or more, so I won't be easily scared. In the end it should save me money, if I live long enough!

Thanks again for your offer. Remember, there is room for two or three at a time here, or for the whole family off season.

Love to all,

Monhegan, Maine
April 11, 1976

Dear Family and Friends:

It doesn't seem possible that I haven't written since February. So much has happened since then I scarcely know where to begin. Ever since I started on a journey on the Ides of March all sorts of things have happened, both good and bad, but the bad seem to have outweighed the good.

Emma had suggested I come to Mexico while she and Hollis were in Cuernavaca and combine a tour to Cancun put on by Mount Holyoke Alumnae with a visit with her and Hollis. Such a combination couldn't be worked out so I decided to go to Mexico City and take day trips from there with an overnight in Cuernavaca, ending a ten day stay in Mexico with three days in Puerto Vallarta. Well, the whole thing got out of hand. My ticket didn't reach me until I went to Thomaston the day before my departure for the south. Then I found I had hotel reservations in Cuernavaca, which I didn't need, a non-redeemable ticket to the bull fights, which I didn't want, and other unnecessary items. I was to fly to Ft. Lauderdale, stay with Woody and Betty Weinhagen for three days and return via Tampa and a stay with Ben and Jean Johstone in Punta Gorda. After the awful winter we had I was interested in looking over places to stay come another year.

Doug and Harry were planning to go to Barbados for a week or so at the same time. I took Laura to Nute Searls' where Paul Horn was to take care of her until a fellow was to arrive and stay at the Mooring until my return on April 2nd.

On Sunday the 14th Doug, Harry and I left on the Monhegan and it seemed that the curse of the Ides of March would be neutralized. Doug gave me four lobsters to take to the Weinhagens, advising me to cook them, not trust to getting them to Florida live twenty-four hours later. Fortunately I took his advice because the next day Security insisted on unwrapping my carefully insulated bundle, instead of putting it through the fluoroscope. Why? By the time I reached the Weinhagens' all the ice in the container had melted but the lobsters were still cold and were a great treat.

I had planned to have Ivan Simmons, who keeps my car in Port Clyde, drive me to Portland but by going in a day early I was able to catch a bus. Zimmie Brackett met me at the bus station and took me to a very nice and interesting restaurant down near the waterfront. It was good to see Zimmie. He is still busy with bridge playing though he is playing less than previously. He plans to come to the island the end of the month, after the water goes on.

The day after I landed in Florida we drove to Everglades National Park, down to Flamingo. In one spot we had to drive by a smoky grass fire which threatened to jump the road, but we got through safely both going and returning. We saw a pair of bald eagles (through a telescope) which were nesting on a mango island, a huge alligator sunning himself just below the catwalk we were on, and many interesting birds including roseate spoonbills. Coming back we stopped for some luscious strawberries and ice cream and then were stopped on the throughway for speeding (all of 62 mph) as another car flew by us at no less than 80! Woody tried to tell the officer he had the wrong car on the radar, but it didn't work.

That night Deedie Riddle Harson called to say that Hollis had died the day before in Cuernavaca in his sleep. When I called Emma later she said that Parker was on his way down from Oregon and that she didn't need me until the following Monday, the day I had planned to go to Cuernavaca, but that she would like me to go back with her to Akron later that week. So, instead of going to the Tampa area on my return from Mexico we drove over, through Alligator Alley (no alligators, tho) on Thursday, had dinner with Jean and Ben at a nice little restaurant in Punta Gorda right on one of the canals. On the way over we drove around Marco Island, stopped to see friends in Naples (very pretty area) and arrived at Jean and Ben's in time for a tour of the house and garden before supper. Their house is ideal for them, is at the end of a side street and on a canal. The garden is full of exotic (to me) plants and trees and must be quite beautiful when in bloom. Jean teaches school and Ben plays in a band, so they are both quite busy people. We drove home that night in bright moonlight. By the way, I saw Comet West early one morning with the aid of glasses, despite the night glow from Ft. Lauderdale. We had had too much cloud cover up here to have seen it.

Friday the 19th I was off for Mexico City. It was cloudy and smoggy all the way and as we approached M.C. we flew into a thunderstorm with bolts coming down all around us. **I was scared.** My seat partner's son-in-law owns Anderson's restaurant, a swank place up near the American Embassy where I hoped to have a meal but didn't manage to with my tight schedule.

Mexico City was interesting, but too noisy, smoggy, and hot for me. The altitude got me, too. I had nosebleeds, swollen feet, and palpitations. Needless to say I didn't climb the pyramids - didn't even walk from that of the Moon to the Pyramid of the Sun, and felt like an old lady for the first time in my life. I met a mother and daughter from Denver and a couple from Minneapolis who were on the same tour with me and at the same hotel, which made it nice. We went to a restaurant on the 42nd floor of a nearby hotel for dinner one night. The city spread out in all directions as far as we could see, even crawling up the sides of the mountains that ring the valley. There are more than 11 million people in Mexico City, mostly in one and two-story dwellings, so you can imagine how big it is. The murals, both inside and on the outer walls of buildings, were marvelous, the pyramids awesome and the Anthropologic Institute amazing in its contents and excellent in the manner the mass amount of artifacts were displayed and labeled. I found I could read quite a bit of the Spanish, though it would have been better to have had a guide.

When I left the Anthropologic Institute one of the guides offered to drive me back to the hotel. He was a student at the University and had lived in the States so his English was quite good. He led me to a rather battered car which had been recently spray painted a brilliant blue, including the windows! First he had to adjust the motor with a pair of pliers. Then, in the midst of the Reforma (the largest thoroughfare in the city) we had to stop while he filled the radiator with water!! It was with profound relief I alighted at the Hotel del Prado.

The ride to Cuernavaca was over a beautiful new road. Bouganvilla and roses were in full bloom along the roadside. We climbed up over a mountain range and wound down the other side about 3000 feet below the elevation of Mexico City. Emma and Parker were waiting for me in the cafe in the center of town, thinking I would arrive at the time I left Mexico City. Somehow our wires got crossed. And I thought I was early!

Emma has a lovely apartment in a complex closed off from the street by a high wall and an iron grillwork gate which was closed each night at six. This made Parker feel he was cloistered while Emma and I felt nice and safe. There were lovely gardens and a large swimming pool which Parker and I made good use of. We had the use of a car belonging to a retired Goodyear couple in Cuernavaca who were in Houston temporarily so I didn't get to meet them and missed them last year when they came to New England. Parker drove me over to Popocateptl and up the mountain to 12,000! We were still in an area of tall trees but the snow wasn't far above us and it was cold enough for coats. We had a hard time getting any pictures because of the clouds which kept the top of the mountain obscured. On the way we stopped at a small village where they were setting up their weekly market and ate a tortilla (not a hot one). On the way back we stopped for dinner at Cocoyok, an inn converted from an old sugar mill. It was a beautiful spot with swimming pools, a small pond with geese, ducks and swans, horseback riding trails and a golf course. I ordered the chef's special only to find to my dismay that it was beef smothered in a very hot pepper sauce and had to cool my mouth off with a gin and tonic after each mouthful. Another night we all went to a beautiful restaurant right in the heart of Cuernavaca, again hidden behind high walls. There we had drinks in the garden and our dinner on the patio.

Cuernavaca was beautiful but again too hot and noisy for me. There is no kind of muffler or horn control in Mexico and racing, honking cars are everywhere. There must be more VW's in Mexico than in Germany and ever present diesel motored busses. I guess I'm a snow bunny and had better make up my mind to stay here in the cold and quiet during winter.

Parker went back to Oregon and Emma and I took off for Akron. The Mannings' chauffeur drove us to Mexico City airport, which was a great help for Emma had both her luggage and Hollis' and I was laden with mine plus all I had acquired in Mexico. We were due in Cleveland near midnight so Emma had arranged for us to stay there overnight at the Hilton. Well, our plane had to stop in Columbus and pick up 65 stranded passengers of a disabled plane so we didn't get in until nearly one AM. Our room was about as far as possible from the elevator and when we finally reached it there was someone in it! The next room had only one three-quarter bed. Finally, close to 2 o'clock, we were put in a two bedroom on the top floor. When Fran Ackles called for us the following morning she was told we hadn't registered. She was just leaving the lobby when we found her, otherwise we'd still be waiting for her!

After we reached Akron we went directly to the Post Office to pick up the mail. Winnie had told me I had a lot of first class mail so I had asked her to put it on Wednesday's boat. It wasn't in the mail that Friday. In fact there was no first class mail at all and the post office had not then (the 26th) received Emma's request to stop forwarding mail as of March 10th. So my mail went down to Cuernavaca with Emma's and is somewhere between here and there on burro back. Among the things in that packet of mail are my bank statement, my federal refund check, deposits from people renting the cottage this summer. I'll probably receive it sometime in May, judging from how Emma's **airmail** letters came through. One took over a month to reach Monhegan from Cuernavaca. Meanwhile I have no idea how much or little I have in the bank because I have automatic deposits. Let's hope they got deposited. The bank has not been able to furnish me with a copy of my statement so far. Yikes!!

John arrived from San Francisco soon after we did. That night we all went out to dinner and John had a great attack of nostalgia as we drove around the part of Akron in which he and Parker grew up. Emma had a different kind of an attack during the night, and I followed the following morning so we couldn't go to Winnie and Vern Wolfe's for brunch, etc., but John went after he had gone to Cleveland to pick up John and Susie. Leslie, Hollis' daughter, came in from Montana that morning, too. Emma recovered on Monday but by then John was sick and I was still sick. Where and what we picked up we don't know, but it was violent while it lasted.

Tuesday was a beautiful spring day. The memorial service for Hollis was held in the Lutheran Church that afternoon. It was quite impressive. The only flowers were those around the base of a globe, symbolic of Hollis' life in Goodyear Export. Many of Emma's old friends attended. We went from there to the Akles for dinner and a reception. It seemed as though Hollis was there with us - no sadness.

Rue and Nancy arrived that morning from San Francisco and Deedie Harson from Reading along with Jay Baisly Allen. Jay's son Mitchell was there, too. Everyone stayed up at the Hilton Motel just up the road. Deedie and John moved in with Emma when I left for home on Wednesday.

When I finally reached Monhegan on the 2nd I found that Laura had grown much worse and was passing bloody urine. I had to go back to Rockland on Wednesday to go to the dentist so I took Laura with me and left her at the vet's to be put to sleep. It was an awful decision and I am more lonely than I everthought I would be. Poor Laura, I should have put her out of her misery before I left but she didn't seem that ill.

When I called the island on my arrival in Thomaston I found that the Phalarope had taken a load of people in for Manville Davis' funeral. Manville had died on the previous Sunday following a heart attack on Saturday night. This Monday we learned that George Anthony Palmer had died that day, also after a heart attack. He wrote under the name of George Anthony.

Last Tuesday was a gorgeous day with spring peeking around the corner, (It is snowing hard right now) so I decided to wash bed spreads etc. I had both washers full and the generators wouldn't work! Bob Boody came up, burned out a battery charger he was trying to test the lines with and left me a new battery of his. Both of mine were defunct. Last night I couldn't get the little generator to work so called Doug (Bob had gone in on Friday) and it started like a charm, but soon refused to go: George Chasse came out on Friday with three new washers for the store and said I needed an adjustment on the washer in the cottage. The motor demands far too much power for my system, so he will bring out something to adjust it when he returns in three weeks. Let's hope so, for that machine has loused up my whole system, what I didn't do when I shorted everything with the pail handle I used to catch the oil the last time I changed oil.

Elva and Chris Nicholson are here now. Doug is fishing with Chris. The Doogon has been sold to someone in Castine and the Monhegan is in Tenants Harbor for repairs. Both the Sprouts and the Sprouls are expecting. Barbara Stanley is recuperating from surgery. Dyan's mother is here, also recuperating, and Dusty has had several heart attacks and may soon join Laura in Dog Heaven.

Not enough for another page. Jim Jackson is working on plans for a windmill for me. I hope one can be made to work.

Love to all,
Alta

Monhegan, Maine
May 5, 1976

Dear Emma:

How stupid can I be! Last night it never occurred to me to say I would come out and help you convalesce from your operation. I'm still rather overwhelmed by the thought of clearing out the apartment. But by 14th I could think of getting out to Akron. That would be about the time you would be getting out of the hospital, I should think. I should be here on the 22nd, and the weekend of the 29th. After that there will be no one here until June 23rd. The 22nd is a big day. About 100 people are coming out for the day to study island ecology and I'm supposed to be the leader of the discussion groups!

After the initial discomfort of any operation, I'm sure you'll feel so much better you'll want to tango by July 4th. Dot Stanley had the most serious involvement of her arteries of anyone Dr. Root had operated on and she was walking up Horn's Hill in a few weeks after her operation. So take heart. I'm so glad you are under competent medical care.

There are two gals in the Mooring this week - my vintage and same educational level. I've spent so much time with them I've done **nothing** about the house. One is a friend of Libby Magneson on Vinalhaven. She feels I should take my letters, columns, articles, etc. to Libby, who was a Time-Life editor, to see what could be done to get them in shape for publication. That sounds good to me, if I can ever get things back in order again. At seventy-one a new career?!?

Yesterday on our way back from White Head I found five violets in bloom and have them in a jigger on the table here. I also dug a mess of greens later. Doug, Harry and Rita will be up tonight for a ham and greens dinner in celebration of the day.

Phoebe is calming down and turning into quite an obedient dog, though still very lively and in constant demand for recognition and loving. She licks the deer blood off all my flowers so I'm not sure I'll be able to protect them from deer incursions. She will probably do that, she thinks they are loads of fun to drive away. I hope she won't learn to run them in a pack.

Forsythia and daffodils are finally out, but my magnolia looks as though it would never bloom. It doesn't seem possible that we are so far behind Akron. Five weeks ago everything was out there.

Let's hope today's mail will bring that packet of much travelled mail. All of ours for yesterday was sent to Rockland and we got a Rockland pouch! The higher the stamp cost the worse the service. Earl Field got the mail contract again for another three years so we have no hopes of better freight or passenger service for that time. D--n

I've just cleared out the porch and moved the plants outside in real hopes for warmer weather. Onie, Rita's dog, went through the screen door the other day so I had to have it repaired. Ernest Lewis spent all yesterday afternoon on the job and now we can't make the door slide all the way shut! Ho hum, that doughnut that didn't get turned over is **me.**

Monhegan, Maine
May 19, 1976

Dear Friends:

After three days of rain the sun has come out this afternoon in bursts of light between dark clouds borne on 35 knot winds. The ocean is a mass of whitecaps and the air feels as though snow was in the offing. Forsythia has bloomed and faded but daffodils and tulip's are still bright and the ground is covered with violets - more than I have ever seen here. Strawberries, both wild and cultivated, are in bloom and give promise of a good crop. But my magnolia, which was so beautiful last year, looks as though it has been winter killed, except for one small portion containing two buds. The winds we had all winter must have been too much for it.

The last time I went in I exchanged my ditto machine for another - smaller and simpler to run. Here's hoping it lives up to its reputation and will turn out good clean copy for you.

Tom and Irene Martin are staying in the cottage for two weeks. Poor Irene fell out of bed soon after they came and has been laid up with a broken rib ever since. But she did get out Sunday and walked with a group from Mid-Coast Audubon who came out 65 strong on the Laura B. It was a perfect day - sunny with no wind and temperature about 70. The birds out-did themselves, particularly in my yard where

Tom had spread large quantities of seed all around the edge of the lawn and under the blackberry bushes. Even though people came and went all day the birds stayed around avidly feeding. We saw indigo buntings, gold and purple finches, rose breasted and evening grosbeaks, white crowned and white throat sparrows as well as the more common birds. One birder saw a blue bird and many saw cerulean and perula warblers in the Meadow. It seemed that to a man (woman) the birders are regular readers of my column, which is most gratifying. I've never done anything that has been as rewarding as the continued pleasure that little column brings to so many people. I receive letters about it from all over the country - even received several condolences after I had Laura put to sleep.

Speaking of Laura, after spending a week without her I was so lonesome I called the pound and on April 16th Phoebe Starling came to live with me. Mary Burton picked her up and brought her out on the boat. She is a three year old (?) GOK, yellow with black snout and spectacles. She is extremely loving and full of more pep than I can reckon with. For the first week I was up at dawn (before 5) to walk her. Daylight Savings Time has helped a lot and now I let her out alone when it rains. If she stays away from Jasper she is no problem, but from him she was learned to go on the "deer run" - not so good. Her name, as many of you know, it that on the oldest stone in the cemetery. It is such a delightful one I couldn't let it pass into limbo.

Sandy Dickson just brought Phoebe up from the Spa - all 26 pounds of her - where she apparently had run into a sharp object and punctured her shoulder. She seemed to be in shock but is moving about again okay. She tears around and runs so fast she must be part whippet. This probably won't be the last accident to befall her.

Last weekend the family came up to claim what they wanted of Mother's furniture and family lares and penates. Elizabeth and Priscilla arrived Friday, I went over on Saturday, delayed by overnight guests in the cottage. Cindy and Scott brought Tommy and Bruce up late on Saturday, too late to arrange for a U-Haul so C, S, and B stayed over until Monday, picked up and packed the trailer and Scott's wagon. Elizabeth had taken a lot of stuff and I had a half-truck load of stuff which is still mostly in cartons littering downstairs. I guess I'll have to have a lawn sale as soon as I can check on how many duplicates I have. Had I brought over all I wanted to keep I would have had to put on an addition to the house and may do so, anyway. I do so need storage space and a place to work in. If only I could figure out how to put the old office downstairs to use. I'm glad the furniture is going to stay in the family. Two chairs were my Grandmother Ashley's and now Elizabeth has them - her great granddaughter. I closed up the Thomaston apartment. There didn't seem much sense in keeping it now that I go in shore so little, and then only because no one else can do what has to be done ashore.

The packet of mail which went to Mexico is still in transit or sitting in the Cuernavaca Curreo (not a very descriptive word for the service!). Possibly some of your letters are in it. When and if it ever arrives the news conveyed will read like ancient history.

Emma had arterial surgery at the Akron General Hospital on the 8th and is progressing well, though slowly. She was due to leave the hospital early this week but will not go directly home until several more days, then will have someone come in a few hours each day to help her during her convalescence. This has not been her best year so far. Let us hope all the ill wind came the first part of the year.

Doug and Harry feel as though '76 has been a poor one for them, too. First there was the loss all the men suffered during the Great Feb. 2nd storm, then major repairs to the Monhegan and the fishhouse. Next, they had to replace the three machines in the laundromat. Then the oil tank on the dock sprang a leak and a new tank will be necessary. The last blow was the collapse of the Store truck body which Bob Boody spent two days welding together. Bob has been a busy person since he retired from the Coast Guard. About every truck on the island has been up at his place being repaired in one way or another, my golf cart, Apollo, included. What did we do before Bob came?

Mace Carter was supposed to come over late in April or early May to repair the dock but hasn't arrived as yet. Probably on the Fourth of July he'll come and rip off the slip and all the ladders and people will have to go ashore by skiff as they did prior to 1910 when the first dock was built. (This machine is jumping again, thus the odd spacing). By the way, Earl Field was awarded the mail contract for another three years. He has gone up on his freight again, but the boat fare and schedule remain the same. The first Sunday boat will be June 27th. The Yew opens next week, I believe, and the Inn and Gull June 16th. I don't know when Jeanie and Victor are opening the Monhegan House. They, not June, will be running it this year.

Tomorrow Gib Strout, our school agent, will be here for a School Board meeting to which everyone is urged to go. The state is demanding 29 thousand dollars from us, giving back 13 and we need a minimum of 19 with which to run the school. Sparks will fly, I know. I say, give the state $10 thousand and keep the 19. Even so, it will mean more taxes. Just how does this system guarantee equal education to all when we can't run our school on what is allotted?

My freezer is full of hake and cod. Doug, Harry and Rita were up for halibut last night. Fish are becoming more plentiful. The price of lobster, though, is now less than that for fish. $1.60 for lobsters, 1.80 for halibut.

Now to try out the new machine. Jay and Donna Sprout are going to VT where he has a church. No other news.

Lot to all,
Alta

May 19, now June 15.

You guessed it, the new machine doesn't work. I took it back when I went in last week for the Maine Medical Association meetings and ran through a retyped first page and a second page, then one of the rollers fell off and I can't see how to reattach the spring which came loose. Meanwhile the lavatory downstairs where I have the machine set up is knee deep in ruined copies.

The new oil tank is in place on the wharf but the dock is in worse condition than ever. It **will** be July 4th by the time anything is done. Last week I took Apollo to the dock to carry up the load of stuff I had brought over. In front of the Spa a kid came flying by on a bike. I stepped on the accelerator, instead of the brake - the brake is applied by the left foot - and we collided. The boy got only a bruised shoulder, but Apollo's windshield popped out and shattered. Thank goodness it was no worse.

The island is filling up. Jackie and Julie have returned, Vi Sturner is here, too. She will be curator of the Museum again this summer and plans to stay until October. The Inn opens on Saturday. Also that day Dint's art gallery, which he has built opposite the Periwinkle where Virgie's barn used to be, will have a gala opening.

Tax bills haven't come out yet because of hopes that our school tax will be lightened. The town voted against raising $5000 needed in excess of what the state plans to return to us and we are making many papers for our stand. We are paying in $29,000 as it is, which is almost twice our entire tax levy of a few years ago. I hate to think what my bill will be.

The new edition of INTRODUCTION to MONHEGAN just came out and the Store and the 1784 House ads on the back were left off. I don't know whether it was my error or the printer's, but I didn't catch it when the gally proof came so I'm out what they would normally pay - and I ordered 15,000 copies so as to not have to have a reprint for two years! This has not been my year so far.

My mail finally came back from Mexico in two lots, one without evidence of forwarding and I was confused about being notified of events long since past until I saw the postmark dates. I wonder if any more will stray in.

Emma has come through her operation and will be going to the Farm the end of the month with a companion who will do the driving for her. I hope to get over there in July when I have this place rented. I'll be going down to E. Falmouth to visit the Peters then, too.

Jim Jackson is coming here on Friday to see whether or not it will be feasible to erect a windmill. I don't have really enough land for a high tower, but he may find the wind currents won't require one. Here's hoping.

Garbage collection starts today and I have a full boatload accumulated. Bryan Hitchcock was awarded the contract so Stevie Rollins is going to start building his house. I hope Carl picks up from the house as he did last year. The Sprouts are leaving so we have been interviewing teachers. More than forty people have applied for the job and the selection is going to be hard. Must stop.

Love to all,
Alta

Monhegan, Maine
May 25, 1976

Dear Emma:

Where has the time gone? I wrote a long letter and tried to put it through my new machine and ruined the master, so now I have to take this one back. I guess I'll have to buy a new, rather than second hand copier.

The mail from Mexico arrived yesterday with two rubber bands holding it together. There was nearly $1000 in checks in it, all safe. Also there were two applications for rentals of the cottage, one for

April, fortunately no bills.

No, I didn't send flowers and am very ashamed not even to have sent a get-well card. This moving out of the Thomaston apartment and readying the cottage, between renters, for the summer season have just overwhelmed me. I'm beginning to think I'm growing incompetent, it takes so long to do things. And then, there's no place to put anything. Yesterday I got rid of my day bed, making room for the hide-abed I had in shore. I have it out on the sunporch. I also have a new slip cover for the one I have up here in the apartment. They used the inshore one for a model and the thing fits beautifully. But it cost almost as much as the bed itself. Phoebe thinks it's just the place for her, though she is now sleeping happily in her basket under the table. Laura soorned the basket.

Phoebe is still taking a lot of time. When it isn't pouring (and when has that been these past ten days?) I go for a walk with her before breakfast. This regime is beginning to affect my girth. My skirts are growing loose and I no longer can see my tummy below the upper structures. But the weight still hangs around 150 pounds, clad.

During the week I have my apartment rented I'm going down to Nancy's on the Cape. She is delighted we can get together. I hope I can get in a lot of swimming. I'll also see what can be done about the Aunt Helen-Franklin situation. He was all for having her live with him, now he is wailing about the situation and driving Elizabeth mad. I still think Helen should take some action.

The kids were delighted to take most of the furniture. They went off with a U-Haul stuffed to the doors, as well as cartons in Elizabeth's car, furniture and cartons filling Scotty's station wagon and a box spring tied on top of that. Even so I had more than ten cartons, a bookcase, TV and stand, a folding cot and the hide-abed. We tried the TV out and the reception was awful except for channel 10. Then I found the rabbit ears weren't even connected. Now it works perfectly. Imagine good color reception with only rabbit ears.

Let me know if you want me to help either in Akron or N.H. I hope by now you are feeling much better.

A new publication, New England Offshore, lifted my article in toto about the Arizona and the Flying Cloud from the Courier-Gazette. According to a conversation I had yesterday with the Courier sparks are about to fly.

I must get this to the mail. I thought you would like to know about the mail from Mexico. I have written Dot Manning too. A week or

Lawn Sale

Military island funeral

so ago I wrote asking her if she could see what had happened and I guess she wielded some magic.

Write when you can. And give my best to the Wolfes and Ackles.

Love,

Monhegan, Maine
July 7, 1976

Dear Friends:

Am I EXCITED! Ever since the Fourth I've hardly been able to settle down to do anything. The Monhegan Bicentennial Committee asked me about two weeks before to give the welcoming address at the end of the parade and after the raising of the flags. I thought of saying about a paragraph of "welcome and don't forget to pick up your trash!" Then Rickie Boehmer, our Bicentennial chairman, asked how I was coming with my speech and if it was written! That set me to thinking and about the end of June I began to work it up.

Well, it turned out to be eight pages long and took about twenty minutes to deliver. At the end my knees turned to water and my legs began to shake, making me want to flee the scene. Then all sorts of people came up at the end of the ceremonies and asked for a copy of the speech. I had made just **two**! I think at least fifty people have asked for copies since. Bill Boynton has taken it to Washington to see about having it copied or printed. I sent him a cover design today and must wait to hear from him. He also thought **60 Minutes** might be interested!!!

The whole day was wonderful. We awoke to a thick foggy morning that looked like rain. But by the time the parade had started down at Lobster Cove at 9:30 the sun was out brightly. There were four floats and a lot of people were dressed in period costumes. Several of the little girls were adorable in their Kate Greenway dresses and bonnets. I finished my skirt on Friday night and didn't step on the ruffle and pull it off until the ceremonies were over. I finished typing my speech at the time the parade started, at the last minute as usual.

Nute Searles, dressed in a uniform of the Monhegan Cornet Band of 1914 (they played at Monhegan's tercentenary celebration that July) raised the flag because Chris Nicholson, somehow was told the wrong time. Rickie Boehmer read a Declaration of Interdependence and Bill

Brittain, the resident minister for July, gave the benediction. It all went beautifully, mainly because of Rickie's good organization and enthusiasm. I forgot to say, the crowd sang several patriotic songs to the accompaniment of a flute played by one of the Northern Valley Boys who were here for the weekend, giving concerts and playing for dancing at the schoolhouse and up at the ball field where there were also games and a box lunch. We took lunches and exchanged them, rather fun.

Sunday night we held a Patriotic Service at the Church and Hannah Gross sang some negro spirituals and the Twenty-Third Psalm in Hebrew - very beautiful. She has a **magnificent** voice. We even had a nice sunset to greet us as we came out of church. All the next day people went around saying how beautiful the whole day had been. The one thing we neglected to do was ring bells at two o'clock, and I had a nice new one mounted by the door, too. Bill Brittain's sermon was superb and he has promised me a copy for my historical files. It looks as though I have started on a new career.

My nephew Jim Jackson (Priscilla's husband) came out here in June with a retired engineer from G.E., Arnold Briggs. They had a busy twenty-four hour stay looking over everything, measuring, inspecting the one functioning windmill on the island. The wind didn't blow until they left and it hasn't blown much since except for a few hours now and then when it has gone to thirty or more knots. No news from Jim since so don't know whether or not wind power is feasible for my set-up. Houses are so close together it would be hard to put up a tower without endangering a neighbor, unless that neighbor wants to share. I may get a bank of batteries to use as auxiliary power late at night and during the day for short periods when the generator wouldn't normally be running. It seems awful to generate 3.5 KW and use only a 60 watt bulb for a reading light on a bedside table.

This has been a strange summer. We have had almost constant fog and haze. Only occasionally have we seen the mainland, yet we have had no rain. We got a 1/2 inch in June and only a trace since. I'm watering with dishwater, water from small washings of clothes, etc., and so far the garden has done well. My rain barrels, filled in the storm two or three weeks ago, are about exhausted. Yesterday I took my first bags of lettuce and Swiss chard to the Store; tomorrow I hope to take more. The strawberries have been luscious and now raspberries are coming in, but they are puny because of the drought.

Monday my tenant helped me put in three cedars (red) which I hope will form a screen between me and the Pitkin house where the spruce have died. Also I hope I have at least two sexes so there will be berries for the birds. The magnolia was almost completely winter killed, had only two blossoms and now has a fringe of green coming from the bottom. Last year it was about fifteen feet high and would have afforded some shade to the house this year.

Today is a hot one. It was 70 degrees at six when Phoebe got me up and now at boat time is nearly 85 degrees with not a breath of air except down under the big elms by Ida Proper's. Will it be another August 2nd with temperatures over 100 degrees?

Next Friday, I'm going in for a week or so while this apartment is rented to a couple of gals who begged me to let them have it last year. I'll stop with Elizabeth Gardner in Concord, then go to Falmouth to visit with Nancy and get in some swimming, I hope, then back to New Hampshire to see Emma and return here on Saturday or Sunday. My tenants say they will be glad to keep Phoebe, who hasn't calmed down enough yet to inflict on my hostesses. I hope they won't regret a rash decision.

Wednesday the Laura B broke her drive shaft as she backed into the dock and had to be towed by the Phalarope back to Port Clyde. People waiting to come out were brought back by the Phalarope and Hap Wilson, so no one was stranded.

Sherm had to take Gramp Murphy in yesterday. I guess attending the festivities on the Fourth was too much for him. He apparently had a stroke and is now at Pen Bay Medical Center. He was 99 in March. Freda and Alfred went in with him and are on the mainland in all this heat. Betty Merefelder is coming along well after surgery and will know next week whether or not she will need some chemotherapy, poor girl.

Julie and Jackie Hudson and Norm and Bette Richardson have gone to Boston on special invitation to attend the reception for the queen - J&J because they belong to the English Speaking Union, I believe, at the Richardsons because Norm wired the tower of the Old North Church. He got a replica of the lantern for a souvenir.

Better close rather than start a new page. The Rockwell Kent show at the Museum seems to be attracting a lot of interested people.

Love,
Alta

Monhegan, Maine 04852
Sept. 30, 1976

Dear Friends et. al:

It has been a long time since my last letter, chiefly because I have been waiting for a new duplicating machine. On my last trip in I found one but it was in Bangor and had to be shipped out sight unseen, plus not being tested by me. Well, yesterday I finally had time to try it out and failed miserably even with Anne Hubert's help. Then Ed Hubert came by today to make a phone call and he, with the help of the Rockland Business Machines over the phone, got the thing going. So here I am at the typewriter once more. I took this machine in for an overhaul and to see if it would stop jumping spaces but, as you can see, that fault wasn't corrected. It wouldn't jump when they tested it - like a child at the dentist!

In my last letter I was working on my costumes and speech for the bicentennial. The skirt was finished on Wednesday and I went to work in earnest on my speech whose last page was taken out of the typewriter at 9:30 a.m. on July 4th, just as the parade was forming down at Lobster Cove. Many pictures were taken of the parade which included about every child on the island, including summer visitors, plus five floats, two of which were about the Educational tax which is causing so much difficulty in funding our school. Sandy Dickson and Louisa Winchenbach had identical colonial dresses and caps and rode in the pony cart behind Sam. It was the prettiest float. The Boynton's float was a clever parody of the American Dream which I felt was not in the best of taste for the occasion. Many others felt so, too.

We all gathered at the schoolhouse where Lee Court's cannon was set off to start the ceremonies. Newt Searls, dressed in the costume of the 1914 Monhegan Band, raised the flag - Chris Nicholson was to do it but was mistaken about the time. The Northern Valley Boys led us in some patriotic songs and then I gave my speech, a mixture of history and a plea for continued self determination which has been characteristic of Monhegan over the years. It was long and I was amazed that people stayed to the end. So many asked for copies I made inquiries and found that it would cost nearly $1.50 for a printed copy, not allowing for any profit to the vendor. Maybe I'll try to put out some copies on my new machine, if it proves itself worthy. The day was a great one and its success was due chiefly to Ricky Boehmer's leadership, organization and enthusiasm. She was the chairman of our Bicentennial committee. Bill Brittain gave a marvelous sermon that evening in the church. Celebrations such as the bicentennial are so unifying we should have them more often. Even the weather cooperated after an early pea souper which threatened to dampen the parade (this machine is jumping worse than ever) but which blew away and burned off just in time.

In mid-July I rented my place to a couple of gals and took off for a vacation in shore. I stopped at Elizabeth's in Concord, then went on to Nancy's on Cape Cod and finished up with a short stay with Emma in New Hampshire, having sponged on relatives all the way. It was fun and fortunately not too hot. I should rent my place more often. Then, maybe, I could keep it picked up!

The summer bumped along about as usual for the first half. We had fog and drizzle but no real rain all July. The town water supply held up but the laundramat which depended largely on cistern water had to be put on short hours, subject to the production of the two wells, by August. So far, so good. The first page came out well after the first few which didn't print on the sides, corrected by putting more fluid in the bottle.

Early in the summer I made the remark that in all the faces that came off the boats there was never a familiar one. Well, I had to eat my words in August. Not a week went by that at least some friends didn't come up the slip and later up the hill for refreshment, or at least a talk. Among those coming were Anna Newbert from Augusta, Pete and Robby Pawle known through the Family Physicians group, Helen Mann who was one of my last patients in Wellesley; Doris and Charlie DeCarlo who used to rent the cottage and Helen Gunn Webster and her husband neither of whom I had seen since Cincinnati days back in the early thirties. Two different couples on the island came from Center Tuftonboro, N.H. where Emma lives and we all got together a couple of weeks ago over there when I went to visit Emma again.

My golf cart, Apollo, now nicely housed in the new tool shed has been there all summer because of two accidents. When I returned from my July trip I met a boy on a bicycle at the corner in front of the Spa, tramped on the brake, which was the accelerator! (the brake is applied with the **left** foot). The boy fell off his bike and the bike knocked out the windshield, otherwise there was little damage, but it unnerved me and I didn't try to drive the cart until weeks later when I found the battery was dead. We took it up to the generator shed to recharge it from the generator. It short-circuited and the whole place nearly burned down had Harry T. Bone and Alfred Stanley not seen smoke coming out of the cupola. I lost a pile of burlap bags (thus the smoke) two wooden packing boxes and a two foot square piece of shed floor. Another few minutes and the generators and nine cyclinders of LP gas would have gone up. Thank goodness for neighbors.

The Nelsons weren't so lucky. Their generator caught fire in September and had to be all rewired but again, no serious damage.

A couple of weeks ago Norm Brackett's seine boat capsized while four of them were pulling in a purse seine loaded with fish. Don Cundy and Steve Rollins were fishing nearby and went to the rescue. The accident occurred about 11 p.m. Fortunately all four men got out of the water and onto the Candy B II without being tangled in the seine. Although the seine boat dislodged a mooring when she was towed upside down to Middle Beach the seine was not badly damaged and the men were out fishing again in a day or two. A crew got the boat righted and off the beach at high water the next morning.

Then last week a beautiful red-hulled sloop came into the harbor from Portland with two men aboard. They were given a mooring which turned out to be one for a lobster car and not a boat, unfortunately. Because, during the night we had a gale which blew the harbor. The boat dragged the mooring and she fetched up on the rocks at the south side of the dock. All of us were pretty surprised in the morning to see the thing high and dry as if placed there by a huge hand. The men were able to jump onto the rocks when she hit and again no one was lost. High water was just before noon so the Phalarope and Laura B were able to haul the sloop off but not without one broken line which snapped and wound itself around the sloop's mast, missing all the people crowded at the edge of the dock as well as the people on all three boats. Just as the sloop was afloat the Coast Guard came in from Boothbay Harbor, having been called the night before and again in the morning, and took her in tow to Boothbay Harbor where the men's wives were waiting. The two men were long in praise of the Monhegan men who did such a good job particularly Craig Sproul and Ron Stanley as well as the crews of the Phalarope and Laura B, but had little to commend the Coast Guard for.

The biggest event of the year was the Rothstein-Dolan wedding. Robert and Joan tied the knot, with the help of 300 or so invited guests, down on the rocks in front of Mostels' with Bobbie Ives officiating on Sept. 11th. Wrenna tried to see that all island residents, summer and winter, were invited. Temporary visitors were quite surprised to see long skirts, creased trousers, tailored shirts, neckties and polished shoes as people swarmed down the road to the festivities. It was a beautiful day with a brisk SW wind which sent great rollers crashing on nearby rocks. The ceremony was impressive, especially when Dan Rothstein asked the Seder blessing and distributed the bread. Joan's father was here to give her away and Angela acted as flower girl with her collar decorated with wild flowers. Since this was a Monhegan event it was not entirely conventional and Jeff Dolan caught the bouquet! Robert and Joan are now living in the Rothstein cottage which they are winterizing.

That just about brings us up to date. It is now Sunday, Oct. 3rd, and a beautiful day. As yet my garden has suffered no frost and I hope it won't until the last plantings of peas and beans are harvested. I've eaten from it since July first when I harvested my first peas. This is the first time I've had a really good yield, thanks to Sam's stable dressing put on last fall. My raspberry bushes are still blossoming as are my magnolia and forsythia! Wild strawberries are blooming here and there, too.

Just about all seasonal houses are closed and this is the first day we have had no boat. But yesterday's Laura B was crowded with people. Four dogs came on and only one left. The school teachers went to a convention and the Boodys are ashore so this part of the island is unusually quiet this weekend. The teachers, Dorcas and Jack Butler, and their two children are staying this winter in the Mooring. I shall miss the many people who have come out during the winter, but am glad not to have the trouble of changing beds etc. My guest room remained idle all summer, but I'm hoping one or two of you will use it another year.

No more progress about the windmill. I'm still waiting to hear from Jim Jackson. With the price of LP gas expecting to escalate almost any investment in other types of energy should pay off in short order.

Lee and Jack Dana from Gardiner were here in August and Marion Chapman whom I knew through KVAA in Augusta came while I was away at Emma's last week.

Just to put matters straight. My cottage is the Mooring and is attached to this part of the house, the Pillbox or apartment. I've given up any place in shore. Right now I expect to stay here much of the winter doing more work on my writing, which never gets done in the summer. Maybe I'll even clean the apartment one of these days!

More than thirty of you are interested in receiving these letters which probably won't all be addressed for three or four more days. I was going to ask for a subscription but an increase in my pension and S.S. checks make me feel flush at the moment. Keep your letters coming and goodbye for now.

Alta

P.S. There are several items I forgot to put in the letter. One in particular concerns Jim Thompson who, I believe, came to Monhegan every year for at least overnight or longer since 1907. He and I joked about our celebrating our 70th anniversary of our arrival on this hunk of rock when he left the island earlier in the season. It was quite a shock to hear from Mary Baker that Jim had died in his sleep on September 12th.

Knowing Jim's interest in the Monhegan Museum Mary has set up a Jim Thompson memorial fund to benefit the museum. If there is sufficient money and it is feasible to do so the Manana bell might be set up on its stanchion - just as long as some prankster can't set it ringing at any time of the day or night. If anyone is interested they could send their checks to me, made out to the Monhegan Museum and designated for the memorial fund. Seventy-five dollars have already been donated.

My earliest recollection of Jim is our sitting on the floor of the old Monhegan House putting on our shoes and "stonkins" as his twin sister Helen used to call them. It will seem strange not to see him come off the Laura B at odd moments all during the time the Inn is open.

Some gourds I bought in Wolfeboro over four years ago went to seed in my compost pile and came up as the most vigorous plants in the garden. They put out fruit every three inches along their stems. What to do with them. The children are going to be offered them instead of candy at Halloween, with the enthusiastic consent of the parents involved.

I forgot to mention that Win and Louise Robbins from Hayford Heights dropped in one day as "trippers" on the Laura B. It was good to see them. One of these days I hope to get to their camp in Rangeley.

The fringed gentians are more beautiful and numerous than I've ever seen them. I'm mixing up the ecology of the island by taking seed pods and scattering them over my back hill and along the trails to White Head. I found several witch hazel trees on the Red Ribbon trail. They are now in bloom and have a lot of seed pods ripening from last year. Now to find a sassafras tree.

Doug and Harry Odom and possibly Jamie Wyeth will be up for supper, a big fish chowder, tossed salad, all from the garden, and an apple crisp taken from the new Monhegan cookbook "Finest Kind" the profits of the sale of which are supposed to go to the Breakwater Fund. We have cleared expenses and now may be able to buy a few pieces of granite. It is a much needed thing to stop hurricane seas from tearing up the harbor and everything in and around it. We were lucky in not being hit by Hurricane Belle but one never knows when one will really hit us such as the one we had last February.

I think I hear the Store truck so again, Goodbye.

Monhegan, Maine 04852
December 30, 1976

Dear Friends:

Today is the first day in ages I have had time to do the many things that have had to be done over a period of more than a month, that is since before Thanksgiving. The tree came down today and all but the swag at the front door has been laid away for another year. I consolidated the decorations into two large and three small boxes which are now waiting for someone to come along to help with getting them up into the attic. Once more I can draw the curtains across the big window in the living room to keep out the cold - in the nick of time, too, for sub-zero weather is predicted for tonight.

Phoebe and I went to Concord for Thanksgiving where I came down with a terrific cold and chest congestion and spent the entire week on the barcolounger. No mittens or socks were knit as had been planned. The first Saturday in December I picked myself up and drove to a medical meeting in Portland. That night I stayed with the Pawles in Falmouth Foreside. He is a family physician whom I have known through the association for a number of years. They came out to Monhegan last summer and suggested I stay with them on any trips to Portland, which makes it just great. They have a little suite on the top floor and loads of room for the car and for Phoebe to run in. That Sunday we attended the State Street church where Mr. Wilson, the new Portland City manager, an ordained minister, took the pulpit and also held a discussion meeting after the service on what each one can and

should do to help the city solve its problems. It was a very interesting hour. That afternoon there were two interesting musical programs to attend but I fell by the wayside again and just stayed by the fire all afternoon and evening. The next day Phoebe and I were off bright and early to catch the Laura B. It was cold but calm and the trip out wasn't too bad. It took me another week, though, to get on my feet, so my Christmas letter was both short and late in getting out.

The middle of the month a couple from Georgia, interested in the geology of the island, came to occupy the guest room down stairs. I thought they would be out all day with picnic lunches, but Nature dictated otherwise. It was so cold, snowy and WINDY the whole two weeks they were here it was difficult to stay out more than an hour or two at a time during the whole two weeks they were here. Also, the severe weather made it almost impossible to keep the downstairs living room warm. Consequently all three of us plus Phoebe and their Irish setter-golden retriever puppy (five feet long and three high) tripped over one another most of the time in my tiny living room. I even had to close off my bedroom because of the cold, and each night moved chairs etc. to open the hide-a-bed. The rough corners of my disposition poked out on occasions, but in general we all had a good time and I miss them a lot now and Phoebe is very lonesome without Maude to romp with.

The Santon creche was set up in the closed-in porch and Jack Butler, my Mooring tenant, fixed up a small light to go on the 12V system to light the stable. I finally cut the spruce tree in the garden and set it up in the bay window upstairs and got it trimmed a day or two before Christmas Eve. We had a capon for Christmas dinner and was it good! Island cranberry sauce and a pumpkin pie from my pumpkin were on the menu, too. I fixed stockings and hung them by the Franklin. It was fun! The day after (Sunday) we took advantage of the Butlers' being away and had an Open House, using the Mooring living room. Thirty people came despite sleet and snow. All hors d'oeuvres went but I'm still drinking up the punch.

It is now January 3rd and Trap Day is slated for tomorrow, if all goes well. Bryan Hitchcock and Robert Rothstein had trouble with their boats and had to take them to Boothbay on New Year's Day. Carl injured a rib before Christmas and wasn't eager to lift traps, Dougie Boynton was down with a sore throat so they didn't try to set on the first. That gave Doug time to go rocking today with Marty Moses to help. Harry didn't go because today was a boat day and he expected a fairly large order. Doug still has a lot of pain from the shingles he came down with the end of June and is pretty discouraged. But it is easing up a lot. Marty plans to stay and go with Doug until all his traps are set and he has things pretty well under control. With so many people here Harry has too much to do in the Store to go with Doug.

This is the first time I have known the men not to get their traps down on the dock the day after Christmas. In a "lobster meeting" yesterday they agreed to wait until today to do so. Also, they all but Don Cundy agreed to a 600 trap limit per boat. He held out for 800. Before dawn today trucks began carrying traps to the dock. By boat-time the dock was so full there was no room for the mail or Store truck and no lanes left open for anyone to handle the lines from the Laura B. Neither Carl nor the Odoms had taken any traps down themselves so you can imagine how happy they were. At another "lobster meeting" today the decision was to start loading the boats even though it was low water.

Now it is Tuesday. All night, it seemed, trucks kept going to the dock. Robert Rothstein and Steve Rollins worked well past midnight to take their traps down. A few boats were loaded. Lights were on in many houses before six. By seven-thirty the dock was a mass of people, trucks, dogs and children as the men struggled to load their boats. Steve's traps were piled on top of others and some of Doug's old traps had been crushed. Happy Day! Several people wondered where Monhegan's spirit of cooperation had gone.

By 8:16 everyone was ready and at the sound of a boat horn they started for the fishing grounds. Don Cundy, Carl Wincapaw and Rusty Court put over traps immediately. No one seems to know whose trap went under first. Just as they started off a few snow flakes fell lazily to the ground. It has continued to snow all day, just enough to make the roads and the slip slippery. I got my mug-up basket down to the dock just as the Monhegan came in for their first return load but they had to wait for three boats to be loaded, including Don's which carries 135 traps! Tonight I'm taking the soup I made from the Christmas capon and some corn muffins down to feed the three weary bodies. Harry went out with Doug while Marty drove the truck and handled traps on the dock. It is nearly dusk so operations should come to a close for the day in short order. It is legal to lobster only in daylight hours.

At a Planning Board meeting last night it was voted to deny the Hudsons a permit to build on the land they have owned since 1898, even though their building would not violate environmental protection specifications! What should one suspect from those who have gotten theirs recently and feel there should be no more people and homes on the island? If I sound bitter, I am.

Now that most of the traps are under water lawns are appearing once more and we can see along the road past the piles on the Curtis' lawn which made it so dangerous for coasters on the road. By the way, the young people have formed a cooperative with headquarters in the Curtis cottage. They are buying food by the hundred weight or so. Doug and Harry may close down the Store in the winter. But who cares? As long as we have what we want why bother about the Store? Well, Happy New Year!

Alta

SEASON'S GREETINGS
1976
from Monhegan

Here it is the second week in December and I am just sitting down to send you my annual greetings. The delay cannot be laid on the weather. We've had winter off and on since mid-October and today we are in a deep freeze. It must be laid to the completely disorganized state I've been in for the past year and to creeping senility - Heaven forbid!

Although the sun is shining brightly the temperature here in my living room hasn't reached sixty, even with a fire in the Franklin, one Aladdin heater burning downstairs and the big stove turned up high. The wind is blowing pretty steadily out of the NW at 30+ knots and hitting close to 50 every few minutes. The sea is covered with vapor and white caps, and the gutters rattle and groan periodically. Night before last it was 50 (today 9) and the inside temperature reached 85 with all fires out or turned low! Such is our New England weather. How boring to be unchanging. This is what keeps roses in our cheeks.

This Bicentennial year was a busy one. First of all I spent several weeks preparing six articles for the Bicentennial edition of the Rockland Courier-Gazette, of which they chose five to publish. The sixth required too many photographs. I'm still mulling over where to send it for publication. I thought of the many pages and photographs which were accepted and almost calculated my returns as a "stringer" when I found this was considered a bicentennial contribution. Thank goodness I didn't spend the prospective funds. Sid Bransom asked for a regular column for The Pine Tree F.P., the organ for our Maine chapter of the Am. Assoc. Fam. Physicians. So far I've gotten off one, but have promised to do better. Writing a weekly column for three newspapers continues.

In March I took off for warmer climes, starting with Fort Lauderdale where I stayed with Woody and Betty Weinhagen. We visited Jean and Ben Johnstone (all cousins) in Punta Gorda on the west coast, stopping in Naples on the way. In this way I got some idea of what Florida is like in the winter. I also had a short visit with Lucile Cheney with whom I lived when first arriving in Augusta. She has a house in Margate. It was all lovely but too hot for me. From there I flew on to Mexico City and then to Cuernavaca where Emma and Hollis Riddle were staying. Unfortunately Hollis died a few days before my arrival so further plans for Mexico were cancelled and I went back to Akron with Emma until after the memorial service held for Hollis there. Mexico, especially the Anthropologic Institute in Mexico City, was interesting. But the weather and the altitude were not for me. Maybe the next place to try should be Puget Sound or the islands off the Georgia Coast.

We had a big celebration on the Fourth of July. The Northern Valley Boys were here for a concert and square dance the day before. On the Fourth we had a four float parade, flag raising at the schoolhouse with the firing off of Lee Court's toy cannon! Bill Brittain, our visiting minister, gave the invocation and I gave the welcoming address. So many people wanted copies of it I mimeographed 100 copies, after much labor and waste paper, to be sold to help the Breakwater Fund. A fife and drum corps and the N.V. boys met the incoming boats at noon and at night. Bill Brittain led us in a Patriotic Service at the church. It was hard to come down to earth the following day. Another Bicentennial effort was the cookbook "Finest Kind" which the entire winter population has a hand in. Proceeds will go toward the Breakwater Fund, too.

Speaking of the Breakwater Fund - 1976 has been a year of storms.

The first one, on Groundhog Day, was the worst in anyone's memory and did a lot of damage to the dock and houses down on the harbor as well as to lobster cars and fishing gear in general. The need for a breakwater was most apparent during that storm when huge rollers coursed through the harbor and right over the dock and the Burton's house. Boats dragged their moorings and four lobster cars broke loose. Several storms since have washed out the dock repeatedly in places. On trap day it was so rough the men had to come to the dock with their skiffs rather than row to Fish Beach. The little mole which was built five years ago is almost completely flattened now, giving almost no protection. Summer and fall storms gave visitors a chance to photograph spectacular surf at Gull Rock, Lobster Cove and even in the "Harbor."

Laura B's health failed early in the year and she finally had to be put to sleep on my return from Akron. After one dogless week I could stand it no longer, called the shelter, and had Phoebe Starling sent out under the care of Mary Burton. The original Phoebe was a month old baby who was the first person to be buried in the Monhegan cemetery in 1784. It seemed only fitting to perpetuate her name in this bicentennial year of memories. She is a pure American GOK Dawg of unknown age, tan with dark brown accents and a propensity for jumping and lapping faces. She is convinced she is a lap dog in other ways, despite her 27#, a great companion. She loves to go walking and chases anything that moves. She travels well, too, which is a help, and barks very little, but almost talks.

My garden was a huge success this year. I'm still eating from it. The very cold weather while I was in Concord for Thanksgiving, froze in my carrots and killed lettuce in a cold frame. I'll have to wait for the January thaw to salvage the former. Pumpkins furnished jack-o'-lanterns and pumpkin pies for Thanksgiving and another for Christmas to come. There are rutabagas in the cellar, potatoes in the frige and frozen beans still. The garden is now covered with about a foot of rock weed and pony dressing in hopes for a good crop next year.

Many people came to the island during the year, both old and new friends. The Mooring (my cottage) was a busy place all winter and now houses our teacher and her family. Among the many people visiting in the cottage or the Pillbox were the Pawloes from Falmouth Foreside, the Peaslees from Augusta (Ed held dental clinics here in the twenties!), the Websters of Cincinatti (Helen was in our crowd there in the thirties!) Ashley Day who knew me through Elizabeth and also Marion

Watson, Joan Mondale - Fritz was too busy to come, I wonder why! - the Huxtables and the Drownes, both of Center Tuftonboro who didn't meet each other until they both came to a cocktail party at Emma's in September. Ted Drowne is descended from Shem Drowne who owned all of Monhegan in 1749-58. He gave me a pewter seahorse he fashioned. Shem was the artisan who made the grasshopper weather vane for Fanueil Hall. Like g-g-g-g-grandfather, like son.

Jim Jackson and his friend Arthur Briggs came out in June to study the place for wind power. The result of that visit is that now the place is wired for DC current furnished by batteries charged up by the generator during the evening. We now have lights as needed during the day and after eleven. I hope to be able to get along with only two hours of generator power. The Mooring has an insulated roof and two skylights so the loft should be more comfortable.

All the Jacksons, Sutherlands, Phoebe and I gathered in Concord with Elizabeth for Thanksgiving. David was in Florida and Bruce in St. Thomas. No more paper! Best wishes to all.

Alta Ashley & Phoebe

Monhegan, Maine 04852
February 27, 1977

Dear Friends and Family:

Two months have gone since I have written, at least since I started my last letter. So little has happened I've had a hard time to write my column, let alone letters.

January and most of February have been cold and windy, but the past week or ten days it has been moderately warm. In fact, I've been able to turn off the stove downstairs and keep comfortably warm with just the Franklin. Even the open fire has been too much at times.

Our main problem this winter, other than persistent high winds, was that each storm ended in heavy rain followed by a drop in temperature. A month ago it was 41 degrees at 10 p.m. and 14 degrees by morning. The roads were awash with two inches of rain which turned into a solid sheet of ice. Even now this hill and the one by the Yew are covered with ice after several days of thawing. Each night they freeze up again. Bob Boody has put so much sand on the roads he says he won't have to put any on in the spring! But the ice has been so cold and the winds so strong it hasn't done too much good.

We really fared better than those in the mainland and especially those in western New York state and Ohio. For a while the roads were so bad the oil truck couldn't move at all and some of us had to get our supplies in five gallon lots daily and pour it into our tanks. Then the Coast Guard inspectors wouldn't allow the oil boat to come out. Some calls to the powers that be did let it come out as an emergency. Then, on the next trip Doug was down to only a few gallons in the truck and the boat didn't appear. Fortunately she did get here before anyone was completely without fuel.

Yesterday Doug and Harry brought the rig up this hill all right, but when they went down to Ice Pond road to Sandy Dickson's they skidded and slammed into a tree bashing in the right hand door. Somehow Harry didn't get hurt although it was his side of the truck.

Lobstering is slower than ever, but the price is higher than ever, too, which helps a little. I guess the water is too cold for the lobsters to go seeking food.

Last year I bought a CB but it has never been of much use because I have been able to hear only two or three people and only they could hear me. I bought a walkie-talkie for Rita for Christmas and we couldn't hear each other! Well, today Courtney came up with a broom handle wound with a copper coil for an antenna and now I can hear everyone, including the boats. Robert Rothstein has just had to go out by Gull Rock buoy to tow in Bryan Hitchcock, whose generator broke down. When they get in Danny Bates is going to have to dive to see if both Robert and Rickie Cundy have warp in their wheels.

Kathy Wincapaw is having a birthday party this afternoon with all the kids, except Lance Burton, in attendance. She has a new puppy, a cocker spaniel named Charlie and is one delighted little girl.

Friday the ice storm of the night before fouled things up so Earl first thought he would have to cancel the trip. Doug, Teco, Martha and Richard, Bobbie and Ruth Ives, and the pilot Billie Abbott were all waiting to come over. The Butlers (including Pewee), Bob, Mary and Lance Burton; Billie and Cathy Payne and their guest Jean Hayes, Rickie and Peter Boehmer (going in to pick up their generator) were all waiting to go in. Out here things looked quite good at boat time. Finally Earl decided to start out. He had to call people from various places - the Ives from Wiscasset, Teco from Rockland, etc. By eleven everyone was aboard except Martha and Richard who didn't get the message and Teco, who everyone thought was aboard. Between eleven fifteen and eleven thirty the barometer took a nose dive and the wind shifted from south to west at 45-50 knots just as the Laura B came out through the islands. Rather than turning back Earl came all the way. Even Doug admitted it was rough and poor Ruth Ives was sick four times. Danny Bates' sister and husband were aboard with their two dogs, both of which were sick and whined all the way. By the time the Laura B rolled into the harbor all those going ashore chickened out. Jimmie Barstow was aboard, as he has been in every stormy day. Going back they went through the harbor and out to the eastward of the island, to stay in the lee as long as possible. I suppose they also went outside of Burnt Island, too. When the boat reached Port Clyde Jimmie called up to say they had landed safely and the trip was worth the fare!

By the way, you may know now that John Rothstein and Jimmie Barstow have bought the Laura B, the pier, house, and parking lot. Jimmie has also bought the other hotel in Port Clyde, so will have that area, too, in which to park cars. I believe the fare is going up a dollar round trip. Earl will probably stay on until Jimmie completes his required hours at the wheel. Ivan Simmons may help, too. If they can, they will make better passenger accommodations with plexiglass enclosures etc.

Wrenna Rothstein had another massive pulmonary embolus and had to be taken in on a terribly cold and windy day in January. She has been home now for three weeks, but is unable to see anyone now. Joan Bernstein fell and broke her hip this winter, then found her disease had spread to the neck. She is now in a nursing home, I believe in the new one at the Camden Hospital. Mort has light house keeping arrangements at the Trade Winds in Rockland. All very sad.

JoAnn and Bob Mersfelder are expecting in June. Orca, Dwight and Jason are all growing like weeds, as is Kole. Kris Boehmer is on the Kent's Hill School wrestling team and doing well. **He** is the one that is growing like a weed. And is **he handsome**!

Yesterday I looked in my cold frame and found several lettuce plants with four or more leaves. You just can't kill lettuce seeds, I guess.

Speaking of boats, Craig has a new one, Doug is having one built, Don got his new one just in time for Trap Day. I can't tell whose boat is which, for someone seems to buy the discarded boats.

Harry fell and broke or cracked a rib the other day and Doug is still having a lot of trouble with his post-shingles neuralgia. The apartment just shines since the arrival of Georgia Anarker who is cooking for them. Doug is slimming down and Harry looks far less harried (!) now he is relieved of the housekeeping.

Bob, Mary and Lance went skiing this week. I forgot to say that Alison Wilson came out yesterday, bringing Martha and Richard and taking back all those who wanted to go ashore. The Boehmers cancelled out because they couldn't get their generator on a Saturday and Teco apparently went back to Portland after being left on the dock the day before.

My nephew Tommy Ashley and his friend Sally Cox were here for the long weekend and had a great time. They went all over the island despite the snow and ice. Needless to say I didn't show them the way but Phoebe did. She was in seventh heaven and came back from each walk one exhausted dog. She is changing her coat. Daily there seem to be more black hairs. She has developed a ruff and some stripes, as well as long hair on her legs and belly. Maybe I'll have a collie before long. The vet has finally decided that she is only two years old, not four.

The Butlers, who have been in the cottage all winter and quite comfortable even in the worst weather, are going to be living in the Orne house over on Horn's Hill. They are working like beavers fixing it up, insulating, tearing out and putting up partitions. The Mooring will be available once more for rental, short-term, during the winter. The Butlers plan to leave when town water goes on in mid-April. Jack wired both of my houses for twelve volts. Now I have the generator on only from 8-11 p.m., charge up my batteries for an hour or two. The reason I use the generator then is because I want to use my color TV which won't run on 12 volts and I haven't a big enough converter to do the job. It is great to be able to switch on a light in the night without starting up the generator. Only the pump, toaster and vacuum require the generator - oh yes, and the washing machines. We try to vacuum and wash after the generator goes on, however.

The school is still running. The state changed its method of collecting the school tax so that we keep our share rather than having to wait for the state to send it back. We will end with a shortage of about $500 as things stand now. Town meeting is the 14th. Heaven knows what the taxes will be.

Bob Burton is going in after returning from skiing for his operation on the cervical disc which has been causing him so much trouble. I think the operation is going to be done in Pen Bay, but I'm not sure. I'll find out tonight when I go to the Store for dinner, and let you know definitely.

The Ives have been coming out once a month for church services and Sunday School. They have asked me to sing at each service. Tonight I'm going to sing the tear jerker "There is a Green Hill Far Away" of Gouod. Also, in Ruth Scully's absence, I'll be playing Mendelssonhn's "Consolation" for a prelude and Bach's Prelude in C as a postlude. I don't know how to soften the organ without either having it cipher or omit notes so everything will be double forte!

The Haussers have bought the Twining property at Lobster Cove, so will no longer be tenants of mine in August - just as I fixed the loft with insulation and skylights!

Bob is to go for his operation on Tuesday March 1st. The address is Penobscot Bay Medical Center, Glen Cove, Maine 04841. Apparently I am to put up the new School Superintendent tomorrow so heaven knows when I'll get these letters addressed. Sandy Dickson also came back yesterday after attending a hearing on school taxes. A new plan, if passed will give us a break.

As ever,
Alta & Phoebe

Monhegan, Maine
March 20, 1977

Dear Friends and Monhegan Taxpayers:

Happy first day of spring! Here we are with a white world and more whiteness predicted, though we may miss this next storm. Friday's storm dumped about six inches of the white stuff on us, then whipped it in to huge drifts which were too much for the plow. The roads, which had nearly cleared of ice, are a mess again. At least this mess won't last long.

Last Monday was Town Meeting night. There was so much apathy that only 27 voters showed up. Sherm and Barbara, Craig and Judy had taken off for their spring vacations, Joan Rothstein stayed in to arrange for the shipment of the cooperative. (How can they outdo the Monhegan Store when one person has to go in for two days to repackage the stuff in small lots, then bring it out for 50 cents a carton?). Several others on the island just didn't bother to show up and several others left early.

The officers stayed the same except that Bob Boody gave up Fire Chief and Ron Stanley took over; Craig Sproul gave up Harbor Master and Shermie was elected. Also Robert Rothstein and Craig resigned from the planning board. I ran for it but lost out by one vote to Sandy Mason Dickson. Victor Lord was also elected.

As far as the articles went all were carried as written except that $2,000 for the wharf had been omitted, but was carried, the Air-Pack measure was defeated as was the appropriation for traffic signs. The appropriation for re-valuation was upped to $2,000 in order to employ two persons to do the job. In all $16,223 were appropriated over and above the education and county taxes. That for education should be less than that for last year, so our taxes should stay about the same until after re-valuation, which will help some of us and up the taxes of those who have been paying on unrealistically low assessments.

There was much discussion about the gasoline pump, smoking around it and leakage. The cooperative has found it can't make ends meet. They had gladly taken over from Doug thinking he had made a big profit from it, which was far from true. He was glad to get rid of a losing proposition. The upshot was that signs will be painted by someone - there are no officers or directors of the cooperative - and people will be urged not to smoke around the gas box.

Trash cans came in for a lot of discussion. The "trippers" were still accused of being the chief litterers, although Bryan stated that things were thrown on the dock after he cleaned it up each night by six o'clock. Trippers?? There will be trash cans for use on the dock, by the Store and Periwinkle and possibly the Spa. No mention was made of the museum.

The Butler's are remodelling the Orne house and will move out of my place when the water goes on. I'm afraid it will take me until I leave for Nashville to get the place back in order again. But I would like to rent it for the month of June, if possible. Any prospects? I'm hoping to go to my 40th and 50th reunions in Nashville and South Hadley by Amtrack in May but have no information as yet. Must close, no more paper.

Joan Bernstein died Thursday and Mort has had to go to NYC because of another detached retina. No service planned as yet.

As ever,
Alta

Alta Ashley, M.D.
Monhegan, Maine 04852

October 21, 1977

Dear Helen and Robert:

I have just ordered gold medals embossed with emeralds for you two and your efforts toward the welfare of Franklin and Aunt Helen. While most of us have sat wringing our hands you two have **done** things. In the haste of a long distance call the other day I failed to say that Emma thought we all might try to get together and devise a plan of action. What we may have to do is move them bodily to a place, if we can find a place where they can have supervision. If only we could find a place something like the Methodist Home with a bit more supervision and diversional activity. Unfortunately this type of facility has a long waiting period.

This fall has galloped past without much accomplishment. I have had a family in the cottage on a working vacation for the past two weeks. Today is the first really clear day they have had and they are taking the noon boat! On the really rainy days Mer finished insulating the cottage and putting up wall board. He also made under eave storage areas, so the loft begins to look more like a room than an attic.

Tuesday I am off to Vinalhaven to spell Dr. O'Keefe for five days. I thought that time would be good, with the garden already put to sleep. But we haven't had a frost and the garden is loaded with green tomatoes and flourishing turnips, even a few string beans.

Emma and I will get in touch with you to find out when would be a good time for us to get together. As for the mattress, what about taking one from Aunt Helen's apartment.

Must get this in the mail. Thank you so much for all you have done.

As ever,
Alta

THE MOORING * MONHEGAN ISLAND
Alta Ashley

The Mooring is a story and a half cottage on the road leading to the Lighthouse and overlooks the Meadow. It is about a quarter of a mile from the dock; it overlooks the waters to the north and south of Manana and the Harbor.

On the first floor is a living room, dining alcove, kitchen, bath (with shower but no tub), double bedroom, enclosed sun porch with single hide-a-bed, and open porch - very private. The second floor consists of a loft with a double bed and a single bed, also a folding cot. All bedding is furnished. The pantry is stocked with staples. All that is asked is that you replenish whatever stocks you exhaust.

There is a gas refrigerator with small freezer space, gas stove and gas hot water heater. There are a few gas lights for emergency.

Water is furnished from a deep well. The quality of the water is excellent but the well reconstitutes itself at only 1/2 gallon per minute. Water is a scarce commodity in the summer, so care is requested in its use. Also, the pump is run by electricity so when the generator is not running it is requested that you limit water use as much as possible, especially after 11 p.m. and before 7 a.m. When water needs to be pumped into the pressure tank the pump comes on automatically. It is noisy and is said to disturb guests in the Golden Gull Inn next door.

Lighting. This is furnished by two systems.

1. DC 12 volt system from storage batteries. Total capacity 500 AH. This powers the Zenith radio and B&W TV and certain lights. Do not move appliances or lamps to other outlets unless you know what system is involved. DC outlets are silver. AC 110 are ivory or brown.

2. AC 110 volt system. Any light except the fluorescent kitchen light will start the generator. I prefer not to start the generator before 8 p.m. in summer, 6 p.m. in winter, so think before you turn on a light. Cost of generator time is at least $1/hr at 1977 prices for fuel.

Firewood is available. This, also, is expensive ($100/cord) so it will be greatly appreciated if you can find drift wood or hard wood in the woods.

Waste disposal. Rubbish and garbage is picked up three times a week. Ask for the day and time, it varies yearly. All plastic must be bagged separately. Non-meat non-fat garbage may be placed in a compost pile. Combustibles may be burned in the fireplace or incinerator after a rain.

Flowers and vegetables are available to you from the gardens in season.

When leaving please strip all beds used, gather up towels etc., including kitchen towels, and soiled table cloths. I would like to be able to get into the cottage by 10 a.m. of the day of departure, or by 3 p.m. if the following renters are not expected until the following day. It will be greatly appreciated if you can leave the place as clean as you found it. Ask about vacuum cleaners.

Charge for the cottage is $ /per week. Special rates for short term off season. Winter rates include heating which is furnished by a kerosene space heater, so are not less than summer rates. A 25% deposit is required with a minimum of $50. This will ordinarily be forfeited if reservations are cancelled within 4 weeks.

If you have any questions you may reach me at 207-372-9610 or Alta Ashley, M.D., Box 87, Monhegan, Maine 04852.

Although there are several beds it is requested that a maximum of 5 persons reside in the cottage. Exception can be made for short-term stays. The reason for the restriction is the limited water supply and sewage absorption field.

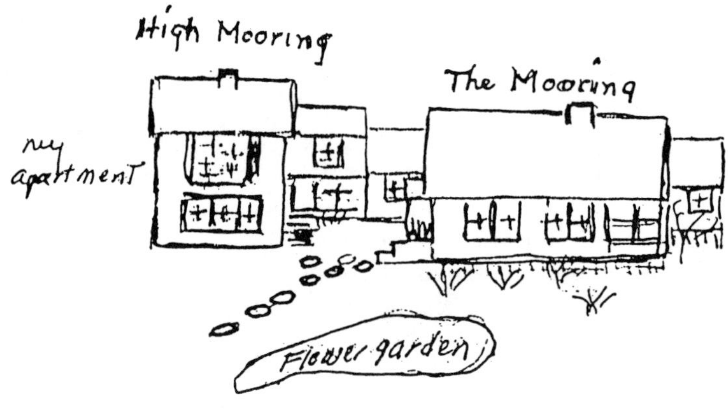

SEASONS GREETINGS
1977
from Monhegan

This is the first general letter to come off my typewriter since last February so all you who have been looking for the "Dear Friends" letters can rest assured you haven't been dropped from the list because no letters have showed up since then. Where the time has gone and how it has been spent, I can't figure out, even after going through a whole year of my newspaper columns to get information for this letter. Phoebe demands a lot of time. She begs to go on walks and when we do go I catch sight of her tail occasionally about fifty yards down the trail, but she won't go off on her own except to supervise things at the dock in the early morning when the garbage boat is in operation. Then, we have to have a "quiet time" once or twice a day during which she sits on her haunches beside me in the rocking chair and dozes blissfully as long as I keep scratching her neck.

All spring I labored on my Bicentennial speech, putting it into booklet form. The task was greater than my ability with typewriter and ditto machine but I did get about thirty copies suitable for sale, the proceeds from which were added to the Breakwater Fund, which now stands at about $2,500 through sale of the cookbook "Finest Kind" and donations from interested people. Maybe we'll have that breakwater before the Burtons' cottage "The Breakers" is washed away. This fall I've been working on albums of my news columns, in duplicate, one copy for me and one for the Library. What a job! Back in spring I did '73 and half of '75 with library paste and scotch tape, only to find that is the wrong way to preserve records so all those pages have to be soaked or treated to remove the paste and tape. That will have to wait until spring! Meanwhile '71, '72, '74 - '77 are done, and four fat albums are sitting on the bedroom floor ready for filing. This time the material has been mounted with a minimal amount of tape and slipped into full page plastic pockets.

On my last trip in shore I picked up an electric typewriter and have had a great time with it. Here's hoping it will cut good stencils. It makes beautiful carbons. One problem is that the power return doesn't work and I have to return the carriage by hand and turn the platen, thus the poor spacing on occasion; the right hand margin hold doesn't work and the shift doesn't bring the letters up quite high enough. Otherwise it is a joy to use and I find I can use it with my inverter so don't have to turn on the generator each time I want to type.

That brings me to the major change in the house. Last year the teacher's family lived in the cottage for a few months during which time Jack, her husband, wired my house for a 12 volt system, using two 220 ampere hour marine batteries and a recharger from Sears used by garages. Already I have used 12 fewer cylinders of gas for the generator this year since the system was made operable in February, than I used other years. At about $20/cylinder this is quite a saving and the system will pay for itself in a couple of years or so. Now I have a big inverter which when put into the line will run the water pump as well and should account for more a saving on the gas. Then, when I get my windmill and solar cells (still in the dream area) I won't need much gas at all. Then we'll have to start thinking of how to conserve on kerosene.

This was a particularly nostalgic year. Many familiar faces were found among the "trippers" coming off the Laura B and Balmy Days, as well as other boats coming to the harbor during the summer, among them Myra Luce from Camp Low days back in the twenties, Virginia Hill, Ed and Evelyn Peaslee from Augusta, Audrey and Bill Kennison of Augusta Players days and now living in Thomaston, and Charlie and Doris Dickason DeCarlo who used to summer in the cottage. But the most nostalgic of all was the 5,000 mile trip Phoebe and I took in May which carried us to almost all the states east of the Mississippi and to the homes of many old friends as well as relatives.

This was the year of reunions - fiftieth for Mount Holyoke, forty-fifth from Cincinnati and fortieth from Vanderbilt. At first I thought of going by AMTRACK, but the inconvenience of that, plus the need to buy a new car and Phoebe's poor accommodation to being cared for by others made travel by car the better means of transportation. Then up came an invitation to a family wedding in Milwaukee early in the month. We left the island on the Monhegan on May 3rd and didn't return until June 3rd. Phoebe proved to be a very satisfactory travelling companion and a fairly good guest, even in the homes of other dogs.

We were in Milwaukee for the wedding and also for the dedication of the fellowship room at cousin Frances' church, a very touching occasion for she died after a long illness a few weeks later. We went on to Midland, Michigan and visited Ed Dunn and saw the Irishes and Edgar Adams, all fellow graduate school acquaintances from Cincinnati. It was a delightful visit. En route to Nashville we stopped overnight with the Websters who gave us a tour of the city, most of which was entirely unrecognizable. Eleven of the fifty members of the class of '37 M showed up for reunion in Nashville. Here again the place was quite unfamiliar. I got lost trying to cross campus to the med school and hospital. Most of the group and spouses had dinner together after a cocktail party at Campbell Haynie's so we had plenty of time to reminisce.

From Nashville we went down to Bainbridge, Georgia to visit old friends of Wellesley days, Bill and Marjorie White. Day trips took us to Thomasville and Tallahassee. Enroute to Savannah we had to stop for some tortoises crossing the road as we passed the top of Okefenokee Swamp and were held up 15 minutes in the broiling noonday sun for a freight train in Waycross. We could have spent days in both Savannah and Charleston, but hurried on for a visit in Catonsville, MD with the Nesbitts (summer tenants of the cottage) and in Arlington, VA with Stuart and Anna Barber. Next came a two day visit with Corrine Hogden Robinson in Newtone Square, PA, my Cincinnati roommate whom I hadn't seen for almost forty years. That reunion was particularly enjoyable because of the tribute she received at an AARP meeting on her retirement as chapter president. From there we drove to Naugatuck, CT where my grandniece and two great-grand nephews live. Phoebe expects to return because she buried a luscious steak bone in the neighbors' yard.

The Mount Holyoke reunion was a bit of a disappointment because I was separated from all my close friends and hardly got to see them at all, especially Dot and Allan Johnston who had come from Arizona. But the class did itself proud, had the most and the highest % return and was runner-up in class gifts to the endowment. We are now looking forward to 1982!

After a visit with Emma Riddle in Tuftonboro, NH we returned to the easy life of the island, a dried out garden and a mountain of mail. The trip was marred by my leaving my purse in Thomaston at the start, being stopped for speeding in Maumee, Ohio, and dropping my driver's license on the floor of the Websters' garage in Cincinnati. It was made comfortable by the car air-conditioning, which seemed such an extravagance in New England.

(I'm having an awful time with this machine but it does make good masters when I type them on the right side!). Summer brought Fred and Scotia Waite to the island following Fred's 50th reunion at UNH. Then in August Emma brought over Parker and the two boys Aran and Kevin who had come east from Oregon. We had a family dinner for Aunt Helen at Concord Inn when Elizabeth John came on from Wisconsin in the fall. But the biggest celebration, as far as the cottage was concerned was the ten member Thanksgiving dinner in front of the fire place. Elizabeth Gardner, Priscilla Jackson with Mary Alice, Susan, Stephen, Libby and Amy from Baldwinsville, Tommy Ashley, Sally Cox and I sat down to a dinner of capon and a turkey which had to go back to the oven for more cooking, home grown squash, island cranberries, home grown pumpkin pie and many more fixings. The weather was drizzly but the ocean behaved itself for both crossings. If everyone enjoyed it as much as I did we all had a great time. The Jacksons had a bad time of it getting back to New York, I have yet to hear how bad.

Many things happened concerning the island. Danny Bates and Amy began bringing out animals for their place, once owned by Ray Phillips, on Manana. They now have several goats and are making cheese (very good, too), ducks, geese and a donkey who spends his time braying, especially in the early morning. His former owner, Joanne Henderson and a former colleague of mine, came over to visit him during the summer.

The Laura B was sold to Jimmie Barstow in the summer. He and Ivan Simmons are at the helm with occasional help from Earl Field. We have another water tower and new wells so there was no water shortage this summer. The rig Bob Boody devised to raise the water tower was used this fall to suspend the Manana Fog Signal bell onto its stanchion at the Museum this fall with funds from the Jim Thompson memorial fund.

The big island event was the marriage of Jackie Boegel to Billie Boynton in July. Alice Becker and Barry Berg were married quietly during the summer and Paul Horn and Barbara Ulrich were married off island last week. We all gathered on the dock yesterday, armed with rice, to greet the newlyweds and they didn't arrive!

The end of October I climbed back into harness and went to Vinalhaven to cover for Dr. O'Keefe for a week. It was great to be back there with so many good and old friends, but both Phoebe and I were glad to return to Monhegan's quiet life.

Many old friends of the island passed away this year among them Leo Meissner, Nick, Luisi, Zero Mostel, Ray Barker and Betty Ebert. There was a memorial show for Luisi at the Plantation Gallery early in the summer which was well attended. There were several of his paintings on display at the museum, too.

You probably wondered what life was like here last winter. It was tough, but not as bad as it probably was in shore. Our chief problem was ice which piled up more than a foot deep on the road because each snow storm ended in rain and then froze. Going was awful and the oil truck and most other vehicles were immobilized. Also, the oil boat couldn't get here for a while in January and we got perilously low in fuel. Then came a welcome, early spring. Creepers and shovels were stored and forgotten.

Best wishes for the blessings of Christmas and a wonderful New Year 1978.

I forgot one of the most important items. My eight year old night-blooming cereus had two blossoms a night apart, duly admired and toasted with champagne by over twenty admirers. What beauty - too delicate to have been caught on film. Here's hoping we don't have to wait another eight years, for a repeat performance.

Alta Ashley

January 4, 1978
Monhegan, Maine

Dear Friends:

Christmas is well behind us and Trap Day is imminent, so I think I should at least get a start in writing a "Dear Friends" letter before the year grows too old. I hope I'll manage more than two for this year. Also, let's hope this electric typewriter will do better than the previous one and the old manual one. It is hard, though, to keep from activating the keys, as witness above.

Now to back pace a bit. Phoebe and I went in shore early in December to attend meetings and do some shopping. However, the biggest storm of the winter so far struck Rockland and immobilized everything so we came back after two days and found this place in a shambles with Phoebe's bag of kibbles strewn across the floor, a dish of paper white narcissus in disarray. It was bitter cold so I decided to sleep in my hide-a-bed in the living room, as I had done at Thanksgiving time. When I opened it up I found a rat's nest inside the sheets, the sheets and blanket gnawed and another nest and pile of kibbles on the floor! The next day I went to do a wash and the washer didn't work. That night there was a lot of gnawing and I found the electric cord to the washer eaten through! Jack Butler came over to repair the cord and when he moved out the washer he found a rat sitting on a pile of kibbles back of the washer. I had borrowed the cat next door and she stared at the rat who immediately jumped down a hole in the wall. We found about two pounds of kibbles inside the washer down by the motor. No wonder the thing had been working hard!!

For some time I had been smelling something musty, which no one else seemed to detect, but which now I know is a typical ratty smell. We set traps, which got sprung, got three different cats up here to go hunting and set out Decon, which was gobbled up. Evidently something worked because no more traces of rats have been seen even though

Phoebe and I went over to Emma's in Center Tuftonboro for Christmas. But the ratty smell persists and probably will until the washing machine can be taken out and thoroughly cleaned. Phoebe was of no help in chasing our intruder. Every time she heard it she jumped into my arms, a mass of quivering dogflesh.

We had a nice time at Emma's. Jezebel and Phoebe made their peace with Jezebel maintaining the upper paw, well armed with claws which Phoebe objected to with blood curdling screams, even on near misses. We had freezing rain on Christmas day so we did a minimum of driving. Of course, I got stuck in the driveway and needed the assistance of three attractive young males to get me out. We had no snow on the ground here but it was piled high in New Hampshire and I didn't realize what a mess I was driving into when I tried to get up to Emma's house on my arrival.

Misfortune seems to dog my steps. When I returned after being in shore in December the gas line to the house froze and I ran out of gas just at suppertime. The refrigerator went off and so did the water heater. It was bitter cold and Carl, bless his heart, came up and fixed things early the next day. When I returned from Emma's neither generator would work. We thought the gas supply had run out, but I guess it had frozen, too. Carl came up again after dark and put on two tanks but the big generator refused to go, the little one would keep going only if the restart button was held in. Finally it did go, but the big one just sat there. Next day Bob Boody found a gear in the magneto stripped and we have been waiting for parts ever since. Island Living! Between the trouble with generators and rats I feel like throwing in the towel and moving to warmer territory come another winter, but probably won't, especially if I go thru with plans for my extra room off the kitchen with solar panel and windmill.

The island has been full of visitors since the weekend before New Years, many to help with "setting", others to observe and write stories and take photographs. A couple were in the cottage all this past week, but had to leave today. Several of the "helpers" also left today and still we wait for Trap Day. At 6:30 Monday morning Don Cundy started moving his traps to the dock. By evening the entire dock plus both sides of the Dock Hill road were piled high with everyone's traps. This time they left a wide alley for trucks, but plugged up the sides so that a whole row of traps had to be moved when the Laura B came in yesterday in order for the lines to be secured on the pilings. Tuesday it snowed all day and this morning it blew a gale until after noon. About 2:30, though, the wind let go and they began to load their boats. Steve Rollins was the first to load. Something happened and a few fell in the water but were promptly retrieved. According to conversations on the CB they plan to set by 8 am tomorrow. All the wives will be happy. The men are like caged lions when they are ready to go and have to wait for weather or someone's ill health. Carl Wincapaw, Dougie Boynton and Doug Odom have all been under the weather but should be fit by tomorrow. Alfred Stanley won't be fishing this year, at least for quite a while. Sherm, I think, will be going tomorrow. Both had knee operations this fall. There will be fifteen boats in the fleet, counting Bobbie Mersfelder's dory and Victor Lord's skiff. Rickie Cundy is fin fishing with Norm Brackett out of New Harbor, so he won't be lobstering here. it will seem funny not to be fixing up a lunch and mug-up basket for Doug and Harry after so many years. Now that Georgia is living down at the Store I am out of a job. They said, however, I could share the first catch in a dinner at the Store, which will be nice. I may make up some sandwiches and coffee for the crew working at the dock, just to keep my hand in. As soon as lobstering starts this island will be just about devoid of males from sun up to sun down for the next six months.

A photographer, Ivan Masser, interviewed me today and took steen pictures, including one of Phoebe and me and her "quiet time". He is a free lance photographer and sends his stuff to many magazines. I'm sure he'll forget about my story, so don't go hunting it up. He could make a nice story around Phoebe, though, if he only knew. I must leave enough room to tell about Trap Day. Thank you all so much for your cards and notes about my letters. I hope the copies of this will be of better quality. So much for now.

January 5

Today was the day. Both Tuesday and Wednesday the wind blew pretty steadily at 25 knots or above, except for a short time yesterday afternoon. Then last night it howled for a while and there was so much surge at the dock early the men decided to wait until after a "lobster" meeting at eleven, when they set the time when all boats were loaded. Few had already done so, one being Rusty Court who carries close to a hundred when fully loaded. Bobbie Mersfelder, fishing from a dory, was the last to be loaded. As he pulled away from the dock with his load a horn blew and off the boats raced to the setting grounds. Sherm

Stanley has decided to go this year, but Alfred will not, at least until much later in the year. Jim Balano and Bobbie Mersfelder are the new ones, Kris Boehmer at fourteen is the youngest and Doug and Harry the oldest, having obtained their licenses in 1946. They are also the oldest in age, though they don't want to admit it. With Mort Bernstein off the island as of Sunday, I'm the oldest person here. Casey Anne Court is the youngest, just under a month of age. There was such a scramble when the boats took off I don't know who went first or who dropped off the first trap.

The sound of boat motors once more fills the air and Monhegan is once again a fishing island.

HAPPY NEW YEAR TO ALL!
Alta and Phoebe

Emma:

D & H got only about half their traps down by last night so our feast won't be until Tues. or Wednesday. I was going in on Monday but doubt it with winds up to 50 knots predicted. Why don't you keep the panty hose - very warm & comfortable - if big enough in the fanny or tummy!

Monhegan, Maine 04852
January 15, 1978

Dear Friends:

Before my letter about Trap Day had cleared the Post Office we were struck by a storm which rated even greater than that of Ground Hog Day of 1972.

Rita came over to dinner on Sunday evening. I had a nice fire going, which is an unusual accomplishment for me, and as the evening wore on we began to open windows. The room temperature reached 80 degrees and that outside 50 degrees. By the time she went home the wind was a bit blustery, bumping up to 49 knots frequently. All night the house shook, gutters screamed and shingles played tattoo. In the morning the temperature was still up to 49 and the wind indicator swinging between 25 and 30 knots. At 11 o'clock it was screeching at 57 knots. The Laura B, naturally, didn't leave her mooring. Luckily, too, for the dock in Port Clyde came apart. The newly repaired end was floated up on the tremendously high tide augmented by the coincidence of the height of the wind. The rest of the dock accordion pleated and slammed into the freight shed. Fortunately they got all cars and gear out in time.

About 7:30 that morning Mary Burton called over the CB that water was washing over the dock (nearly three hours before scheduled high tide) and traps still on the dock - chiefly Doug's and Jim Balano's - were in danger of being washed and blown overboard. Men dashed to the dock and moved about 200 traps to safety. The slip had been lifted by the tide, its sides washed away and the cement footing for the iron plate at the top lay at a crazy angle. John Murdock fell waist high into a hole by the slip and was pulled out just before a trap was washed into the same hole and was never seen again.

While they were working on the trap removal they saw a wall of the big fish house now owned by Shermie Stanley and Dint Day was torn off and barrels of bait and other lobstering gear was being washed into the harbor. Everyone then congregated at Fish Beach and tried to salvage what they could from the fish house. But the water was so high and the force of the waves so great little could be done but watch the destruction proceed and take pictures of the grim scene. Ron Stanley was knocked down by a wave, my boots were filled at least once. Bait and broken wood was tossed all over. The men working to salvage gear were soaked to the skin. Carl Wincapaw's fish house floor was carried away but he was able to get his generator and some other equipment out before much was lost. Shingles were torn off Fish houses around Fish Beach, the Burtons' cottage and the Barnicle. The stone wall at the Days' held but Bob Burton lost part of his sea wall and the deck, he so hopefully restores each year, was damaged. At low tide the beaches were covered with dead lobsters of all sizes and many a "shot" claw. Many snarls of traps have been found, some smashed traps. Estimates of damage are $75-100,000 to the dock; a like figure for the two badly damaged fish houses plus replacement of posts and shingles for some others. I doubt anyone has been in the woods or on the back side to see what trees are down.

Bob Boody worked on the dock, got the slip pinned together with bobby pins and gum, hauled what gravel he could to fill in the worst holes. Jim got the Port Clyde dock so freight could be carried to the boat at high tide so we finally had a boat on Wednesday. My story of Trap Day and the storm lay in the Post Office until then. The Courier printed three of my most uninteresting pictures of Trap Day, actually all of the days preceding it. Fortunately Ivan Masser, a visiting photographer, had taken my films and story in on Friday before the storm.

If that storm wasn't enough, we had another hit us yesterday. This was a nor'easter, so damage to the harbor was minimal. But it dumped about six inches of snow and sleet on us. Even Rita decided it was too bad to go to the Store in the afternoon. There was little need anyway since Doug and Harry were there and few people ventured out to do any marketing. The wind reached something over 45 knots during the night. My indicator stuck there and yesterday it hit 48, this time with temperatures in the twenties so it was a bit hard to keep warm. Last night Sherm and Shermie had to go out and pick Billy Abbott off a ship. Wow!

We had our lobster feast Thursday night topped off with a coconut-rum liqueur Dougie and Debbie brought back from Haiti and gave me for Christmas - very good, too. The lobsters were superb. Ever since the storm a spray of forsythia I cut after Christmas and a bowl of paper whites have been in bloom and one amaryllis is ready to blossom - a nice touch of spring. The snow which is piled high around the house is doing a good job keeping the wind out of the cellar. With no wind today I have put out my fires, even that in the Franklin and am warm! Now I must answer a letter from a man who wants me to send any information (on Monhegan) to enable me to prepare a factual paper of that historic area''!?*?!

It is now the end of January and we have had what the men call ''trap buster'' storms at least one a week, three in all. If last winter was bad this one is proving to be AWFUL. I had scarcely run off the previous page than we were hit with a rain storm that soon changed to freezing rain on the 18th. There was no boat that day. The temperature fell from 33 degrees to 18 degrees by evening and we picked up about six inches of snow on top of ice, making walking just dandy. Then on Friday noon, just at boat time, it began to snow. The wind hit 48 knots as the Laura B went back to Port Clyde. Ivan, thank goodness, called on the CB to say they made it in safely. it was cold so the snow was dry and was piled in huge drifts by the winds which blew all night long. That weekend all the men and some women spent shovelling. Our plows were unable to handle the drifts.

Trucks were immobilized by the snow so many of us invested in plastic sleds to help with mail and groceries. A load came out on last Wednesday's boat. Then Thursday arrived with temperatures in the fifties and torrents of rain on 45 knot gales. The roads became rivers of slush. I went in ankle deep several times between Cundy's and the Spa. Friday the wind was so bad the boat didn't come. Temperatures fell turning the Meadow, now full off water, to a skating rink. The outlet stream had filled the passage to Middle Beach with water the day before.

Ever since the storm of Jan. 9th, which damaged the pump on the big storage tanks on the dock, Doug and Harry have been waiting for a day when there hasn't been rain, snow or spray so they could work on it. Finally on Friday they got it repaired late in the afternoon and tried to bring the oil rig up this hill, but couldn't on account of the ice. I'm down to just above a quarter of a tank full off kerosene so have been running my heaters only a few hours a day, relying on my Franklin and on an Aladdin in the cottage. I have to keep that warm on account of the water. Twice this month I expected renters who have never showed up or even written. Maybe they are still snowbound in Pennsylvania and Maryland.

Three of my pictures about Trap Day appeared in the Courier, one in the Portland Press Herald and a picture of the snow storm landed on the front page of the Courier this week. If only I could rig up a dark room here. Then I could send prints to all my newspapers. As it is I send the film to the Courier, they pick the ones they want and file the rest. Grrrrr. I got them back and sent scads of negatives off to be printed. Now have a sheaf to be filed here! I have a zoom camera now so will use that for B&W and my good one for color - B&W for the papers.

Saturday was Doug's birthday so I had the Store gang up here to dinner - Doug, Harry and Georgia, plus Rita. John Murdock and his girl, Winnie Neilson, came for drinks. I didn't have room for them here for a meal and felt too lazy to tote everything to the cottage where there is more room. We had roast pork, home grown beets which I had stored in sand, and Doug's favorite cake - a Washington pie with raspberry jam filling, jello and ice cream. Wedding bells are to ring in November for John and Winnie who is a delightful girl.

Much of the island has been felled by a severe cold this past week. So far I've escaped and hope to continue to do so. Maybe that flu shot in the fall which Doug, Harry, Rita and I took will do the trick.

Dot and Alfred Stanley certainly showed good judgment in taking this winter off. Alfred decided to give his knee a good rest after his operation in the fall. They are in California, I believe. The rest of the men have lost up to half their gear in this month alone. The Bajupa is here today buying for the second time only this month. The dock is littered with huge snarls of buoys and warp and smashed traps as well as a lot of Doug's traps he never got to set and are now frozen together. Bob Boody has hauled loads of sand and gravel to fill the worst holes in the dock, but major rebuilding will be necessary. Much of the underpinning has been washed out, leaving huge caves. We expect one of these days to see a truck go down through. Don't expect a drop in your taxes!

Today is beautiful - no wind and the temperature in the twenties under bright skies. Everyone is out hauling. It is supposed to stay like this until the weekend. I suppose we'll have another storm by Monday, since I am planning to go in, the first time since my return after Christmas. I wish the walking was better so Phoebe and I could go off somewhere. We just had a sonic boom and she came racing home in terror. Yesterday we had one at boat time when she was at the dock, a no-no. She took off and fairly flew home so we've decided to order a boom each boat day. Without one she refuses to stay in the house unless I go out and catch her and carry her home - a thirty pound load these days.

Ruthie Stanley is in shore waiting for her second baby. Joan Dolan Rothstein is expecting in June. Do Stone Wheeler had a hip operation last week. The Boehmers and Mort Bernstein are off to Europe for three weeks next Monday, Ruth Scully (Dyan Fitzpatrick's mother) left yesterday on the start of a three week stay in Florida. That is the extent of my island news.

Best wishes to all,
Phoebe S. & Alta Ashley

Feb. 12, 1978

Dear Friends, Family and Fiends:

Enclosed is a copy of the story I sent to the papers on Friday. I decided it was easier to make more copies than write a new story to you. I'm a little afraid to mail it. The past two letters have been immediately followed by a horrendous storm and one is in the offing, now.

Last week Emma called to say that Franklin was in a nursing home and we should go down and help Helen and Robert with plans for Aunt

Helen. So I left on Friday. The oil boat came that day but the Laura B tied up to her so we didn't have to wait for the oil delivery and got over to Emma's in good season. Saturday we drove to Elizabeth's in Concord where Helen, Robert and Aunt Helen came over to dinner. Emma, Elizabeth and I had spent the afternoon looking at retirement, rest, and nursing homes around Concord. They were for the most part quite nice for "gentile ladies". But we had doubts of Aunt Helen's adjustment to regulated living.

Sunday was a beautiful day. We picked up Aunt Helen at Robert and Helen's in Needham and went over to the Baptist Home in Newton where Gertrude Anderson, once resident minister on the island, has been for several years. She and Aunt Helen struck up an acquaintance and Gertrude spoke glowingly of the life there. We were then taken to dinner in the general dining hall, a very pleasant place. The meal was delicious and well balanced. I hope it was a typical meal. then we went to see Franklin who is at Wellesley Manor for physical therapy for his arthritis and quite unhappy, poor guy, but his treatments should help.

Monday Helen was going to see about getting Aunt Helen squared away for admittance to the Baptist Home. Emma and I headed back to Center Tuftonboro, fortunately before noon for that was the day of the huge snow storm. We ran into a lot of snow on Rte. 128 near Winchester where we had gone to pick up some boots, but drove out of it shortly. I dropped Emma and headed for Rockland with the wind blowing a gale and clouds thickening. I almost drove to Augusta to attend a KVAA meeting that night but decided not to risk it. Well I did. The storm hit that night.

Snow came in through every little crack. I decided to stay at the Treadway-Samoset where I found Peter and Ricky Boehmer and Mort Bernstein on their way to Europe. The vent to my room heater was open so the next morning I had a huge drift in my room. Snow piled up along the corridors, against windows and doors, sifted down from clerestories. It was wild. One waitress and one chamber maid got to the hotel on Tuesday. No shops were open and there was but one pack of cards. Everyone spent his time staring out at the complete whiteout, then rushing to the bar and dining room early for each meal, all the world like residents of retirement homes!

My car was full of snow when I got to it on Wednesday. The vent on your side had never been hooked up & was wide open! No wonder you were cold. No snow yet (Tues. PM) so I guess we'll escape this one. Any news re Aunt Helen?

Phones on the island have been out since the storm. Telephone men came out on Friday and spent over an hour trying to find the source of the trouble and decided it was around Burnt Island. The Store hasn't been able to put in any orders and the provisioners just haven't used their heads, except for the Hood's dairy. Doug, Harry and Rita are due here tonight for dinner. Fortunately I have the remains of a ham and two huge jars of pea soup and plenty of ice cream in the freezer. I made some lemon cookies this morning and hope they'll shape up into little hearts in honor of the day.

Jimmie brought the Laura B out on Saturday with Bob Boody's back hoe aboard. First they had to wait about an hour for the tide to go, then it took another three hours to unload the thing which tipped into the pilot house, breaking a window, then slammed into a piling and bent the front loader so Bob had to spend time repairing it before he could put it to work.

Last night the Coast Guard came out to refuel Manana - they had a huge oil leak not discovered until they ran out of fuel last month. This morning another C.G. boat brought out the SBA man to talk to the fishermen about the Jan. 9th storm losses. He'll have to come out again about the last storm damages - can't do it all at once! I guess all our lobsters landed down on Marshfield Beach, according to the papers and the emptiness of traps here. This winter has been a tough one for lobstering.

An Army helicopter brought out a C.D. photographer to record damages and a C.G. one came out to survey damages on Manana, so we've had a lot of extra traffic this past week.

Betty Mersfelder died Saturday night or early Sunday. The services are to be today at 1:30 in Thomaston with Bobbie Ives officiating. A pilot had to be picked up this morning so Shermie couldn't go in and stay long enough for people to attend and the others had to be here for the SBA man. Then, we are expecting snow to arrive at any time, only flurries but they often turn out to be heavy falls. I would have liked to have gone in but couldn't stay overnight. Alfred and Dint did go in. Sarah Macpherson died in late December. I got a card with Christmas greetings from her in January, mailed from Haiti by Bob.

Mace Carter is due any day with a crew to repair the dock and I am expecting to put them up in the cottage. If they come while Georgia is away I'll be cooking for them, too! The Laura B docked on the south side of the dock yesterday. She'll have to adjust her schedule until the slip is fixed. By the way, the fare is now $8 r.t. He put an Aladdin in the cabin for us last trip (heated cabin provided!).

I forgot to say that Gertrude Oliver, Llewellyn's mother, died last week at 95. She lived here when he was the teacher. The Butlers are leaving so once again we are looking for a teacher. Glad I'm off the school board. Enough for now. Best wishes to the snow-bound, and all.

Alta & Phoebe S.

Monhegan
Alta Ashley

The Sea Strikes Again

Feb. 10. As I sit here looking out on a peaceful, sapphire ocean dotted with white lobster boats busy harvesting the sea it is hard to realize that only three days ago Monhegan, along with the entire northeast coast, was dealt another body blow worse than the one she received exactly four weeks before, again at the time of the new moon and its attendant abnormally high tides. Cottages and work shops set back away from the shore came through the storm unscathed, 'though some were heaped high with huge drifts of snow piled high by the 50-60 knot gales which blew from Monday evening to Wednesday mid-morning.

For the buildings along the shore, especially the fish houses on Fish Beach, it was a different story. Huge seas, which reached their peak at the times of the abnormally high tides Monday night and Tuesday, welled into beaches and coves, smashed against ledges and sea walls and over the dock, taking with them everything that could be pulled loose by the undertow as they ebbed, only to fling it all back on the next sea.

Following the destruction to Shermie Stanley's fish house in the precious storm he had moved in with Carl Wincapaw. Together they had repaired damages to that building, which also housed the summer quarters of Ann Marie Sullivan and her family. Both men were working in the fish house feeling they were not in danger, despite the high tide, because this was a nor'easter and Smutty Nose would protect the harbor and Fish Beach. After all, most of the buildings on Fish Beach had withstood the ocean's fury for 100 to 150 years. About an hour before high tide a tremendous sea engulfed the fish house sending the two men scampering to higher ground for safety as the newly replaced floor floated loose and the entire building shuddered and began to tip. Almost immediately another sea crashed into the beach and carried the fish house out to sea, never to be seen again. The same seas lifted Chris Nicholson's fish house off its pilings and set it down 12 feet or more closer to the water's edge where it still stands canting at a 45 degree angle. More of the lower part of the old fish house nearly destroyed in the precious storm was carried away although the building still stands.

Windows in the Addison Winship house (once known as the Cabot Apartments) next door to Carl's fish house were broken by floating debris, shingles were ripped off and the house suffered considerable water damage. Parts of the seawalls around the Red (Barstow) House and that of the Seawall Cottage were washed away. Shingles, underpinnings and the entire front porch of the Wik Wak belonging to Natalie and Charlie Johnson were ripped away and the entire cottage was pulled away from its supporting chimney. Minor damage was sustained by other cottages in the Dead Man's Cove area. The Rockcrest Shop on Monhegan dock was damaged by seas which crashed over the dock taking the porch of the shop away as well as lumber, gas tanks, lobster gear and the gasoline pump with them. Fortunately the valve to the pump from the storage tank had been closed. The surface of the dock was entirely washed away. It now looks like a shingle beach. The slip, repaired by Bob Boody after the previous storm, was completely destroyed this time. Fortunately the big fuel storage tanks on the dock weathered the storm. Only the Friday before the oil boat had come out with a new supply. Roads were passable enough for the oil truck to travel on Sunday and all homes had been refueled, just in time.

All boats stayed on their moorings but the winds and tides were so erratic they headed in all directions and rubbed against one another during the height of the storm causing some damage to gunwales and transomes, as well as to paint.

The dock on Manana Island was damaged, the north side of the boat house was washed away and the pea pod driven out through the south wall. All phone service was disrupted and is still out. Communication with the mainland is only through CB and VHF radio and messages the boat men can carry. No Monhegan mail came to the Thomaston post office on Wednesday so the boat trip was put off until yesterday.

Although this small island with its hundred and fifty or so buildings and eighty winter inhabitants has received a high percentage of damage in this storm it has been spared the grueling hardships suffered by areas to the south and by its sister island, Matinicus. Once more the men have lost much of their lobstering gear and must spend many hours disentangling snarls of traps and rebuilding or replacing damaged ones. We now wonder what will be in store for us as a new winter storm sweeps across the Rockies to the central plains and into the east by the beginning of the coming week. In its short life 1978 had outdistanced 1977 in severity by a huge margin.

March 19, 1978

Dear Fellow Tax Payers:

Here we are in the grips of another winter storm with high winds and wet snow which so far hasn't accumulated much at all and shouldn't with the temperature at 36 degrees. I seem to have to have a storm raging to spur me to pounding letters out on the typewriter.

Town meeting was held last Monday with few changes in town officers. Richard Farrell gave up third assessor and took one of the vacant places on the planning board (defeating me once more). Don Cundy took third assessor. When will he be here to go to meetings of the assessors, water company and planning board, all of which go with the job of assessor? Victor Lord took the other planning board seat.

Just under $14,000 were appropriated. This does not include the state and county imposed taxes so our total tax bills will be enough

Interium radio phone — Bud Murdock, Store owner

to cover twice this amount, I'm afraid. But we have had one stroke of good luck. We are to receive a large **federal grant** for reconstruction of the dock and the breakwater. Right now the small breakwater at Fish Beach is strewn all over the place. The new one will be about in the same place, maybe a bit south of the present site to help protect the buildings on Fish Beach. Joan Rothstein is the new school board member.

Bob Boody is still looking for a place to live. If he can't find one the family will have to move off island and we will lose a valuable member of the community. MaryAnn, too, has been an asset during her service on the school board. Too bad they can't buy the Hultberg place down below the Golden Gull.

In his will Leo Meissner donated to the town the lupine field across from the Spa and the Swim Beach property, to be held as a memorial to the Jenney family.

I just read my newspaper columns of this time of year for '76 and '77 and find that the weather this year is just about as bad as it was then, way into April. I thought I was out digging up the garden last year at this time, but it was a month later. So take heart, spring will arrive in due time, not on March 21st.

Elizabeth Gardner and I are off to Monte Carlo on April 8th for a week of sun and fun. Poor Phoebe, I'm afraid, will have to go to a kennel, she is too much to handle for someone who doesn't know her many eccentricities. I hope to be back by April 21st after going to the dentist on the 19th. Our phones are still out!

As ever,
Alta

ALTA ASHLEY, M.D.
MONHEGAN, MAINE 04852
March 29, 1978

Dear Emma:

Last night's telephone conversation illustrated better than I could put into words the difficulties associated with radio-telephones. I heard only a small portion of what you were saying because your voice kept fading and when I pressed my key to talk and found you were still talking. It takes about five seconds for the sound to come through after the key is depressed and nothing is heard by the transmitter during the time the key is depressed. I can talk well with Robert but hesitate to call him except early in the morning - and Prue isn't always up then - because Roger answers the phone when Helen and Robert are out. I haven't tried to call you after making some calls and not finding you home.

I didn't plan to come down during March because I expected renters last week and was told we couldn't get into Aunt Helen's apartment until April. Then, everything seemed under control, according to the latest letter from Helen. I'm sorry so much has now fallen on your hands. As you said, there is no point in my going to the Farm while Aunt Helen is there, she would probably throw food at me, she has always been good at throwing verbal barbs. Possibly the best plan is to put her back in her own apartment and see how things go. She might regain her mental balance back in her own surroundings.

John Manypenny has moved back to Philadelphia. Perhaps we should latch onto him for help with either Aunt Helen or Franklin. I hope Franklin will improve at Mt. Ida with less medication.

There is little hope that our telephone situation will be improved. The Coast Guard is reluctant to repair the cable and the New England Telephone Company wants out. I hope something will be done. Radiotelephone conversations can be heard by anyone with a short wave scanner, so our conversation was heard all over the island and Knox County last night, maybe as far away as California!

Enough for now. I wrote to Helen and Robert Monday so you'll see that letter. Now to write another letter to the telephone company. When next you write would you put a date on your letter. Mails are so erratic I'm never sure which week you are talking about. Let me know if and what I can do. Usually by the time I get a letter it is too late for me to go in to help, with only three boats a week.

Love,
Alta

*They didn't show up!

Monhegan, Maine 04852
April 25, 1978

Dear Cousins et al:

In January you all heard from Helen Woodbury about Aunt Helen and Franklin and the situation she and Robert got into due to their compassion. Now to catch up.

Franklin was discharged to Wellesley Manor, a nursing home hardly worthy of the name. There he apparently suffered some kind of stroke. He was lucid on admission but soon exhibited weird behavior, allegedly attacked a nurse with his cane, and had to be heavily sedated. Helen and Robert with Carol and Peter's help had him transferred to Mt. Ida Nursing Home in Newton Center (?) where he alternates from thinking he is in jail to believing he is in a swank Golf Club and offers you a drink when you go to see him. He is confined to a wheelchair because of arthritis and osteoporosis of the spine.

Early in February arrangements were made to enter Aunt Helen into the Baptist Home, either in Kingston or in Newton. Then came the great storm of February 6th. The necessary pre-admission physical examination appointment for that day was impossible to keep. In fact no one moved out of home shelter for nearly a week, an interminable one for Robert and Helen, already at their ropes' ends after sheltering Aunt Helen and coping with Franklin's problems.

Aunt Helen was finally taken to Kingston, on a trial basis, about the middle of February. In March her trial period was extended because of "an incident in the dining room". She was said to have struck at one of the women at her table. The director, Mr. Bartos, solved that difficulty by moving her to a table with **two men** and letting her sit facing the room so she could see everything that went on. A week ago Helen and I went down and had another talk with Rev. Bartos, this time to be told that Aunt Helen had sworn at someone in the dining room and "We can't have people using such words in the dining room!" He wouldn't repeat the words, they were "too bad", so now he doesn't know whether or not Aunt Helen will fit into the family at Evanswood, but they will keep her until the end of the month. Where to if she fails this trial period?!?**

Meanwhile the apartment at 7 High Rock Way is still full of Aunt Helen's things. The rent is paid until June 1st even though she has not occupied it for about two years. I had planned to go down this week and start clearing out the apartment, but then realized I had no authority to do so and have asked Robert and Helen to get me the name of her lawyer so as to get power of attorney or be appointed as guardian, as the oldest sane member of the Woodbury clan. I hope all this can be accomplished by the 8th of next month so we can have time to empty things and get them distributed or placed in storage.

The other day I made an inventory of things in the apartment, then left it there when Phoebe disappeared and was missing for about an hour. I was so distraught I wasn't functioning very well.

Helen has now written that Robert has had some cardiac irregularities felt due to the strain he has been under during the past three months so he and Helen are taking a vacation on the Cape for a week or so. They deserve it, too, for they have worked day and night on the problems both Aunt Helen and Franklin have presented.

There is very little in Aunt Helen's apartment of real value, but there are a few pieces which Great Grandfather Woodbury made - at least two chairs, a chest and possibly a table. Much of the furniture is in poor condition. Some was acquired from the Dunhams when they all died off, including a diningroom set which Sally Peters, Nancy's daughter, may be interested in. She is to be married on May 20th. This would be a good solution to moving that, if no one objects.

Our phones are still out since the storm in February. They were supposed to start repair work on the cable yesterday but no one has caught sight of the cable boat as yet. Appeals have reached the desk of not only the president of New England Telephone, but of the United States, through Jamie Wyeth who has done so many presidential portraits and met so many D.C. bigwigs. If you have any suggestions or objections to any of the above do write immediately. I'll probably leave here on May 8th or 9th to give me time to empty the apartment by the first of June. I'll stay down until after Sally's wedding.

When Helen and Robert first went to the apartment they found all the silver and took it home for fear it would be stolen, for they found the door had been jimmied. They have no silver of their own and it seemed only fair to Emma, Elizabeth Gardner and me that they keep it, considering all they had done on Aunt Helen's behalf. We took two figurines of Aunt Helen's and her night light down to her. She agrees the apartment should be emptied, but I didn't get this in writing. My phone is 207-372-9610, Eliz. G.'s 617-369-3398, Emma's 603-569-1521. I'll stay with Elizabeth when I go down to Mass. Let me hear what you think.

Love,
Alta

Monhegan, ME 04852
May 5, 1978

Dear Friends and Family:

It is almost three months since I last wrote. Each day I was sure I could write you about definite developments concerning the island phone system and family matters but as yet I have little to report in either case. Being without phone service has been particularly trying during this period of uncertainty about Aunt Helen and Franklin. The only thing I can say is that finally I have gotten my name listed in the phone book - all for naught?

For the past two weeks the cable boat has been here - really in Port Clyde - working on the telephone line to the island. We all received a letter from the Coast Guard Service saying that Manana Fog Signal Station will be automated by 1980 and the cable will be declared "surplus". They said that they will repair the cable this once in hopes that the island will be able to obtain alternate service by the time the cable is removed or it breaks again. Ever since the storm of February 6th we have been trying to get help in re-establishing service. We may be able to get some system going which will be better than the marine radio system we have had to rely on, and possibly better than the pay phones we did have IF there is enough local interest. Keep your fingers crossed!

Because of the lack of phone communication I have lost track of what is happening to Aunt Helen and Franklin. I did receive permission to move Aunt Helen's things out of her apartment, mailed from Evanswood on May 1st. (Later) Today's mail brought a letter from Helen and Robert to say that Aunt Helen's rent at Evanswood is now paid to June 1st, but she apparently is still "on trial". Franklin is in the Mt. Ida Nursing Home in Newton (I don't know the address) and is probably there to stay. Emma and I will go down and clear out the apartment at 7 High Rock Way. They want us out of there by the 15th. I had planned on the 19th, but maybe we can get things packed up by the earlier date and have a lawn sale for things not worth keeping. This all makes me realize I've got to make plans for my care and the disposal of all my accretions. Today is my birthday, too far along toward the 3/4 century mark to be comfortable.

The island is bumping along without phone service. All week we have had no grocery order and shelves are **bare.** Each boat day Harry has gone to the dock with the big truck and returned to the Store with milk and bread, and that's about all. They probably sent the order to Matinicus!

While I was away there was a fire at the Store which started in the laundry over the drier. Rita smelled smoke and Doug went down to investigate. When he opened the laundry door the blast hit him, singeing his hair and eyebrows and burning his forehead. Fortunately his eyes were spared. Had the fire gone on another minute it would have been a repeat of the 1963 fire. Flames were already licking up the sides of the walls in the back room. Town water had been turned on just that afternoon so there was no lack of water. Also, Ron Stanley was going by the Store and saw the flames so he wasted no time in getting the fire engine to the scene. Water was also pumped from the well at the parsonage. Only the laundromat was lost and is still out of commission. Thank heavens for my machines. Peter Boehmer and crew rebuilt the laundromat and Norm Richardson and Normie came out last weekend to work on the electric cables, boxes etc. lost in the fire. We didn't have any ice cream until Wednesday because the ice cream box connections were burned. Larry Leonard came out with Norm. He and Doug have worked since then on trying to recommission the big generator which conked out a year or so ago. That has proved to be a huge job.

A week after the Store fire there was a grass fire at Court's again put under control before it reached the generator, due to CB and quick action by Ron. The following day grass around the Store generator caught fire. This was discovered within minutes and dowsed without any damage. The fire truck has had more action in eight days then it usually has had in a year.

The dock is still a mess. Mace Carter's crew is due any day now to work on that. They got the slip put together in February. They stayed in the cottage and I cooked for them for six meals and nearly collapsed by the time they left. What appetites!

The first, no, the second week of April Elizabeth Gardner and I took off for Monaco for eight days, which turned into only seven because a disabled plane had us spend the first night at Logan. We had a great time, lost all the money we set aside for gambling, took side trips which carried us from San Remo, Italy to San Raphael, France; walked in Snow in Valberg, lunched on cheese and crackers and wine and got chased out of sidewalk cafes for doing so; travelled on the Grande Corniche and all the others, plus the toll road; visited Roman ruins which were everywhere. It was a wonderful week even though it was cold and we had neither time nor inclination to swim. Elizabeth picked up stones in Ireland, Monaco, Ventimiglia, Cannes and Nice. Must close.

Phoebe spent the time in an Acton kennel with a gentleman fellow canine.

As ever,
Alta & Phoebe

Monhegan, Maine 04852
June 6, 1978

Dear Friends:

Some time ago I thought surely I'd meet the increase in postal rates with my next letter, but many things interfered, least of all a game of BOGGLE which Elizabeth gave me. Now the house is littered with pieces of paper filled with lists of words. First I try to beat the timer, then try to find all possible words. This morning I made forty-nine, the highest yet, twenty-three in the allotted three minutes. In the first game I found seven!

Summer finally arrived this weekend and apple trees have responded exuberantly. Beans planted last week popped out of the garden and radishes burst their skins. Lupine on the south slope by the Golden Gull came out last week even in the fog, others are just beginning to show some color. Miners are killing off the squash plants one by one and cut worms got a few plants earlier. I've had two pies from my rhubarb and am about ready to call quits on that delicacy.

First of all, we again have phone service. The cable boat worked more than four weeks and finally found all the broken areas and service was restored on May 19th. But the Stores phone weren't in service until last week; the lines were damaged by the fire in the laundry and the repair man didn't make it out here until then. We sent out a questionnaire about phone service after receiving a letter from the Coast Guard stating that this was the last time the cable would be repaired and that in 1980 when Manana Fog Signal Station is to be automated the cable would be removed, anyway.

We received answers to a little more than 50% of those sent out and fifty people said they either would want phones or would take one if it was needed to maintain service to the island. That seems like a fair number to interest either NET or any other phone company. There were only two completely negative answers - one written and one by word of mouth. I'm sure if any cottagers had been here when there were no phones at all more of them would have said "yes" to a home phone. It really was grim and **expensive**. Each VHF call I made cost $6 or more and none was three minutes long!

I missed the excitement of the Store fire by being in Monaco, and the thrill of having the phones reactivated by being in Massachusetts where I had gone to close up Aunt Helen's apartment. What a job that was. Emma came down to help but arrived with a temperature of 102 degree and was out for five days, sleeping the clock around. Elizabeth helped a few hours, thank goodness. There were about 6 hefty baggies of discarded materials to be taken to the cellar; I filled my car with clothing, etc. for Morgan Memorial and had another carfull which went to the therapy department of McLean Hospital, as well as two carloads of treasurers which I took back to Concord and carted about half to Monhegan. After nine days of working in the apartment I closed the door and went off to see the Huberts in Dover. It was a hot day and I arrived covered with dust and grime. A hot shower, shampoo, steak dinner preceded by a long drink brought temporary revival. It was a pleasant and most welcome visit.

Morgan Memorial truck promised to be at the house between 8 and 2 that day so I stayed overnight in Needham with Robert and Helen the night before and battled early morning traffic to Allston. The truck arrived at 1:59 1/2!!! They refused to take rugs, a cot and mattress so I just left them, plus curtains and draperies which required a tall ladder to remove them. Painters were supposed to have come in on the previous Monday but hadn't come by Thursday. We were supposed to vacate the place by then, even though the rent was paid to June 1st. So I decided the landlord could well afford to pay for cleaning up, since I had no time to arrange to have it done. This didn't sit well with her, however, and she took it out on poor Helen and Robert. Family, friends and neighbors (for a price) had just about emptied the place, which was a help.

After seeing what a job it is to empty someone else's quarters I was all fired to come back here and clean out everything I didn't intend to use within the week. Instead I just added more to the accumulation of the ages. I then sold my bureau to owners of The Mooring Chain who had a matching piece of furniture and am living out of boxes and suit cases until I can get the chest I took from Aunt Helen's to Concord. Now for a good rummage sale to get rid of the accumulations which was in the top drawer.

After being in Concord, Center Tuftonboro and Allston for three weeks I arrived in Thomaston in time to fly out here in Alan Mersfelder's amphibian, freight and all. He was flying out his (Bobbie's) family, so Phoebe and I hopped aboard. It took less than fifteen minutes from take-off at Owl's Head to jumping onto Swim Beach, a very pleasant way to make the crossing. Bobbie is working on a float which will be put in use when Alan gets squared away with red tape, etc. He plans to pick up or take people to Rockland, Port Clyde (6 min. run), Portland or Boston. He has had a lot of experience as a bush pilot in Alaska. Last week he brought out most of the freight for the Rockcrest Shop.

As usual this letter has taken more than a week to write, due to many interruptions. Now for island news. The island is filling up. The Periwinkle opens on the 14th, the Inn and Golden Gull on the 16th. Alan Mersfelder has been cleared for taking passengers but still has to get an office in Rockland and a radio for out here. Bobbie will help with bookings.

Vi Sturner is here and will start soon to ready the museum with Julie Hudson's help (or is it Jackie or both girls?). Jimmie brought out a backhoe on Monday. In the course of unloading it some damage to the slip occurred. As yet nothing but emergency repairs have been done since the February storm. The dock surface is in terrible shape. Probably Fourth of July weekend will see the pile drivers **et al** here to compete repairs. Remember, the afternoon boat leaves at 2:30 from Port Clyde and 4:00 from here this year. The morning boat is the same 10 AM from Port Clyde. Sunday runs start on the 25th. Twice daily boats start on June 23rd and go to Sept. 17th.

At Plantation meeting on Monday we voted to raise $42,385 in addition to the nearly $14,000 raised in March, but $8,000 will be taken from the Emily Thompkins Farrelly Fund and some of our school expenses of last year will be returned, which will help the tax burden.

Pru Cundy went to Washington and met with our legislative delegation about the breakwater. She was armed with pictures of the three big storms we had and felt she spoke to sympathetic ears. The telephone survey results are also encouraging with an overall return of 65% and a total of 51 possible customers. Let us hope the present cable will last until we can come up with a good plan for continued service.

I just learned that the backhoe damaged the Laura B slightly, not the slip. They had to stay all night and catch the right tide early this morning to unload the monster which is being used for a septic system at the Yew. The contractor wouldn't hear of using Bob Boody's backhoe.

Cindy and Sandy Barker were here early this month. Bud died early in May after a massive coronary. Ray died last fall of one, too. Bud was made president of the family company at his father's death, a position he held barely six months. Happy and Peter Estey are due here shortly.

The Sullivans are here and hope to get started on building their cottage. Meanwhile they will be in the Square House for two weeks and then at Betty Ebert's. The laundromat is finished. There are now two driers. All phones are back in service until the next cable break. Horse chestnuts are blooming and rugosa roses are budding. Lupines are in full flower. Enough for this time.

As ever,
Alta Ashley & Phoebe S.
(Ann Landers to the contrary)
6-13-78

Monhegan, Maine 04852

Dear Residents and Cottagers of Monhegan:

Recently many of you received a letter from the Coast Guard Service which in essence stated that because the Fog Signal Station on Manana was due to be automated and since there is now no further need for communication with the mainland via telephone, the telephone cable from the mainland to Monhegan has been declared surplus property. However, in view of the fact that Monhegan is now dependent on this cable for telephone service damages done to it during the winter of 1978 will be repaired (work is now in progress) but such will be the last maintenance of the cable by the Coast Guard Service. In the event the cable malfunctions in the future the cable will be taken up and disposed of.

It is, therefore, incumbent of the people of Monhegan to plan for alternate means of telephone communication to substitute for the present cable. New England Telephone feels there is not enough interest in telephone service for that company to take over service to the island.

Many feel this is not true, particularly after three months of the inconvenience of no direct communication with the mainland.

To help determine the amount of interest there may be it is necessary to estimate the number of potential phones which might be supplied and the volume of calls made. It must also be kept in mind that any new system would require more than the five talking lines now allowed by the present cable so that if you want and need telephone service you may not be able to enjoy the luxury of using your neighbor's phone. * An expanded service could mean multiple public phones at the Store, the Inn or other area.

Would you please answer to the best of your ability the following questions and return as soon as possible to:

Box 87, Monhegan, Maine 04852

____ I would like telephone service available to Monhegan
____ I would like a phone in my home
____ I would accept a phone in my home if necessary to the service
I make approximately ____ local calls per week
and receive ____ toll calls per week
Service would be needed for ____ months per year

Comments:

Signed* A substantial need and desire for phone service is needed for the NET to consider rendering service.

Telephone Service Questionnaire
May 1978

During the fall and winter of '77-78 I had occasion to contact some of the top officials of the new England Telephone Company concerning a private matter. Just as these negotiations were ended the Coast Guard telephone cable to the island was ruptured in the Storm of February 6th. Having made contacts already I again talked to NET officials in both Bangor and Boston. During these conversations I asked if it would be helpful to make a survey of sentiment on the island concerning telephone service. It was the belief of many on the island that there now was more interest in such service than the telephone company had been led to believe several years ago. For this reason the questionnaire was developed and sent out. It was thought that a representative from Rep. Emery's office was due to come out and that the Coast Guard Service was also sending a spokesman, thus the questionnaire was developed in haste. For this I apologize but I wish to thank the 76 people who did return their forms and the significant information which was obtained. particularly from comments.

Here are the results:

Length of service desired	More than 6 months	Less than 6 months	Total
Number sent out	34	84	118
Number returned	28	48	76
Returned by P.O.		1	
% returned	82	57	64
Want service to island	26	47	
Not interested in service to island	1		1
No answer	2	1	3
Would like phone in home	16	10	26
Would accept phone if necessary	6	19	25
Potential phones	22	29	51
Emphatic refusal of any service in home	3	5	8

Most people were unable to estimate the number of calls, local and toll, which they made or received. Estimates ranged from 2 a season to $600/month. Since most calls from the island are made on credit cards, to 800 numbers, or be reversed charges the total volume of calls put though any particular phone could not be determined.

The survey was not meant to be slanted towards or limited to service from the New England Telephone by cable, although many thought so by their comments. Many suggestions were made such as radio-telephone, micro-wave, formation of an island company, hooking into a small company on the mainland. Investigations will go on, in view of the 1980 deadline set by the Coast Guard and the negative attitude of NET toward service extended to more islands than they now serve.

6-10-78
Alta Ashley
(with Bob Burton's approval)

MONHEGAN FIBER FESTIVAL
July 17 to 21, 1978

All week Monhegan has been treated to an unusual, colorful and exquisite showing of pictures, wall hangings and designs executed with the use of, not paint or ink, but with fiber and thread. The exhibit is being held at Carina House whose shocking pink north wall has served well as a foil to the brilliant pieces which have been hung on it for display. Inside the house there have been demonstrations of the various techniques used by the artists whose works have been on display.

The festival was delayed one day by the rain which threatened and fell fitfully all day Monday but did little else than wash out that day's showing of Natalie Minewski's paintings in wool, an art form which developed from her interest in tapestries seen in the Unicorn Tapestries in France and the Boyeaux tapestries in New York City. These tapestries are produced by needlework, in contrast to weaving which is characteristic of most tapestries.

Nita is a graduate of Cooper Union and of Poitiers Univ., France. She has exhibited widely in France and New York State as well as on Monhegan. Presently she is an art teacher in New Paltz, N.Y. and Art Director of the local newspaper, "Huguenot Herald". She started working in this medium when in La Rochelle, France in the 1950's. Over the years she has developed her technique to create landscapes, still lifes and portraits "using the concept of color relationships". Again, according to Nita "The artist conceives of each work as a painting and the yarn as strokes of color used freely and creatively, not restricted by conventional stitches". Most of the so-called paintings depicted scenes or impressions of Monhegan.

In the workshop inside Carina House Gladys Rosen had on display many examples of her exquisite silk and metal thread embroidery which she has been working on for the past two years. A resident of Manchester, Conn., Gladys is the wife of Dr. Theodore Rosen and the mother of five children, now grown. Up until five years ago she says she was "busy just being a wife and mommy". With time on her hands she took up needlepoint. In an attempt to convey the form of the Temple of Jerusalem on a Matzo cover for Seder celebrations for one of her sons she developed a form of embroidery using silk and metallic thread through which she was able to bring out the perfection of form and ethereal light of the dome of the temple. She was helped in this through courses in color interaction and two and three dimensional design, as well as inspiration she obtained from the work of Robert Semple and his Carina House group of which she has been a part in the past two years. Her matzo covers have as their themes the true values of life which should be preserved and the roots of the Hebrews which must be identified for each generation, brought out in the four questions asked by sons at the Seder celebration. Not all of Gladys' embroidery, however, is of religious nature. She also does exquisite pieces of flowers, birds, and butterflies.

On Wednesday wall hangings created by Robert Semple, himself, were on display. These varied from bargello pillow tops to large wall hangings in libra point, a stitch which he has originated.

Robert Semple, III was born with thread in his veins, apparently, for one of his ancestors founded Clark's O.N.T. Cotton Company. He didn't realize this, however, until seven years ago. He was one of the first psychoanalytically oriented art therapists in America, has authored many monographs and has had one-man shows of his paintings in New York, Philadelphia, Santa Fe, Paris and Zurich.

Seven years ago he quit smoking. Having nothing to do with his hands he looked around for something he could do to busy them, anywhere, at home, while waiting for a phone call in the office, on the train commuting to and from work. Bargello was the perfect answer. He set up a summer studio-workshop in his Carina House on Monhegan Island, renaming it Carina Bargello + Point. Libra-point, the use of random length stitches (from 2 to 12 canvas threads) allows for freedom of movement of color and texture. Robert's knowledge of oriental and Indian designs has given him a rich background on which to draw for patterns he creates. One of the largest pieces he has made is a hanging for Kate Mostel based on a painting by the late Zero Mostel entitled "Monhegan Forms", a montage of rocks, waves, boats, fish, angular island roofs, etc.

Since the inception of the Carina House workshops summer visitors and islanders alike have been inspired by the enthusiasm which Robert has for his medium. Under his guidance they have created their own designs for bargello and libra-point creations. Among those who gathered at the workshop on Friday were Prue Cundy and her daughter Donna, Karen Wincapaw and her daughter Kathy, Barbara Hitchcock, Georgia Anacker, Leslie Whitney (all islanders), Gladys Rosen, Tirza Mailkoff, Nancy Stanich, Peg Holbrook, Mae Cass, Pam Krause and Jean Edison. Isabel Farrell, Marion Wiley and a friend were introduced to the medium that day, too, with the aid of the more experienced workers.

According to Tirza Mailkoff, a piano teacher from Ann Arbor, Mich., who demonstrated the use of the Aran stitch in bargello on Thursday, she became an ardent pupil of Robert's two years ago when she visited Carina House. "After seeing all that color and hearing all that Bach I was hooked!" Robert's one pillow worked in Aran stitch, similar to that used in Aran fishermen's sweaters, inspired her to work with wool and canvas. Soon she was combining Aran stitch with bargello and libra-point to create pillow tops, wall hangings and even a rug. Besides her stitchery she is also teaching one of the island children, Timmy Boehmer, to play the piano through the use of electronic tapes and is greatly pleased by the progress he has made.

Although the festival came to a close on Friday one can still see the work done by Robert and his students by visiting Carina House where Robert is always ready to talk about his medium and aid any serious student of stitchery. In the words of Gladys Rosen "He has the ability to reduce a complex subject to its simplest terms, so that any student can have a feeling of self-confidence and achievement. His sophisticated designs, translated from Pre-Columbian motifs, details of North-African tribal patterns, designs from royal seals, are all part of his repertoire.

"His uniqueness lies in this relationship to his fellow-man. He senses what it is that each student demands of him. Those who have worked with him recognize that he is in love with people and with nature."

Monhegan, Maine
August 1, 1978

Dear Friends:

Blessed, blessed rain is falling. After only four hours the rain barrels are full and the rain gauge measures a half inch, more than fell all of July. In ten weeks we had only 2 1/2 inches, even though for the first four or five of that time we also had almost no sun. Had it not been for the fog things would have been a whole lot worse. I finally resorted to watering my vegetables with the hose last week, a measure tabooed for those on town water.

My strawberry bed produced well, then the entire middle section dried up. Maybe a bale of peat moss and manure poured on it will bring it back to life for next year. Despite the drought greens and broccoli have done well and I find it hard to keep up with them. But my peas were an almost complete disaster. Male corn tassels have appeared even though the stalks are scarcely three feet high, far too small for the pole beans which were supposed to be supported by them, a Jim Crockett.

In mid-July I went in shore to see why I couldn't get down to accomplishing anything and why I hurt all over, was put on some huge capsules - four a day! - and now have a new lease on life. I actually haven't attacked the albums I am making of my newspaper articles, in duplicate, one for me and one for the library. I even vacuumed the rug the other day and began to reassemble things on the shelves in my room. There is still a long way to go toward restoring order. In fact, such really can't be done until I have more space which I can vision only with the addition of that room to the east of the kitchen, no matter what others say about using my old office down stairs.

It is now November, but I'm not going to put any date on until I finish, knowing how many things can push these letters aside. To begin with, the renewed vigor I spoke of was of short duration and I had to go back to Pen Bay Medical and have my medication changed because of a drug reaction. But I've been on naprosyn for several weeks now and really am feeling practically back to normal.

Several things have come up since I started this letter. Nita Minewski showed me a coloring book she and a neighbor worked up for the bicentennial for New Palz, N.Y. We thought we could modify my bicentennial address and make a similar thing for Monhegan. I've been reworking that, have gotten more engrossed in the history of the island and have rewritten and expanded the speech to such an extent I now have to go back and extract the essentials if my text is to be used. In the meantime I have been deluged by descendants of Henry Trefethren, and have become fascinated with tracing the lines of the Trefethren, Starling and Horn families as well as that of Shem Drowne. Next I must turn to doing the same thing for the Ashley-Woodbury clan which my niece Nancy asked me to do about three years ago. One project after another!

All summer there seemed to be a lot of extra happenings to write up for the papers. Robert Semple had a week of festivities at Carina House which he termed a Fabric Festival. He and several of his pupils showed their works in bargello, libra point and wool painting (Nita). And there were many more than usual friends showing up as day trippers making for a very busy summer.

The day after I started this Doug came by with the big yellow oil truck. He let the kids out, then backed up the hill to Marion Wiley's, struck a rock and started roaring down the hill, fishtailing all the way. He had shattered the rear axle on a rock, had no control or brakes, and stopped below the library only because the wheels had moved back two feet and the truck body ground down on the rear tires. No humans or dogs were in the road, fortunately. Marion came along and told him she wanted only three-quarters of a tank of oil! and didn't understand why he told her he couldn't give her any. (she was just coming back from the Inn after breakfast when she met up with the truck).

One day I was stopped down by the Spa and asked where the gas station was. I said there was a tank down on the dock, whereupon the woman said "I'm looking for a ladies room"! We'll need a gas station if there are many more trucks brought over. There are now at least 37 vehicles on the island, most without mufflers. Even now it is hard to walk down the road without having to get off and let at least one truck through. And I'm criticized for having a golf cart which I have driven 210 miles in five years, going on six. I've used less than 10 gallons of gas in that time, good conservation.

Alan Mersfelder finally got his license to carry passengers in his plane, but weather and a late start combined to cut down on his business. I came back with him on my last week's trip to the mainland. It was so windy we had to land over by Manana; then the surge at the dock was so great I lost my balance getting out of Bobbie's dory and fell flat on the slip, not in the drink, thank goodness.

In late June Chris Nicholson had a heart attack and the med-evac helicopter was called to take him in. We decided Elva should go, too and be put in a nursing home. Jeanie went along, too. We landed on the pad at Pen Bay Medical and Chris was receiving care less than two hours after they called me at 5 AM. Later this summer the helicopter was back again for another patient. This is a great service. I wish only that patients could be picked up without having to take that horrendous ride to the ball field.

Late in the summer the Troelemans sold their house to Bob Boody. Art was unable to leave the cottage all summer and decided it was time to sell, so now we are assured of continued mechanical help from Bob. I must say it is nice to have him here and also to have the welding apparatus a bit farther away.

To backtrack. Elva was admitted to Rockland Convalescent Center and when there was a double room available Chris joined her, but she died shortly after on August 10th and Chris is back living in the room Jeanie and Victor had made over for Chris and Elva in the Monhegan House. Sidney Baldwin died in Boothbay Harbor this summer. Yolla and Rudolph Niclas both died, just as Yolla was about to arrange for republication of "David and the Seagull" and "The Island Shepherd". Mace Carter whose firm was to repair the dock died a couple of weeks ago. The dock still has not been resurfaced since last winter's storms.

The Museum was very busy this summer as more and more trippers came to the island. Among the visitors were the Gardner Trefethens from California and Hazel Trefethen Frydeen from Minnesota. My mention of them in my columns brought many inquiries about the family, as I mentioned above. The Brazillian owner and international crew of the America were also among the visitors. Nearly everyone who comes leaves with expressions of wonder and thanks and occasional contributions.

The Store has had its problems with the "Bottle Bill". Cases of empties are piled ceiling high in the shed. The chief problems are transportation to a center which will accept them as well as the flies they attract and the space and sorting time they take. I will say the roads and bushes are far freer from litter than usual, though.

My new room will go on in the spring. The carpenters who are building Doug Way's new house will do mine in the spring. John Murdock and Winnie will be in the Mooring this winter. Their wedding is Nov. 25. They moved in a lot of their stuff this week and I had to find room for linens, silver, dishes and cooking utensils and I haven't cleared up after bringing out the stuff from Aunt Helen's. By the way, the oriental rug from her front hall which I nearly tossed in the dump came out bright and beautiful after being cleaned by a rug man in Waldoboro. Nice to know there is such a service so close.

The telephone service is still up in the air. Right now the cable is working and we are all holding our breath. I believe today there is supposed to be a hearing with New England Telephone and the PUC. I sent maps to the PUC and Continental Telephone, after consultations with both organizations, marking positions of poles, existing phones, requested phones, acceptable service if necessary, designated as to seasonal or year round and haven't received an acknowledgement by phone or mail from either, or return calls from NET. Grrrrr. Let us hope something will be done before the cable breaks again. VHF rates have gone way up since last spring along with NET rates for calls within Maine. It now costs 70 cents to call Camden!

As yet we have had no killing frost and I am still eating from my garden. The continued drought has cut down on the size of the root vegetables and the grasshoppers this summer ate all the feathers off my carrots. Although they have regrown the carrots are smaller than my little finger. I got three four inch ears of corn from twelve hills and ten pounds of potatoes from some potatoes which sprouted last spring. Apples have been numerous but quite small and cranberries are more scarce than ever. It was so dry in July that very few of the berries grew. What ones did are very small.

I haven't had a pheasant in the yard since last spring. Mort has seen two this fall. What do you supposed has happened to the flock? By the way, Mort Bernstein has written a book on tennis and nearly finished a second. Ricky Boehmer flew out to Hollywood with him when he was taped for a Merv Griffin show which will be aired as soon as the New York newspaper strike is over, which is preventing the publication of his book. How many more trippers will that bring to the island?

Elizabeth Gardner and I are going to Sapphire Bay, St. Thomas in February. We had hoped to go for Thanksgiving so we could spend the holidays with Bruce and David, at the U. of Virgin Islands and Orlando, respectively, but couldn't get reservations. February should be a better time. I hope both boys can join us for a while, at least. That will be our trip for the coming year. I'll be with Emma for Thanksgiving, then go to the Murdock-Neilson wedding before returning to the island. No Christmas plans as yet.

When we were in Monaco last April one trip took us to St. Paul where the Santons are made. I bought three for the price I have paid here, so didn't buy more. Now I find the Cricket Shop didn't order any for this year because of the rate of exchange and I'm having difficulty in finding pieces to add to those I've been giving the family for the last few Christmases. C'est La vie!

The Sullivans and Carl Wincapaw have built new houses on former Hudson land to replace the house that was washed away. Stevie has added to his shop as a house, now, and built a small shop where his house was to go. Doug Way's house on Horn's Hill is closed in, to be finished in the spring. The Laura B brings out lumber, trucks, generators, it seems, daily. Dyan's studio is being enlarged and the Golden Gull is being turned into efficiency apartments. Many changes. Enough for now.

Phoebe is fine, calming down but miserable when I leave her.

As ever, Alta

Nov. 4, 1978

October 10, 1978

Dear Bob and Shirley:

The collar arrived some time ago but Alta has been busy with telephone surveys and what have you so she has asked me to thank you. It will be so good to have it and not have to depend on scratching my neck to scare away the fleas.

Doug Way's house is going up fast. It is a two and a half story house with an exposed basement on the north side, so is quite a structure. Bushes are to be placed there to break the line.

Bill Seatland, who is the architect-contractor-carpenter is a very competent worker. He came over and talked about my addition and expressed interest in working on your house. We call Jan Wampler on Sunday and he felt it would be best for Bill to do the entire job for me, since he is here on the spot. Jan offered to help and said he would be glad to have Bill do the construction of your place, all next spring. This all seems to be working out very well.

Also, we stopped to talk with the manager of Continental Telephone the other day and found that the PUC is still working on Monhegan's phone service and either New England Tel. or Continental will have to pick up the option. Monhegan will not have to fork over any of the construction costs except, perhaps line extension costs, and the base monthly rates will be reasonable. All good news. We hope things will get started before the present cable fails us again.

Winter is approaching. Today the wind has already peaked at 35 knots, but we have yet to have a killing frost and we are still eating out of the garden.

Alta sends her best. Thank you again for the collar and the pats.

As ever,
Phoebe S. Ashley

October 11, 1978

Mr. Bennard B. Perlman
6603 Baythorne Road
Baltimore, Maryland 21209

Dear Mr. Perlman:

Thank you for returning the page from my mother's old album. I hope to see the reproductions some day.

In 1911 we lived in Allston, Massachusetts, part of the city of Boston. We came to the island for the summer, usually by boat from Boston to Rockland, although I remember going to Bath and then to Boothbay, probably by a smaller boat, and over to the island on Cass Brackett's boat.

Ed Brackett, brother to Cass, owned the Monhegan House where we stayed. Our first summer was spent in the home of Cass Brackett, down where the parsonage now stands.

Sorry to say I was not the child mentioned by Henri. There would have been two or three of us. My father would not have been here in September, at least that late. I wish I could tell you who the child might have been.

Also, I am sorry not to be able to remember much of that time. I do know that Mother enjoyed her association with both Bellows, Henri and their wives. My father had fun playing baseball on the team with Bellows. Both were enthusiastic athletes. I have a feeling Mother stood much in awe of Henri, who was somewhat older than the rest. She said that Mrs. Henri was reserved and didn't join in with their fun and games. I remember there was a lot of laughter when Bellows was around. Henri was kindly, in an awesome grandfatherly manner toward us children, my cousin Emma and me. It is too bad my sister and mother are no longer living so I could ask them questions. Gertrude would have been twelve in 1911, I was barely six.

Sincerely yours,

Bennard B. Perlman
6603 Baythorne Road
Baltimore, Maryland 21209

October 4, 1978

Dr. Alta Ashley
Monhegan Island, Maine 04852

Dear Dr. Ashley:

I hope the photographs of Henri, Bellows and Randall Davey arrived without damage or the page being creased.

Last night I was reading through some new Henri material and found two references which I thought you might enjoy:

Sep. 17, 1911 (Henri letter to his mother) - "Bellows was here (on Monhegan) with us but had to go home because his young wife was about to have a baby (a girl)."

Sept. 22, 1911 (Also written from Monhegan by Henri to his mother) - "There are only a few of us here. One man and one woman and her child besides Davey, Marjorie and I at the hotel. We have rain and fog occasionally but only enough to give variety. Last week got some fine surf."

I think I am correct in assuming that "the child" referred to in the second letter must be you. Henri also received a letter back in 1905 from his student, Rockwell Kent. Writing from Monhegan, Kent said that Mr. Brackett has your fish line put away for you. I expect to write to "Zimmy" about that, since it must have been his father (or uncle), in charge of the Monhegan House at the time, who is referred to by Kent.

It's all quite exciting, and especially after having met you and obtained the pictures. If you recall any other events of the period - perhaps things you remember your mother having said -please feel free to pass them along to me. Incidentally, where were you living in 1911? Did you come to Monhegan for a vacation (as was the case with the artists) and, if so, where was your home?

If I come across any further references to Monhegan similar to the above I shall let you know.

Sincerely,
Bennard B. Perlman

Center Tuftonboro, NH
03816
December 8, 1978

Dear Winnie and John:

It looks as though I shant be back on the island to welcome you; and since I doubt there is a page of instructions I want you to be sure to read this before you turn on lights or water.

The generators are off. Please use the small generator by throwing the switch in the generator house **down.** The water pump switch is also off. It is situated to the right and back of the space heater in my part of the house. If and when you leave the place for any time (off the island) please throw each switch **off.** This prevents a flood, should the pipes freeze or the place be struck by lightening.

Do not switch lights or appliances, TV's or radios. Last summer both the Black and White TV and the Zenith radio were plugged into 110, instead of 12 volt system and both burned up. The TV was replaced, the radio is still being repaired. Twelve volt lamps will also blow up if plugged into 110 volts. The bulbs cost $1 and $2 a piece, also they could cause injury when blowing up. 12 volt plugs are silver, 110 are ivory or brown.

If the TV picture on the 12 volt system begins to shrink it means the batteries need recharging. Turn the charger timer on for the length of time you will be using the 110 system. It will stop spontaneously.

If the water pressure falls there may be trouble with the light which controls the pump. It is located in the cellar and connected by wire to the pump guage by the pressure tank. If you have any trouble CALL BOB BOODY.

Be careful not to overload the small generator, but it will run better initially if it has a fair load. I usually set the recharger and then turn on a light. Do not run the charger with more than one appliance on such as an iron, vacuum cleaner or washing machine. If you plan a heavy load switch to the large generator, making sure there is plenty of antifreeze in the radiator.

My gas supply is on red. Would you mark the date when the other supplies go on red and record the number of cylinders you set up, should you need to do so before I return, probably not until after Christmas.

If you need to reach me you can call me here at 603-569-1521. There may be enough money on top of the phone, if not, I'll reimburse you.

I was rather distressed to find that Katy used the cottage when I expressly said ‘‘NO’’. There are just too many things which can go wrong, as you can see, when someone doesn’t understand the basic workings of the place. Also, I found Tugboat inside with no litter pan. Please take her with you the next time you both leave the island. I think it only fair that you bear any the costs of any damage the cat may do or the result of mishandling of the electric system.

Would you mind checking the space heater in my section, turning it up or down as necessary. Probably a setting of 1 should be high enough except in severely cold and windy weather.

The wedding was beautiful and the reception most enjoyable. It was nice to be seated with the ‘‘older’’ Monhegan group. We should have more such celebrations!

Good luck!
Alta

MERRY CHRISTMAS AND A HAPPY 1979!

Monhegan, Maine 04852
December, 1978

Dear Friends:

Here I sit in the Trade Winds (Motel) using a borrowed electric typewriter set on a chair, the only surface large enough to hold it. I can't find the gadget to change the type spacing or release the carriage, so bear with me.

This year I was going to do everything ahead of time. I left Monhegan to spend Thanksgiving in New Hampshire with Emma. We had a nice capon (and still have several meals to go on it). Saturday we went to Concord while I went to the Neilson-Murdock wedding. Next day we drove to Kingston to see Aunt Helen, who is a happy resident of the Baptist Retirement Home at Evanswood. Emma hasn't been feeling up to par of late, so I decided to plan to stay with her for a while. I am going over to the island tomorrow and pick up a few much needed things.

Last winter many of you know Monhegan was clobbered in both the January and the February storms and were without telephone service from February 6th to May 22. I was already involved with the phone company about another problem so kept on with my contacts. After all these months I learned that the Continental Telephone men were over on the island looking for me today! I'm hoping to catch up with them tomorrow on my quick trip over. It looks as though we will be having an expanded service before long. Sorry I won't know before I have to turn this stencil in to be run off. As soon as I know something definite I'll write those concerned.

The past year has been a busy one. First there was the problem of getting Aunt Helen situated. Bob and Helen Woodbury did most of the work, but I had the job of closing her apartment, a job that had to be done in one week. I was knocked for a loop, somehow, and didn't get back to doing anything until late summer. Every time I thought I'd get down to working on my Monhegan history something else would come up. I finally brought my newspaper clippings album up to date, though there is still '72 and '75 to soak off another wrongly prepared album. What a job! There seemed to be a lot of special articles to write, but that doesn't account for the lack of production (or housekeeping!) or letter writing all these months. It is just that things take a lot longer to accomplish than they used to.

Alan Mersfelder's amphibian plane service to the island has been quite an innovation. I flew out twice but have never returned by plane. Fifteen minutes from Rockland is all it takes. The only problem is wind, particularly from NW. I hope to use the plane for at least one way tomorrow. There are so many things to find and pick up I doubt I can do it while the boat is at the dock. By the way, the Laura B has been off the run for nearly three weeks for a multiple of repairs, unfortunately, not for winterizing, as we had almost dared to hope. Some Bass puff boots should keep my feet warm on the crossings this winter.

There has been a lot of building on the island this year. First of all the buildings damaged by the storm, and the one washed out to sea had to be repaired and restored. Now the Sullivans have a house and Carl Wincapaw a separate fish house. Doug and Bobbie Way have a new house on Horn's Hill. The Mooring has a new porch and many houses are sprouting new shingles and roofs. I am really going ahead with plans for my new room and hope to be able to set up a work room downstairs for the typewriter, which spends a lot of time on my bed when not on the dining room table, my ditto machine, the scads of albums and bulging files.

Last April Elizabeth Gardner and I spent a week in Monaco and Southern France. I visited St. Paul, Provence where the Santons were made and bought only three. When I went to get more at Damariscotta I found they hadn't ordered any this year because of the fallen dollar. This February we are going to Sapphire Bay, St. Thomas for a week with the Sutherlands, my great niece and nephew in a condominium. It should be great fun. I hope to get in some swimming. Monaco was too cold.

Better say Seasons' Greetings to all from Phoebe and me.

Alta Ashley

Monhegan, Maine 04852
January 4, 1979

Dear Friends:

Happy New Year. Today the men (and gals) started moving their traps to the dock with the expectation of being able to load their boats and set traps tomorrow. At a "lobster meeting" yesterday they put a limit of 250 traps per fisherman (or maybe per boat) to be put on the dock. By 10 AM today it looked as though the piles were as high as usual and it didn't look as though the limit had been adhered to. However, with fifteen boats this year and two to three fishing on some boats, 250 x 20 or more makes at least 1000. they were all through by early afternoon despite horrendous traffic jams on the dock and on Dock Hill Road all morning. Alan Mersfelder was to bring out some people this afternoon, but the winds, absolutely calm at dawn, hit 35 knots by mid-morning and have been gusting to 30 all afternoon, so all flights were cancelled.

Trap Day has been put off because of three factors. On New Year's Eve the wind blew over 40 knots, the following day it hit 47 and we were enveloped in dense fog which continued the following day. Barbara Stanley's grandmother died on Sunday and she and Sherm went in on Tuesday on the Laura B, then were socked in in Rockland. No planes flew in or out of Maine all day and driving was slow and hazardous because of fog and slippery roads. Bryan Hitchcock has been trying to get a new engine installed in his boat for weeks and even yet hasn't made it back to the island.

We were all surprised to see Sherm and Barbara come in on the Laura B yesterday, thinking they had gone on to New York. With them back and with no knowing when Bryan could get here Trap Day was agreed on as the first good day after today. It is the weather now that may hold them up even through the weekend, although they may decide to start tomorrow even though no boats have been able to be loaded because of the rough water in the harbor. This year there will be three gals fishing. Debbie Bates will go as usual, Leslie Whitney is planning to go, at least for a while, with Shermie, and Jeanne Gibson will be going with Steve Rollins. She helped him all summer with the garbage and seems to thrive on rough weather. Two hours taking pictures is enough for me. How they can stand bagging bait is more than I can understand. I nearly pass out whenever a bagger comes into the Post Office or Store fresh from their labors. Someone came in here and made a phone call the other day before changing her clothes and one made an hour call at the Store before washing. She won't do that again, if Doug and Harry have their way!

Phoebe and I didn't return to the island until December 27th. We gave up trying to come over for the Christmas party because we might not have gotten away in time to go with Emma over to Hanover on the 20th. From all reports it was a happy time for the many who did attend. Christmas day brought a real blizzard. Emma and I were invited out to the Dows' at Tuftonboro Corner, just five miles away, but

I didn't dare try to drive, for the first time in my life. We were picked up and brought back home and spent a very pleasant day with the Dows, their friends and Sandy Dow's 92 year old mother. (I wondered if she might be 75!) On Tuesday I decided I had better head for Monhegan and not get caught in another storm, also took my chances and spent the night at the Samoset, very pleasant. It was there we were marooned last February. Alan had a flight out to the island early on Wednesday so we went out with him. As we landed it began to blow. Before he got Phoebe and me ashore and JoAnn and Sarah on the plane and took off the wind had increased so he had a hard time taking off and all future flights for the day were cancelled. He is now based at Port Clyde, which makes it better for parking.

We are still waiting for Trap Day. Bryan brought his boat back late yesterday and the men began loading their boats with rock, but no traps were put aboard because the weather report was none too good for today (Saturday, Jan. 6th). this morning the whole island was awake and preparing to load their boats. There was an almost complete calm although a surge by the dock remained. It was cloudy, but visibility was 7-8 miles. I got up and made sandwiches for Doug and Harry and put on a pot of coffee about 6:30.

Just before another "lobster meeting" at seven it began to snow, one of those fine snows that so often turn into huge storms; visibility dripped in fog. Weather reports were conflicting, so the decision to "go" was put off until another meeting scheduled for noontime. Meanwhile, Shermie Stanley was tentatively planning to go in because a dental problem, so we may be waiting over the weekend. News must be scarce because we are being bombarded by reporters. Right now (11 AM) the sun is trying to come out even with snow still falling.

The problem with setting just before a storm is that dry traps, even weighted with stone, are buoyant and don't settle well on the bottom. Rough seas toss them about injuring or killing any lobsters inside. Also if they can't haul their traps and rebait them for several days the lobsters are weakened and may die or be cannibalized. That is why the fishermen are so particular about conditions on Trap Day.

It is now after noon. The decision now is to wait for another "lobster meeting" tomorrow morning. Many of the men are superstitious about setting on Sundays so they may delay until Monday. But a storm is racing toward us, once due on Tuesday, now on Sunday night or Monday. So it is anyone's guess what day they will get started. Right now there isn't a ripple on the water and not an inch of surge at the dock - a perfect situation for setting, but the sun will be setting soon, before the boats could even be loaded.

While the adults are impatiently idle the children are having a great time skating in the Meadow which has been pretty well flooded by the heavy rains in late December. They tried flooding the Ice Pond last week by pulling water out with the fire hose and spraying the surface, but there was too much snow on the top, which floated and froze making for a hubbly surface.

I learned today that someone thought it was I who had died Tuesday, apparently getting me mixed up with Barbara's grandmother or Lota Hogg, the wife of Bishop Hogg who has filled the pulpit several summers. I for one will miss her a lot. She was a wonderful musician as well as person. She coached me and made me feel my singing added to the church services and I found great joy in singing to her organ accompaniment. We did Bach, Mozart, K.K. Davis and other anthems.

Alan's plane is the Monhegan Seagull, not the Starling. Only Monheganites would understand the significance of the latter name. Others might not want to fly with a Starling, but would feel like soaring on the wind with a Seagull.

Emma's next trip to Hanover is on the day Elizabeth and I are due to take off for St. Thomas. On the phone she said she felt much better and would drive over herself, which is very good news. I'm looking forward to that week in the sun after this week on the island. The fuel line to the cottage froze so they had minimal heat for several days before they told me. The little generator quit charging the battery because of a short. It drained the battery to the big generator so that this morning neither worked and we had no water pressure. While I was gone my stove went out, then leaked kerosene on my rubber-backed wall to wall carpet downstairs. This all happened the coldest day of the year and about half my plants died. I haven't had the heart to look them all over and hope my night blooming cerius made the grade. The hoya took a beating and a ti plant gave up the ghost as did most of my herbs and all the tomatoes I had hoped to grow through the winter. I just ate the last cherry tomato today, taken from a dry stalk.

No plans for the new room have arrived as yet. If none is forthcoming I'm seriously thinking of giving up renting the cottage after this year and moving my stuff over there, keeping this place as a workshop and guest house. The little profit I make isn't really worth all the bother. Most of the profit goes back to Uncle Sam and dear old Maine, anyway. I could then drain the water and close up the place in December and go where it is warm until mid-March. A bit of swimming ought to help get rid of some of this thickening middle!

There will be a lot of changes on the island in the coming months. Larry and Judy Hubbard won't be returning next fall so we must start another search for a teacher. Doug and Harry are hoping to give up the Store come spring, Bud and Sally Murdock will be taking over, as plans are now. It will seem strange not to have Doug and Harry at the Store after all these years. I don't remember anyone else there.

Sunday AM

In the wake of a "Winter storm watch" the men have decided to set the traps now piled on the dock. For a while decision was held up because neither Carl Wincapaw nor Craig Sproul had helpers. I guess those they had counted on had had to leave last week to get back to work or school. Now I must get out and take pictures and write up my story for the papers.

May the new year bring health and contentment.

As ever,
Alta

Later: Well, they finally found helpers for Craig and Carl and loading of traps began about 8:30. Then, while Don was being loaded and Doug, Carl and Bryan were waiting to get into the dock, Vernon took off and a horn was sounded at 9:25. I'm not sure who set the first trap. I saw one go off of Rusty Court's boat just at that time. And so two precedents were broken this year - setting on Sunday and starting before all boats were loaded. I believe they had agreed to this at the meeting this morning, because of the long delay and the need to empty the dock, as well as the storm watch still in effect.

I forgot to mention a real innovation - Rita has a new door. No longer can we tell whether she is in or out by seeing the newspaper stuck in the door or the hook spanning the latch. Billy Paine and Peter Boehmer effected the transformation this fall while I was gone.

The Portland Press herald has arranged for the Coast Guard to come out from Boothbay Harbor to pick up my film and story - a fine way to use our tax dollars, I'd say, and told them so!

A & Phoebe

MONHEGAN TRAP DAY 1979

Jan. 7. At eight o'clock this morning the fishermen on Monhegan began rowing out to their moorings to begin the lobstering season for the year. At a "lobster meeting" which began an hour before they agreed to load their boats with traps already on the dock since last Thursday, even though a winter storm watch was in effect. Ever since yesterday afternoon the sea has been glassy smooth and the tidal surge at the dock, which had delayed trap setting, had flattened out completely.

Some helpers who had come over to the island had to return to work or school during the week's delay, and a further delay this morning was occasioned until others could be found to work on two of the boats.

Loading from the dock and Middle Beach began about nine o'clock. At 9:25 a horn blew and the first trap slipped into the water opening the lobstering season for another year. In all the confusion of the moment it isn't clear just who set the first trap. Vernon Burton was first to set off for his favorite fishing grounds but Rusty Court may have sent the first trap over the gunwale of the Annie D.

Two precedents were broken this year. Ever since some years ago when several calamities occurred to the fleet after a Sunday Trap Day the men have avoided scheduling Trap Day on a Sunday until today. The long delay this year and the possibility of a storm by afternoon broke the precedent of waiting for all boats to be loaded before the starting horn was sounded. Two boats were being loaded and three more were waiting their turn at the dock when the starting horn sounded, as per agreement reached at this morning's "lobster meeting". Now, scarcely two hours after the men started for their boats, dark clouds are gathering in the west, the temperature is rising and the glass is falling, heralding the approach of a storm which may drive the boats back to their moorings by early afternoon. Meanwhile they are busy going to and from the dock, and the hum of boat engines once more fills the air.

This year there are fifteen boats in the fleet, including Bob Mersfelder's dory, and three women lobsterers.

Monhegan, Maine 04853
Jan. 25, 1979

Dear Friends:

This is just a quickie because there is little to report since I last wrote. We've had stormy weather much of this month but have managed so far without dire consequences. Last week we had about a foot and a half of snow dumped on us which tied up everything, but Sunday's rain washed most of it away and the ice on the roads finally got down to manageable size for the oil truck. But the oil boat, expected for more than two weeks, put off their visit until tomorrow! We are down to 1/4th a tank, about where my stove usually goes out. Today's storm, expected to bring mountains of snow, just brought slop and the temperature was over 32 degrees so I've kept warm by burning six logs in the Franklin and no oil. The seas kicked up by the storm, though, aren't supposed to subside, so heaven knows when the kerosene supply will arrive. My enthusiasm for spending winter in the north is waning fast, especially since I've done all the shovelling so far this year - young people don't seem to realize the importance of cleaning up after each storm, or they don't know how to shovel.

In my last letter I apparently jumped the gun and could have ridden the wrong horse. Rumors were so numerous and persistent that I mistook them for fact. The only fact is that Doug and Harry **hope** to be free of the Store in the future, but no define time or decision have been reached.

A week from now I'll be off on the first leg of my journey to Sapphire Bay. Weather permitting I'll fly with Alan Thursday morning, giving me time to do some shopping, drop Phoebe off at a Kennel in Warren, visit the Mitcheners and join Elizabeth on Sunday. We fly out on Monday for eight days, all too short to make up for the snow and rain we've had so far. Alan has been busy flying back and forth when weather permits and he doesn't have engine trouble. He made four or five trips yesterday, even bringing out the fellow who captains the Laura B on one trip, who said that one nice thing is that your feet don't get cold on the plane! I haven't seen Jimmy Barstow for ages. His grandmother died last week and he and Magee had to go to New York for the funeral. Magee was here over the previous weekend. Jeanie Lord was here, too, and is feeling much better. She, Victor and Kole are staying in an efficiency at the Trade Winds this winter, Kole is in nursery school and loves it. They have found a medical regime which is doing wonders for Jeanie.

I hope this letter will quell the storm raised by my last. Forgive me for disturbing you.

Best wishes,
Alta & Phoebe

Monhegan, Maine
March 8, 1979

Dear Friends:

Although Town Meeting is only four days away I am going to start this and add a paragraph or two for those interested in Monhegan events after that momentous occasion.

Many things have happened since my letter in January, and the second one some of you received the end of the month.

Phoebe and I left the island a day earlier than expected because I learned that Emma was at Huggins Hospital in Wolfeboro, having under-gone surgery on the 29th (no malignancy, thank goodness). I didn't have time to do more than drop Phoebe off at the kennel in Warren, even though we went over by plane. We were on the third trip that morning! We flew over the oil boat on its way to the island for the much needed fuel. It had been unusually warm all January so we managed to get thru the month on a quarter tank of oil, but a cold wave was expected, so the oil boat was a welcome sight.

Everyone was so low in fuel that the new supplies were exhausted by the time everyone was serviced, even with the delivery truck being filled twice at the dock from the boat. The cold wave was a real one, and one no one wants to reexperience. Pipes, pumps, wells, drains and septic distribution fields froze in the fifteen days of zero and below temperatures and gale winds which accompanied much of the cold. Even the water in my pressure tanks froze. Poor John Murdock, living in the Mooring, spent many an hour thawing out everything in both houses day after day.

I wonder how Phoebe, who is used to snuggling under the covers beside me, managed to survive in the kennel. It was plenty cold in New Hampshire and Concord but both houses had central heat and thermostats.

Emma was still receiving IV's when I had to leave for Concord. On Sunday Elizabeth and I drove (she drove) down to Kingston, Mass. to see Aunt Helen. We got there in time for tea and had a nice visit with her. She didn't want to go out in the cold, so we stayed at the home, a very pleasant and comfortable retirement place. People are free to come and go, may have their own cars, but are looked out for as necessary. Aunt Helen looked her youthful self, as usual. She is incredible at 93. Will any of her nieces and nephews do as well?

Monday, Feb. 5th Elizabeth and I flew off to St. Thomas. We drove to Belmont where Tommy Ashley drove us to Logan. For once we got off without a hitch and landed in St. Thomas in the afternoon and were met by Bruce Gardner, Elizabeth's son who is studying at the College of the Virgin Islands, Scott and Cindi Gardner Sutherland who were sharing our condominium at Sapphire Bay.

What a week we had. Bruce had only evening classes so was able to chauffeur us around the island. He even went with us on a day trip to St. John where we swam at Trunk Bay and enjoyed a buffet at fabulous Caneel Bay. Scott had a Rotary meeting at Frenchman's Reef Hotel one day so we all went to the buffet there, then swam on that beach.

At Sapphire Bay I met Dr. Sam Belknap from Damariscotta, who spent each early morning finding treasurers on the beach with his metal detector, and Bill Berenberg from Children's Hospital whom I have known for many years. Sapphire Beach Resort is a great place, one can get in only by a pass. The condominiums are in a semicircle parallel to the beach. There is always a breeze off the water, the beach is powdered coral and interesting fish can be seen among the coral reefs, if one snorkels. I didn't do it right and got two big mouthfuls of water on my one attempt. There is also a restaurant and bar right on the beach and food stores not too far away. There are tennis courts and a salt water pool. One scarcely needs to leave the area for food or entertainment. But we did.

Bruce drove us all over the island, which is nothing but a series of tropical mountains. The roads are a series of S and double S curves, hairpin turns and blind driveways. Traffic keeps to the left, which was mighty disconcerting to us as we rounded curves and entered intersections. The week was all too short, except for my skin, which developed an itchy rash which is still with me to some extent.

When we left on the 12th it was still zero or so in New England and we nearly froze before we retrieved our winter coats in Belmont.

The following Wednesday, no, I'm jumping ahead too fast. The day before we left for St. Thomas my niece Nancy Peters entered the hospital for surgery by none other than Bob Barker whose family's phone I share here on the island. Elizabeth and I went over to see her at the N.E. Women's Hospital over on Jamaica Pond. She looked pretty, but fragile with her new Shirley Temple hair-do and was all excited about a forthcoming trip to Italy in April with Mary Pierce her college roommate at Wheaton, can it be thirty years ago? On the way back to Concord we stopped to see Nan McKeen Stanley, an erstwhile Allston playmate many more years ago than that, now living in Chestnut Hill. We topped the day off with a visit to Katherine and Ashley Day in Concord. They summer in Five Islands, Maine whence they can see Monhegan on clear days. We were pretty well visited out by the end of the day, but it was good to see everyone.

Emma was still in the hospital so I went back to New Hampshire, detouring to Warren, Maine to pick up Phoebe, who couldn't have been happier when she spotted the car. There was no chance to fly out to the island because of wind and ice in Port Clyde, so we went back to Center Tuftonboro.

Emma left the hospital a couple of days later but was housebound by the snow which was two to three feet on the level and mountainous along the roads. Although the plow had broken a path from the barn up to the house it was too slippery going with snow-covered icy patches for her to navigate safely outdoors. Parker came on from Oregon, hoping she could go back with him, but that didn't seem wise at this time. He and Deedie took Emma down to a boarding home in Wolfeboro to stay until she would be able to get out and around and drive her car. After seeing her settled there Phoebe and I came home, having been away five weeks and more. We caught Alan just as he was loosening the lines of the plane for a flight to the island. I don't know whether Phoebe or I rejoiced the more on being back home. But after hearing the tales of woe concerning the BIG FREEZE we were thankful we had been cozy in Emma's house during that time. After Town Meeting we will be going back to get Emma settled in her house again. Let us hope the mounds of snow will be well gone by then or shortly after. There is no snow here but PLENTY of M U D!

The winter has been strange. All of February there was no snowfall until the last day. Since then there has been almost nothing but rain which has fallen as snow up in Tuftonboro, of course. The Meadow is a huge lake right now and side road culverts are raging torrents. If only we could impound some of it for next August.

Doug and Harry have been in Barbados for the past week and are due back sometime today, though not to the island until tomorrow or Sunday, in time for the next cold spell, 'tho not much below freezing this time.

This winter we have had three unusually high tides, but, unlike last year, they occurred on windless days, so no damage was done. I was here for the January tide. Water covered the entire slip and one third of the surface of the dock which, unfortunately, hasn't been resurfaced since last year's storms. An article in the town warrant, however, asks for a appropriation for $3000 for gravel - oh, that is for the roads. Wharf repairs already have allocations but no one to do the work since Mace Carter died last summer.

My plants took a beating, but most survived. The Ti plant is putting out more shoots, the cerius is only half dead, some tuberous begonias are putting out shoots, so things aren't as bad as I had feared. it must have gotten mighty cold in the breezeway, though.

On the happy news column: Georgia Anacker spent much of the winter in Canada where she met someone who will become her husband on April 28th. The wedding will take place on the Anacker family farm in Maryland. We are all very happy for Georgia.

Yesterday during a pea-souper the Manana horn wasn't working. It was at boat time when the thing stopped functioning and the Laura B blew and blew hoping for an answering horn as she made her way slowly through the Duck Rocks. Today they had a C.G. inspection over there, but the fog had cleared by the time the helicopter with the brass arrived. Let's hope the horn was fixed.

Plans for the new addition finally arrived last week. I've spent hours trying to fit them to the actual conditions. Work should be starting by the end of the month or sometime during April. Do you think I'll find room for this typewriter and the fancy dishes now piled on top of each other or just out on my kitchen working space shelves. I can hardly wait for the room to be available.

If all requested funds are appropriated at Town Meeting on Monday we will be raising around $55,000 in taxes this year, up about $3,000 from last year. Some of this will be covered by new construction, so there shouldn't be too much of an increase in the tax rate.

This just about winds up news for this time, except for results of Town Meeting, which I'll write in for those who are interested. I plan to return to Emma's on Wednesday. Probably will be there for a couple of weeks and then will be back here pretty much all the time, once work is begun on the addition. If you want to write the address is:

c/o E.C. Riddle
Center Tuftonboro, N.H. 03816
603-569-1521

As ever,
Phoebe & Alta

Kila Bates arrived on Manana in January and Betsy Lynn Mersfelder is now with her parents at Green Point. We creep steadily toward 100 year round inhabitants. By the way, the winter people have formed a craft shop whose wares will be on sale in the summer. I guess I'd better get my knitting needles reactivated. As yet I don't know where the shop will be.

Monhegan, Maine 04852
April 24, 1979

Dear Friends:

Phoebe and I have just returned from our first walk to White Head for the season. Last week we went through the Red Ribbon Trail, over Long Swamp Dam and back through Cathedral Woods. On both walks we met so many blow-downs it was almost impossible to get through. According to State Forest Service people we have the worst fire danger conditions of any wooded areas in the state. If a fire should break out before the trails are cleared there would be no way of getting to it with either the pumper or the new rescue vehicle recently acquired. It becomes increasingly scary as each dry August rolls around.

After Town Meeting Phoebe and I returned to Center Tuftonboro and got Emma back to the Farm. Ten days later she had improved so much it seemed time to return home and let her handle things for herself, especially since the place was to be repainted - at least the kitchen - new linoleum laid, etc. We got back to the island on March 26th, this time by Laura B, the first ride on her since early fall. The following week we went in for two medical meetings, expecting to fly home on Thursday, but weather dictated otherwise. Since I had two medical appointments the following Monday I decided to spend the time visiting.

We found Marge and Paul Mitchener at home in Andover and spent a pleasant overnight with them and another night with Elizabeth in Concord. Our arrival at Elizabeth's was followed in minutes by a wild wind-snow storm, complete with thunder and lightening which drove Phoebe quivering into my lap. The storm lasted about an hour and then the skies cleared and the sun shone brightly for the rest of the day. Thank goodness we weren't caught en route. We drove out to Maynard to a salad and entre bar restaurant, took enough of the entres to eat for supper!

Emma seemed fine and the refurbished kitchen bright and cheery. The Huxtables, just home from Florida in time for snow which fell on Saturday, came over Sunday noon and made us drool about their newly acquired condominium in Ocala, Florida.

Instead of flying back to the island Monday after my appointments we battled snow and sleet and high winds, so holed up in the Samoset and made good use of the pool. The storm lasted until late Tuesday, so we didn't get back until Wednesday, then by the Laura B. There was a huge flock of Canada geese on the Samoset golf course which we watched from the restaurant windows. Phoebe went daily down to play with them and returned like a golden streak after a few ganders hissed at her. We met Jim Schweier on the dock, missed Ben Odom and didn't see Larry Leonard who had been here working on the 75KW generator for weeks because we didn't get to the Store because a team of four fellows interviewed me all afternoon for a documentary film. Doug took him in early the following day.

Jim Balano has sold out his boat, gear and truck to John Murdock and on May 1st (I can say it now!) Bud and Sally Murdock will be taking over the Store. Doug and Harry will move to the fish house while work is done on the Underhill house. Bill Sweetland, who will build my new room, is doing that work, too, so mine won't be started until early June. It is scheduled to be finished by July 5th.

While I was at Emma's I visited a movie producer, David Quaid, who lives on the top of the highest hill around. He has a new house with solar panels and a wind generator. Michael Phelps, across the road from Emma, helped with the installation of the generator and is coming out here to study our situation. He is to meet with the committee that is studying the feasibility of central electric power for the island as well. With the increasing cost of fuel everyone is thinking wind and solar energy. One like the wind generator on Cuttyhunk would be quite a boon to Monhegan.

For weeks the island buzzed with mysterious and cryptic messages over the CB radio. Meetings were held at Mary Burton's. Then posters appeared about a Spring Fling for "sometime in April" and "after the water is turned on". Rita White was about the most excited person thinking about a big party and "with a band, too". the entire island including the children, but not Rita, were privy to the real purpose for the party. With the impending transfer of Store ownership everyone wanted to make a fitting gesture of appreciation for all she has done, all the little favors she has done, particularly since the advent of CB radios, filling orders for husbands to carry home, setting some outside the door to be picked up after the Store closed, watching laundry and letting people know when the machines were available, etc.

The plan was to hold the party a week before her birthday, April 21st, then push it forward a week, to keep her off the scent. I invited her up here for her usual birthday dinner with Doug and Harry, then we all would go over to the Inn for the Spring Fling. Several times we thought surely the secret was out. Doug and Harry left here in the jeep to get their dessert at the Inn. Rita tried to stay and help me with the dishes and pick up. I finally convinced her we ought to get over to the party, but she did have to stop at her house to pick up some cigarettes. That gave me time to dash in and tell everyone she was on the way and absolutely unaware that the party was for her.

The diningroom was decorated with streamers and balloons, all red and white. A throne covered with red velvet was set up. Karen had made a lovely crown with rhine stones and blue crystal. Presents were piled by the Throne including a TV, Battery and charger and everyone had signed a card painted especially for Rita by Jon Bogdanov. Rita carried the whole thing out with dignity and composure as if this happened often, yet letting her pleasure and gratitude shine through. it was a great party for all, but especially for Rita. The band played or records were spun well into the wee hours, and Rita was one of the last to call it a day. Even Doug caught the spirit and danced.

Mary Burton made and decorated a huge cake to serve the nearly one hundred guests. Cameras clicked and flash bulbs popped through much of the evening. We need more such parties, though two probably would do us each year, but none will come up to this one of April 21, 1979.

A spray of forsythia came out just in time to decorate my table. June had just given me some red table napkins and I found a paper table cloth with Happy Birthday. Rita obligingly wore her red outfit, which was just the thing to go with the decor at the Inn. Such fun! Georgia Anacker's fiance, Doug Laver, was one of the guests. Their wedding will be held in her family's garden this Saturday.

Barbara Taylor is back with two of her children, Sam and Doug. The latter is in school here as is Steve Whittit's oldest child. Steve is building the Hudsons' studio down at Calf Cove. Steve Rollins is going ahead with his house and the Schubert-Chase cottage will go up on Horn's Hill in the fall to replace the one burned in 1961.

That about brings us up to date. My garden is partially planted but nothing is up. It is still cold, even dandelions are scarce.

As ever,
Alta & Phoebe

Monhegan, Maine
June 1979

Dear Fellow Monheganites:

More than a year has passed since a telephone service questionnaire was sent out. On the basis of the returns neither New England Telephone nor Continental Telephone feel that incorporating service to Monhegan into their companies would be economically feasible and both companies have cancelled any plans for future service.

Although we now have limited service from New England Telephone over a Coast Guard cable, NET states that Monhegan is outside its franchise area and as soon as the cable breaks again or is removed by the Coast Guard the company will not be able to continue service to Monhegan.

The Coast Guard Service says no further repairs to the cable will be made and, furthermore, the cable is scheduled to be taken up in 1980 when the Fog Signal Station and Lighthouse are due for automation. This would leave any communication to and from the mainland to CB and VHF radio services both of which are limited and far from satisfactory. VHF radios are confined to boat owners and service through the marine operator is quite expensive. Thus, if Monhegan people are interested in maintaining telephone service, we must search for a willing company or form an independent company for the island.

Recently someone has shown an interest in forming a Monhegan Telephone company, using the latest sophisticated "trouble-free" and operatorless system. However, between 50 and 75 phones would be necessary to make the system financially sound. So we again ask you to fill out and return the enclosed questionnaire, keeping in mind the following:

1. Wires and cables could be concealed, if desired, in areas where overhead wires and poles would be offensive, though with some added cost.

2. Phones in private homes would not be available to the general public for making and receiving calls, as now obtains, but

3. Public phones could be augmented, their location dependent upon the willingness of owners for maintenance.

4. Direct dialservice would be available

5. Monthly charges to subscribers would be set by the Public Utilities Commission and would be in the neighborhood of $7 - 10, at present rates.

TELEPHONE SERVICE QUESTIONNAIRE JUNE 1979

	12 mos.		Seasonal	
1. Would you like a phone?	Yes__	No__	Yes__	No__
2. Would you want more than one phone, if so how many?	________		________	
3. What is your estimated volume of calls, in dollars made from a Monhegan phone?	Cash ______		________	
	Credit, etc.__		________	

Remarks ______________________________

Please return to: Monhegan Telephone Survey, Box 87, Monhegan, Me. by August 1, 1979. Thank you. The Monhegan Telephone Survey Committee.

Monhegan July, 1979

Dear Friends:

Where has the time gone? My last letter was begun three months ago just after my last trip to the mainland. Each week I think "I just must go in to get garden supplies, clothes etc." And each week there is something expected that requires my presence here, but which hardly ever comes about.

First of all - the Store changed hands on May 12th when Doug and Harry passed over the keys to Bud and Sally Murdock. Doug and Harry stayed on for a month, helping with details which can be learned only by experience. Rita is still going in late in the afternoon to help with the five o'clock rush and the ordering of milk, bread, sundries and produce which she has done for D. and H. for many years. Having been at the cash register for so many years and helping to refill shelves after hours she has developed an uncanny knowledge of what people need and want. The truck strike and the gas situation have united to keep shelves unusually bare, making it look as though the new owners haven't learned how to order, unfortunately. The produce bins, however, have been overflowing with gorgeous vegetables from a farm in Tenant's Harbor.

Barbara Taylor and Dan Rothstein were married on the 15th. Her father, a rabbi, conducted the Jewish ceremony, Winnie the civil. After the ceremony the family went to the Inn for the regular buffet supper where they filled two large tables. Then came a fabulous spread up at the house for **three hundred** invited guests. Unfortunately, after days of warm sunshine we had a pea soup fog all day and a drizzle by evening so all three hundred piled into the house and **stayed!** There were grilled mushrooms, smoked mackerel (which Jim Schweier had done in a refrigerator set up in my yard) various kinds of sandwiches and, of course, wedding cake. It was quite a do, even by Monhegan standards of mass feedings.

This leads me to the third event. They began working on my addition on June 28th. Bob Boody worked a day with the bull dozer but couldn't do much because of the angle in which he had to work, so Jim Schweier (Juliet Odom's husband) picked and pounded for three weeks at the boulders and rotten rock they uncovered, rather than finding nice solid ledge. This appears to be an area covered with glacial till. Rubble was finally cleared down to ledge on Thursday and forms erected on Friday. Saturday Jim poured cement all day, making 2 foot by 2 foot footings for the foundation. That ought to withstand any earthquakes or landslides from Lighthouse Hill, I hope. When people ask what is going on I say I'm building a twenty floor condominium!

Bob Burton has bought Marion Wiley's house but she will use it as long as she wants to. It is increasingly hard for her to make the hill especially over the rubble from my excavation which has split into the roadway to her house.

I started this last week little realizing I would be reporting a major happening. It is now Wednesday July 25th. About 12:30 yesterday morning I heard a scream and thought it was another party on a boat, of which we have had our share this summer. Then I heard the fire siren and saw the fire engine turn the corner to the dock. People began running past the house so I dressed - a suit coat over my nighty (I couldn't find my clothes in the confusion) and dashed toward the sound of voices. I met a group coming back and found that there had been a fire in Uncle Henry's halfway up Dock Hill, but that Paul Krause and Kit Hudon, living there, had put the flames out. Paul had gone into the bathroom to check a drip with a lighted candle and had gone back to bed with the candle still burning. This had ignited a plastic showerstall. After two hours the fire was deemed out and the building was abandoned. Kit and Paul moved over to The Breakers, the Burtons' cottage. At quarter of six Billy and Cathy Payne, just up the hill in the other side of Rita's house, smelled and saw smoke pouring from Uncle Henry's. They alerted Vernon who put the alarm out on the CB. Once again the fire truck was put into action and the auxiliary truck was brought to the scene. The heaviest smoke I have ever seen, made eerie by a pea soup fog, towered over the building, then flames burst out of upper floor windows. The pressure from the fire engine wasn't enough to make a good flow of water even when the intake was dropped off the dock. Pressure pumps couldn't be started or wouldn't draw water. A hose was connected to a town hydrant by the Inn and garden hoses were brought into action, all to no avail. I finally called the Coast Guard to stand by in case we ran out of water.

Ron Stanley called off the C-G when they were en route. About seventy-five people formed a bucket brigade, lifting water up from the slip. Of course it was less than an hour after dead low tide. About the time the bucket brigade went into action the three pumps were in working order. The fire was brought under control in minutes and was out shortly after. No one was hurt, thank goodness. The candle holder, according to Pam Krause, was one made out of a cork float which burn-

Plane service ad — 1979

Wharf Repairs — 1979

ed when the candle burned down. We should warn everyone, since they are such popular items. Caroline Steiner will be out Sunday to see what should be done with the building which Vernon feels will have to come down. Too bad; it is such a quaint-structure and one of the oldest on the island.

Nothing more on my wind generator which is still on order. Meanwhile Bob Boody has reconstructed the one Dougie Boynton had and couldn't manage to use. His will probably be up and running while mine sits somewhere on a siding waiting to be delivered. Just as well, though, because I'm nowhere ready for the wiring and placing of new batteries. I hope I won't have to wait too long. My little generator, which I hope to sell, is becoming increasingly troublesome.

Work on the dock is nearly finished, except for carpentry needed on the slip, Also, the fishermen's breakwater at Fish Beach has been erected and should be finished soon.

Although we have been wrapped in pea soup for days we are drying up - only 1 1/2 inches of rain in June and 1 inch in July so far, while the rest of the country is being deluged. The Store laundrymat has been running on short hours because of lack of cistern water. I have a reel full of towels getting wetter and wetter from the fog. Now we have grasshoppers which are marching thhrough my garden reducing everything to lace row by row. Why don't the seagulls help us out as they did the Mormons? We have plenty. I guess they prefer fish to grasshoppers.

Doug and Harry came back today on the Laura B after a week in shore, their first summer vacation in their lives. They had started working at the Store as teenagers in 1930. They look great and are enjoying their retirement to the hilt. They are still living in their fishhouse while the family is enjoying the Underhill house and its new kitchen.

The gas shortage has had some effect. All hotels were completely booked in June but have had a few cancellations, most of which were soon filled. The Balmy Days has brought only a trickle of people from Boothbay but the Laura B and Alan's Seagull have had full loads most days. Fog has been a disaster for Alan, though, this week. My aches and pains are better and I'm off that old naprosyn. Have even screened dirt from my excavation and gotten a lot of good top soil.

Regards to all,
Alta & Phoebe

Monhegan, Maine 04852
September 22, 1979

Dear Friends:

Be reassured, I am still around and active. Tomorrow is the equinox when fall will come tripping in on cool breezes, the Laura B will make her last two-boat-a-day run and also her last Sunday run for the year, so it is time I caught you up on news of the summer.

When I last wrote the end of July I expected that by now I would be moved into my new addition and utilizing wind power, of which we have had plenty in the past two weeks with David, Frederick and now Henri flicking us with, not their skirts, but with a flick of their great coats or ponchos. But no. The wind generator, first promised to be erected in mid-June, is to be shipped "at any moment", and the room, scheduled to be finished on July 5th, has finally gotten a roof on it and outside sheathing is complete. There are several lines of shingles on one wall and red lines on the floor to mark out closet spaces and the dimensions of the laundry room. Windows are supposed to arrive this coming week and the slick little Jotel stove I found in Owls Head a couple of weeks ago will be set up soon. I **thought** I had two cords of wood ordered for delivery next month and now find I had spoken to "the wrong person". Wood in this land of forests is as scarce as hens' teeth because everyone is turning to wood stoves in view of the cost of oil and kerosene. The price here will be about a dollar a gallon! Needless to say, I am not planning to rent the cottage this winter, which means keeping heat on to prevent the water freezing, particularly where it crosses the sunporch between the two houses.

This has been a particularly busy summer, even though I made but one trip in shore and didn't stay overnight even then. So many people dropped by from out of my past or were readers of my column I have five pages of signatures in my guest book. Yet I had no personal guests stay overnight in this part of the house. In fact, the last family member to stop over was Elizabeth Gardner more than a year ago!

There were a lot of special events to write up, pushing aside the story I am trying to write about Monhegan stores, concentrating on the Odoms' store and its transfer of ownership to Sally and Bud Murdock. Any excuse to lay aside that blank piece of paper rolled into the typewriter at the beginning of a story!

Speaking of the Store, Doug and Harry are new people since they shed the responsibilities of that enterprise. At present they are visiting in New Brunswick, the first time they have ever been able to leave the island at this time of year. Instead of looking pale and careworn at the end of the summer they are brown and rested. Juliet said she grew to know her uncles for the first time in her life, and enjoyed being with them, out fishing, on walks about the island and just relaxing in the Underhill cottage. Meanwhile Buddy and Sally have enjoyed their first season at the Store and are looking forward to being involved in other things as the population thins out.

My garden, which was almost demolished by grasshoppers and the drought, has come back to life and producing like mad. A plant I thought was an acorn squash has given me three large, green striped creations which are probably pumpkins, or may be so-called squmpkins. One fell off the vine so I cut it open. The flesh is a pale yellow and very stringy. One which is growing on the ground is about two feet in diameter and should be useful for Hallowe'en. Only one acorn squash has appeared and that is barely as large as my fist. But the zucchinis are another story. Out of five plants I have had enough squash to supply the neighborhood and they are still coming. The carrots, whose tops were destroyed by the hoppers have revived. I hope they have time to develop some good roots. Beets have been great and there are enough to store in my new cellar for the winter. I braided my onions and have them hanging in the kitchen. A big bag of string beans is in the freezer and a new crop of them and pole beans is coming along as are two rows of sugar peas. Lettuce was pretty well destroyed by hoppers but new plants are heading up. I've been living out of the garden since early July and hope to do so for another three weeks.

It is now the last day of September. Work is progressing on the addition. All but one window have been installed. That one had to be custom made and should be here soon, none too soon for keeping warm. The door leading to the garden on the south side is still to be ordered. When it rains, usually on a southeast wind, there is a great puddle in front of it. Shingling is about half done and uprights for closets, etc. are now in place. I can hardly wait to move in and get my files off the floor of the bedroom. The wind generator, however, is still en route.

A memorial service for Jennie Whibley was held at the church on Friday. Jennie died in the spring, then Anne's husband had a serious car accident and died six weeks later, so the service was delayed until now. Thirty people were still here who remembered Jennie. Even Phoebe attended; as a Starling she was welcomed as a member of the family! Jennie was a Sterling, descended from Josiah and Mary

Trefethern Starling whose baby, Phebe Starling, was the first person buried in the Monhegan Cemetery.

Anne and Jane sang Whispering Hope, which both girls had often sung with their mother, and there was scarcely a dry eye in the congregation. I had charge of the flowers, most of which I took from Elsie and Glenn's lovely garden. Committal services followed the church service at the cemetery where the ashes were placed next to the graves of Josiah and Mary and not far from those of Phebe and little Mary. A small stone will be put in place shortly.

The cruise ship Viking Queen stops here, weather permitting, en route from Portland to Bar Harbor, via Boothbay Harbor and Rockland. One group of passengers boards at Portland and debarks at Bar Harbor, returning to Portland by bus. The other reverses the course. According to the captain we on Monhegan could board here and return or get off at Rockland on the return if a storm prevented coming out here. I'm tempted to take the cruise next spring or fall. During the summer the boat serves the Isles of Shaols as a ferry. I couldn't make plans for this year because of the building and waiting for the wind generator. By next year **maybe** both will be finished and I can leave for a few days. Tuesday I'm going in for Knox County Med. Soc. dinner and some much needed shopping, leaving Phoebe overnight with Jean Hayes who is in the cottage and is the only person Phoebe will desert me for a walk with in the woods. I'd like to stay longer but don't dare to because of things going on here, including an imminent arrival of a dish washer.

Doug and Harry are back, via the Carib from Yarmouth to Portland. Rita is through at the Inn, the boat is on six trips a week, the island roads are almost free of trippers. Some still come by boat and return by Seagull, however. Tom and Irene Martin are here for bird photography, so we are well into Autumn. No more space.

As ever,
Alta Ashley

September 1979

Dear Friends: This letter, which started following the Store fire of 1963 to four friends, has now grown to a periodic report of island doings to more than seventy people. Because of this increase in volume, as well as that for postage and supplies, I feel I must now send out the letters on a subscription basis, much as I regret having to do so. Thank all of you who in the past have sent stamps, donated paper, etc., as well as helping in other ways. I hope to write ten letters during the year, including the Christmas one, for $2.00 a year. Thank you.

Alta Ashley

I wish to continue receiving the Monhegan letters. Enclosed is $2.00 to cover my subscription for the year Oct. 1979 to October 1980.

Signed
Date

Please include any change of address.

Monhegan, Maine 04852
October 1979

Dear Friends:

What better way to use the hour gained last night as we switched back to Standard Time than to start this letter. Many of my chores are pretty well up to date. There are shingles and shavings to clean up in the yard, root vegetables to be brought in, but I can't do that until the new cellar is ready. There is real progress to report, however.

The shingling is done and about half the insulation is in place in the addition. Michael Phelps, the wind generator man, was over here this past week and went over the wiring with the carpenter, Bill Sweetland, so that is taken care of. We hope that everything but the washer, dishwasher, pump and vacuum cleaner can be run off the twelve volt system thus requiring the gas generator about three hours a week, except when the wind is not sufficient to keep the batteries charged. Let us hope that that will be the case because LP gas keeps going up fifty cents a cylinder each time I order.

It will be nice when I can set up the Jotel. I can keep warm with it and also do some cooking with it, saving gas for cooking and kerosene for the space heater. My last tenants, the Bakers, brought out a quarter of a cord of cut and split, mainly oak, stove wood which will fit nicely into the new stove. The Columbia came out with what was supposed to be 30 cords for the island, but proved to be only 26. I got one cord delivered in 4' lengths which will have to be seasoned, cut and split. I hope Newt comes back by then and can do it. Two of the pieces are about 2' in diameter. Each customer got at least that many.

Hardly had I addressed all the last letters than we had quite a storm with tremendous surf. I went to Lobster Cove and on around to Burnt Head. The going was difficult because most of the trails were filled with water and many with muck. I realized that twenty-seven years ago Mother and I took the same walk during a hurricane and I thought she was "so old". Well, Sunday (the following day) I began to feel strange and by Monday morning (Oct. 7) I was definitely in trouble. I decided to try Doc Rothstein's ultra sound treatment, but got worse and went over the bounding main in a basket stretcher which just squeezed into the pilot house of the Laura B. What a ride! The upshot was that I probably passed a kidney stone - it was never found. After some IV's, millions of x-rays I came home on the next Friday a bit worse for wear but am back to normal spading up the garden, cutting brush, lugging in window boxes and planting about 500 bulbs which were dug up when I tried to get the buttercups out of the flower garden. Don't ever try to tame the wild one, they just take over. One thing I can say is that if you are sick Pen Bay Medical Center is **the place** to be.

The island has lost several former summer visitors this year. Word came that Mo Schulman died this summer. Alex Minewski left the island in August and lived only a few days longer. He had put up a galant fight for about a year. Each time I look at the portrait of Gertrude and me which he restored a couple of years ago I think of him. His passing is a great loss to the island. Last month Joel Boxer was killed in a hold-up and last week Mendell Bogdanov was killed by a truck in his home town. He and his wife had just left here after working hard fixing up their house. Manny even went swimming almost daily and seemed to have recovered well from his heart attacks. Sandy Dickson's mother, down in Mexico, has been ill and Sandy was called down there a couple of weeks ago, but is back.

When Jack Cabot left the island a couple of weeks ago he was given a musical send-off by Muffie on the violin, Mary Barney on the concertina and Kathy Rhoda on the block flute. He looked rather thin and now we hear that he is in Maine Medical Center because an obstruction of the throat. On Friday Jeff Dolan gave Zimmy Brackett a "farewell party" at Newt's cottage and we kidded Zim about its being a "farewell". But by party time he really thought it was a farewell. He was hit by a severe digestive upset which lasted until this morning (Sunday). Never-the-less all those attended had a good time and Zimmy was able to leave the island this noon with Steve Rollins, but looking pretty thin after his bout.

Three houses have come on the market in the past few weeks. Remmingers are selling the Staples place on Horn's Hill. They find it too inaccessible to them in Cleveland. Mike Leow has his place and studio on the market and Ralph Hinchman is selling his place. I don't believe any have been sold, due to the tight money market. I wonder if, since the depression, so many houses have been on the market for so long.

For several days now I have been trying to get Doug to take me in to Burnt Island for cranberries. Andy and Betsy Wyeth came in today on the Seagull and Andy said for us to help ourselves on both Burnt and Allen's islands, the family owns both. Doug has to go to Tenant's Harbor Tuesday, weather permitting, to have his compass checked so armed with many pails and a lunch, I'll be dropped off at the dock at Burnt Island to spend a few hours there and be picked up by the Chris and brought back here. I hope the Chris won't forsake me!

I am back from Burnt Island. The Chris didn't forget me. In fact, I was picked up right on the dot. We started out at seven o'clock and got as far as Western Duck Rock when we had to turn back because of a leak in the water line. Harry used up all the fresh water aboard, pouring it into the radiator as fast as he could go and the last drop went in as we reached the dock. Repairs were completed and off we went an hour late. Doug had an appointment in Tenant's Harbor for nine, which he turned over to Sherm. Meanwhile the tide was going and we had a bit of trouble getting Phoebe aboard. It was worse when we got to Burnt Island - I didn't want to go exploring without her.

Burnt Island is a miniature Monhegan, except that there are only a handful of buildings and it is inhabited only during the summer by the caretaker and his family. There are woods, meadows, and swamps as well as trails around the peripnery and criss-crossing the island. We found the cranberry swamp but it had been picked clean. I got only half a cup of berries, but they were huge and a deep wine color. Blaine Davis, the caretaker, came by at noon to see what was going on. By that time Phoebe and I had walked about half-way around the island and had even explored Little Burnt,-ledges and raspberry thickets and a double curving spit tying it to the larger island. The sand looks like coral but is course clam shell. I found some sand dollars and used one of my "cranberry" pails to gather periwinkles for Rita. I also came back with pockets full of rose hips, so the day wasn't completely lost. A book kept me occupied after lunch until the Chris returned. It was then dead low tide and we had to get Phoebe down about ten feet to the boat. Harry nearly lost his balance on the ladder trying to keep her from climbing back up, but we finally all got into the boat safely.

And now for some island news. The Hallowe'en play will be given at the school this afternoon, followed by Trick and Treat. Some adults, I'm told, may do a bit of T & T themselves! Bud and Sally are having an open house at their apartment, Saturday, which has been completely refurbished since the Odoms moved out. That will be on Saturday. Let us hope the hosts will be able to attend!

Paul and Barbara Horn have moved into the Windjammer, formerly Ruth O'Conner's cottage, and are busy winterizing it with the help of Paul's parents. Mr. Horn is now retired from the ministry and able to spend some time here. Ollie Bolman and Jeff Dolan are both here. Ollie is trucking for Carl Wincapaw and Jeff is doing painting and odd jobs; neither has been able to find jobs to his liking. Newt said he would cut and split my wood for me. That is good news.

Emma Field was due for surgery last Monday, but suffered a shock that morning. Just what will happen to Earl, Sr., now, I don't know. He tried to keep Jimmy from using the dock and parking lot in Port Clyde, but didn't succeed. Young Earl has had or is to have cataract surgery, as is Zimmy this fall. Ron Stanley is to have laser beam treatment to try to save the sight in his eyes.

Steve Rollins is ready to start framing his new house; he'll be installing a clivis moltrum (sp?). The Hitchcocks' addition is coming along slowly, the greenhouse, I think, is functioning. My window one is still standing on my front porch. Billy and Jackie Boynton will finish pouring cement today, the last day for town water. Then they'll work here doing painting and helping with wall board before framing their place. This is a carpenter's paradise with all the building and repairing that goes on constantly.

The Seagull has gone up on its price. It now costs $28 to bring the plane out. If there is no one making one of the legs of the journey and only one person going either way that person bears the entire cost. Otherwise it is $14 per person. Even so, plenty of people are still going by plane as weather permits.

Alice Becker and Barry Berg are back. Barry didn't run in the Manhattan marathon, that's why we didn't see him on TV. Barbara and Dan Rothstein and children will be off to the Holy Land later this winter. I believe a stork has been casing the Tribler Cottage looking for a delivery in the spring. Get out the knitting needles, Alta!

Nov. 1

The school play was most original. The kids wrote it and cast it. The title "What Time Is It?" and the curtain line "time for Treats". It was about the Seaghost Airways, a postal inspector with goblins and monsters livening up the stage, as well as the Becka III coming to the rescue, a marshmallow fog (white sheets on pulleys) and controls of the plane all gummed up with peanut butter and jelly. The lines were very clever, though of meaning chiefly to Monhegan people. The six scenes were announced by Sam Rothstein, the production was by the Monhegan School Playwrites (mainly Lance Bolhmer and Timmy Burton). It kept us all in stitches. Too bad neither Winnie Burton nor Alan Mersfelder were there. Winnie had a postal workers' meeting in Portland and Alan some aeronautics business in New Jersey. The State Agent, Mr. Hurlbert, saw the play on Wednesday and thought it was great, as did Miriam Barteaux, our teacher. After cookies, etc. the kids came around for T & T. My 31# pumpkin was outside and a smaller one carved into a Jack-O-Lantern on the steps, but the candle had blown out by the time the children came - a devil, ghosts, witches, et al. I offered pop corn, apples, gourds and mini-Hershey bars, all of which I still have in goodly supply.

So much to do! Five loads of blankets, etc. from the cottage, the garden to tend, yard to pick up, and then the new room to furnish. I hope there will be a place for everything and everything in its place, but I doubt it, with my track record.

Best to all and
Happy Thanksgiving!
Alta & wag of Phoebe's tail

Thanks for the contribution.

HOLIDAY GREETINGS from MONHEGAN, MAINE for 1979

Dear Friends:

First of all I must apologize for the un-Christmaslike paper I have had to use for this letter. The store in Rockland was down to its last few reams of any kind of ditto paper, and most of it was pink, so I settled for yellow. They had no red or green stencils and the only ones I had are now in the back of my car over in Port Clyde! You can see I am my usual well organized self.

In looking over the activities of the past year it seems as though I had done hardly more than wait for Godot, chiefly for the addition to be started, and now for it to be finished. Plans were submitted to the builder in October '78, but he was delayed in starting by an operation on his knee.

First drafts arrived in March and final drawings in mid-April and a contract signed with expectation of a completion date of July 5th. Then came the need for a building permit, which I had been told in the previous fall was not necessary. That took more than a month so the foundation wasn't begun until late June.

Instead of finding solid ledge on which to build we found the exposed rock consisted of huge boulders resting on old "rotton" rock which had to be chipped away by hand because the bulldozer couldn't work at the angle of the back hill. Three weeks later cement for the footings was poured, two weeks after the estimated date of completion. And so it has gone. Now, one week after Thanksgiving they are just putting up trim and have the floor and hearth for my Jotel stove to be laid, storm doors to hang and a walkway between the two parts of the house built and installed. The laundry room is unfinished, waiting for control panels for the wind generator to be installed and the clothes washer moved when the dishwasher, which is to go where the clothes washer is now, is delivered and the floor laid. I had hoped the family could have come over for Thanksgiving!!

The wind generator is still among the missing. It was to be installed in June! Now we are all excited about the possibility of the use of photovoltaic cells as auxiliary sources of electric power. David Sleeper, involved in auxiliary and alternate energy systems, will be out here on a return visit on the 12th to give us facts and figures. He sounds hopeful of supplying economic electric systems.

This leads me to the article that so many of you saw, written for the NY Times. First of all, the survey showed most islanders in favor of central power because of the high cost and inefficiency of individual power plants which rely on fossil fuels. I had hoped to reduce my generator use from 3-4 hours a day to that amount, or less, a week, relying on it for only heavy demand usage such as for the vacuum, dish and clothes washers, and the deep well pump. The 12 volt system, charged by the wind generator (or photovoltaic cells) would supply power for lights and small appliances such as my typewriter through the use of heavy duty storage batteries. Right now I charge the batteries at night when I have the generator running for lights and color TV and have already cut my generator time almost in half. Every bit we can save will be a great help as fossil fuels increase in cost.

The new room is heated well by passive solar power on sunny days, and the Jotel stove heats the whole upstairs in no time, with the help of my Franklin.

The Pella windows used in the new room are terrific. They can be opened so both sides can be washed from the inside. No excuse now for dirty windows. Maybe some day I'll have the rest changed - none too soon. Draperies have been ordered and a new double hide-a-bed. I can hardly wait for everything to be moved into place.

Although most of my energies have gone into the new room there were other happenings. We managed to get through the year so far without any damaging storms, but the deep freeze in February caused a lot of freeze-ups, including the water system in this house. I had rented the cottage to John and Winnie Murdock for the winter and they had quite a time keeping the pipes from bursting. Elizabeth Gardner and I were basking in the tropical sun of Sapphire Bay on St. Thomas and knew nothing about the extreme cold in the north. We had a great time and a week was all too short.

My cousin Emma Riddle in New Hampshire had emergency surgery just before we left for St. Thomas, so I left the island early to see how she was getting along and returned there to get her settled in her home and later in a nursing home, then went back and got her resettled in her home. After the snow and ice cleared in mid-April I returned to the island and didn't leave it for six months.

Just before Thanksgiving Emma had to return to the hospital so I went over again, before going to Concord and then out to Syracuse with Elizabeth to spend the holidays with the Jacksons (my niece Priscilla).My cousin Elizabeth Weinhagen John was visiting her son Walter's family in nearby Clinton, so we all went over there on Sunday for a visit to their beautiful home. It was a great reunion. We were entertained by two wild horses who were loose and ran around the grounds, cutting up Walter's manicured lawn with their hoofs.

My garden, nearly demolished by a plague of grasshoppers in August, rejuvenated itself with the heavy rains of September and is still yielding broccoli, swiss chard and parsley. I wish I didn't have to go back in shore tomorrow for several days, at least. I just tried to run thru the first page of this letter and succeeded in jamming up the machine completely, so in it goes with me tomorrow and someone else will have to run these letters. Ho, hum.

Early in the year I was involved with someone who wanted to set up a telephone system for Monhegan. He took all my data with him and I haven't heard a word from him since. 1980 is getting awfully near when the Coast Guard says they will remove our cable, so now we must petition the Public Utilities Commission for continuation of service. Here we go 'round the mulberry bush again.

Not to be outdone by my relatives, I had a session at Pen Bay Medical Center in early October lasting five days, apparently passing a kidney stone, though none was found. All I hope is that there will not be another. I'm feeling better than ever. Apparently the thing was grumbling around for some time making me think my back was out of whack.

There is room left only to wish you all the blessings of the Christmastime and the best for the new year. We hope some of you will be able to put a visit to Monhegan on your agenda for the coming year. Phoebe gives a wag of her tail to all her friends and thanks those who have sent her a few pats.

As ever,
Atla Ashley

To Quincy, with Thanks!

A grateful group of Monhegan school children, their parents and friends gathered at the Schoolhouse Saturday evening for dessert and punch to express their thanks to Mary Catherine Iannoni, Althea Sawyer, Aliisa and Michael Ananis - all members of the Quincy, MA school system - who had made our children's spring field trip to the Boston area so enjoyable and who are at present visitors to Monhegan.

The predominating color of the decorations and food was pink: pink napkins and tablecloths, pink punch and pink icing or fillings of the many cakes, and pink candles. Each visitor was presented with a corsage of pink and white carnations, save Michael, who sported a single white carnation.

A large banner proclaiming thanks to Quincy, handpainted and signed by the children present was donated to the visitors.

Mary Catherine and Althea were also given a tote bag decorated with the logo of Monhegan Jewelry and filled with presents gathered by the children - note paper, a bag of candies, magnetized animals for refrigerator bulletins, model lobster buoys, calendars and soaps; Aliissa's gift was a copy of "Finest Kind", Monhegan's own cookbook.

Following this Mary Burton, School Board Chairman, told of how the field trip plans were initiated. After a few words of appreciation from Mary Ann Boody, School Board member and one of the parents who went on the trip, the meeting was turned over to Mary Catherine who had brought with her more than one hundred slides of the trip. Her running commentary was both amusing and informative.

For several years Mary Catherine has been a summer visitor to the island. Because of her position as head guidance counselor for the Quincy's Central Jr. High School, she was contacted to see what plans might be made for an educational field trip for the seven elementary school children on Monhegan. For help she turned to Althea who with her sister, the late Merle Sawyer and former principal of Mass. Field Elementary School in Quincy, had known Monhegan and the Odoms since the nineteen thirties. Althea, Counselor for Broadmoor Junior High School responded with great enthusiasm, 'tho at the time she was on crutches with an injured knee. Before plans could take shape, however, the phone system to the island was disrupted and remained so until after the trip was completed. Despite such adversities the trip materialized in May and proved to be both enjoyable and beneficial for the Monhegan children and the school children of Quincy as well.

During the week the children attended classes, saw for the first time multi-classed schools, introduced their island to the Quincy students, watched and participated in gymnastics, swam in the school's Olympic sized pool, toured historic sites in Quincy and Boston, including Quincy Market and Haymarket, the Prudential Building where many a dime was spent to watch a Red Sox game in progress whose members were identified by seven year old Becka Hitchcock by their numbers. They saw the Aquarium, boarded the Constitution, received honorary citizenship papers from Quincy's Mayor Tobin, went to Logan Airport where they saw the foam machine, climbed onto a huge Quincy fire engine and through the efforts of Mr. Goodridge of Channel 5 (Boston), a regular summer visitor to Monhegan, were interviewed for a segment of that station's newscast.

According to Mary Catherine thanks should go to many people who helped with the success of the trip, including Aliisa Ananis, third grade teacher at the Mass. Field Elementary School and her husband Michael, then special education teacher for Quincy, now for Dedham, as well as Supt. Lawrence Creedon, and other teachers and students who helped with the gymnastics and swimming programs.

At the end of the slide presentation the children filled in with some of their memories of the trip, bringing the evening to a close.

Alta Ashley, corr.

Monhegan, Maine 04852
Christmas, 1979

Dear Emma:

I thought I'd have a letter off to you long before this. Where has the time gone? I hope on this dreary Christmas you went over to Manchester to the Trippettis'. Rita is coming here, but our big dinner will be tomorrow when Doug and Harry get back. Today we are having warmed over turkey I cooked the day after I got back as a delayed Thanksgiving dinner for D. and H. tomorrow we'll have a New Zealand lamb. I can't face more fowl after eating the other turkey for nearly a month!

Gil and Ruth Lothrop are joining all six Murdocks at the Store. They came over yesterday and will leave tomorrow. I'm hoping the rain will let up a bit because Rita and I have been invited down there for a late afternoon wassail.

The day I returned to the island David Sleeper and his friend came, too. David is the solar energy man. They stayed in the cottage but ate all their meals with me. The upshot is that David is bringing over six panels on Dec. 31st to install on my new roof. Sandy Dickson will be getting two and both of us will be demonstrations as to minimal and normal use of electricity.

This past Friday four birders came over to make the Christmas bird count on Monhegan. They also stayed in the cottage and ate most of their meals with me. I had drained the cottage before I left for Thanksgiving so there is no water in there. Thank goodness for the toilet downstairs where they all can have cat baths and take care of bodily functions.

Carpenters have been here working much of the time. I haven't been able to take a shower to go to the bathroom unattended, so to speak, since the wall of the bathroom was broken through into the new laundry. Just the other day the door from the laundry to the new room was hung, but they are still working in the laundry room!

I've been sleeping in the new room and have closed off my bedroom. This makes it possible to keep warm with just the Jotel and Franklin stoves, except when it went to near zero when I had to put on the oil stove downstairs. I can also cook my meals on the Jotel, if I time the firelaying right.

With all this extra activity I haven't finished sending out my cards. I still have about twenty to mail, but at least I have made a dent in the pile of stuff on the davenport which is really waiting for my desk and cabinets to be made so they can be filed away in more or less orderly fashion - what's that?

The only decorations I have put up are tree lights which are hung in the windows, the Santon creche and a ceramic tree Jim and Priscilla sent. It was damaged in the mail but it isn't worth the agony of returning it so a bit of Elmer's glue should do the trick.

While I was gone Newt sawed up my cord of wood and took off for Rochester. Another cord arrived on Thursday, so he'll probably be pretty discouraged when he sees the side of the barn where it has been piled.

I hope you have had a happy day and that you will find interesting associates among the people there.

Love,
Alta

Monhegan, Maine 04852
January 5, 1980

Dear Friends:

There probably is no need to tell you who read the Maine papers that Trap Day was on January 1st - the second time since that date was established as the legal opening of lobstering on the island. By mutual agreement at a "lobster" meeting the night before, the time for trap setting was put at eight AM, even though all boats might not be ready by then. Many of the young people took advantage of the full moon and its high tide to get their rocks aboard and even to load some traps the night before. Just before eight o'clock on Tuesday Shermie Stanley in the North Wind, Rusty Court in the Annie D and either Steve Rollins in My Three Sons or Robert Rothstein in the Joanie D set off for the mouth of the harbor and at the stroke of eight one or all pushed their first trap over the gunwale even though some boats were still loading and some waiting in line to come to the Dock for their first load. In fact, Rusty was back to reload before Doug and Harry in the Chris got to the Dock for the first time.

Clay Kent and Walter Ashe, artists from Florida and summer visitors, arrived on Monday, hoping but never believing they would be here for the big event. They had a great time, even going out with Vernon Burton to haul on Thursday and having their lobster feast that night at the Burtons'.

We were going to have our feed last night but Doug and Harry went to the Store to dinner. I guess we are the only ones who haven't had any lobster yet. All that will be corrected tonight, however, with a spread served in my new dining area. (They were worth the wait)

Believe it or not, the extension isn't done yet. Wiring has been done except for bringing the controls upstairs and setting the inverter in the line so I can use the batteries for my TV and move this typewriter into the other part. The desk and cabinets and the planter are still to be built and some trim to be put in place. There are holes in the floor, ceiling, around wall receptacles etc. and when the wind blows over 60 knots as it did the week after Christmas it is COLD. The storm doors haven't been hung so there is a great wind tunnel in the passageway between the lower floors of the original house and the addition. Things in the new cellar have frozen and the plants brought into the breezeway last fall have frozen waiting for the planter to be built. All you who said I showed remarkable patience are rapidly being proven wrong.

There have been tears and harsh words of late. Then, to cap everything, Doug and Harry didn't get back the day after Christmas as planned, for a delayed dinner with Rita and me, because they stayed in Portland to see the Mariner-Russian hockey game. I didn't hear about the change of plans until about four o'clock and had everything on the stove. Then the storm hit on Thursday and they didn't make it back until Sunday. The poor roast lamb kept being heated up until it fell off the bones and the mince pie crust got a bit soggy by the time it was served. Everything did taste good and the extra room in the new dining area accommodated five of us nicely without the need of an extra leaf in the table. Miriam Barteaux, our teacher, made the fifth at table.

When I returned to the island in mid December after the Thanksgiving holidays and another stay with Emma in New Hampshire, David Sleeper and friend Jeanne came out, too. He met again with the community to talk about alternate energy systems. A high wind storm helped him decide against wind power as such a source for the island, and turn him definitely toward solar power, even for electric power. Solar power for heating water is really not too practical for Monhegan because it takes power to circulate the fluid. Any day now he is to return to install some panels on my new roof to supplement my generator-battery system. Even Bob Boody has become less enthusiastic about his wind generator which is yet to be installed. I'm not too eager to climb a 30-40 foot pole to service such a generator. Solar panels, I'm told, are virtually maintenance-free. So, with all the publicity last year about my wind generator, I guess it isn't going to be after all.

When the sun is out, even with the temperature in the teens, I can let the fires go out and keep warm by the sun alone. Today, though, is another story. It is cloudy, blowing a gale, and in the teens, so both the Franklin and the Jotel are going and pretty soon I'll have to turn up the pot burner down stairs. Thank goodness for the relatively warm December we had. Heating oil is over a dollar a gallon now and due to skyrocket.

After having such a time with my ditto machine in November I carted it in shore where they could find nothing wrong and were able to run off 275 copies with little or no trouble. I tried to get the thing put back together after all the parts shifted places on the way over in heavy seas and can't make it go. If I can't get it to run I'll have to take this in or send it in to be run off or go in on Monday with the machine under my arm. Let's hope the storm subsides by then.

John Riddle came on from San Francisco just before Christmas to help Emma move to Epsom Manor for the winter. She has an apartment there and also has her VW. It's about 25 miles from Wolfeboro so she can see a lot of her friends, weather permitting, and is also close to Concord and Manchester.

Now to go back to Trap Day. I tried to call the paper that morning but could reach no one so I called the Rockland radio station. Mike Gross, son of the editor of the Courier-Gazette, took my call and passed the news of Trap Day on to the stations in Portland, the AP and UP. For the rest of the day and into Wednesday I was on the phone, giving my story. Some talk show station in Boston called me, too. I had never listened to the station and didn't know I was on the air! I guess the interview lasted about five minutes. It's good I didn't know about the broadcast because I get a terrible case of mike-fright and stammer so no one knows what I'm trying to say. As it was, everything went off smoothly, so I am told.

The Laura B has been transformed. Plywood panels have been replaced by plexiglass, so we can see out, and there is a woodburning stove in the aft cabin. Plans are for a water barrier to be placed under the doors to keep the waves from washing across the deck. It is so cozy now people stay in the back rather than crowding forward when the boat reaches the dock. There are also grab bars to help keep people upright. Alan is working at the Augusta airport and has put the Seagull up for the winter. Right now he is helping Bobbie with lobstering. Bobbie and family are now at Harry T. Bone's for the winter.

I'm sorry to have to say that Bobbie Way died in mid-December. A memorial fund has been set up for the children in her name. On Dec. 23rd Joe Rains, also died. On a happier note - we should expect three new members of the community: Richard and Martha are expecting in March, Winnie and John Murdock and Craig and Judy Sproul in June.

The Christmas play and Christmas party, a week apart at the School-house, were great fun. No church service was held because of repairs in progress. It is time to say Best Wishes for 1980.

Alta Ashley & a woof from Phoebe

Monhegan, Maine 04852
February 17, 1980

Dear Friends:

First of all, let me apologize for these awful labels (address labels for the envelopes). I wanted something I could prepare ahead of time so the time lapse between my writing this letter and the mailing of it would be shortened. The add looked good, but I didn't realize my name would be so prominent. I fully expected to have the last batch of letters come back to me en masse, but none did, so the postal service is better than we think quite often.

Today is a crystal clear day after a howling blizzard went thru yesterday - the first real storm of this winter. Everything is covered with a gleaming coat of snow, part of which is crusted due to freezing rain which fell for a while. That makes things a bit complicated because, despite high winds, the snow has stuck on the roof. Even the bright sun hasn't melted it and thereby hangs a tale.

The new addition is nearly finished. I moved into it and have been living in it almost entirely ever since. My bedroom has become the spare room, the front room an icebox right now even with the insulating shade drawn. I don't have the Franklin going today because I have to go out to get wood for it and as yet I haven't had the courage to shower and dress.

The end of January Dave Sleeper and his crew came out and installed six photo-voltaic panels on the roof and connected up four batteries. The inverter was put into the line but he didn't have the right kind of relay so I am using it by way of an extension cord strung from its place in the cellar, through a casement window to this room, then extensions to the color TV and this machine. On bright days the panels start charging even before the sun is on them and by noon are over charging, according to the dials I now have. But yesterday they didn't show any charging and by 8PM the lights had turned to a peach glow, so I went to bed. This morning the snow is preventing all but a minimal amount of charge, so I may not be able to use this machine for very long. If the dials are right a couple of more storage batteries should correct things.

I just looked at my dials and they said the batteries were discharged, so I started the washing machine and battery charger and am typing this with the 110 volt system. It takes a PhD in electrical engineering to run this house! That I haven't.

I was going to start this letter by saying "You can keep your diesels, you can keep your windmills, I've gone solar" with apologies to Proctor and Gamble, but I don't quite dare to say it until David comes back with the correct batteries and dials, switches, relays, and what have you. But what a great feeling it is to run this place by the sun except to pump the water twice a day and use the vacuum, dishwasher and clothes washer all very occasionally. Until today I have run the pot burner only at night, the Jotel morning and evening, and relied on the sun for warmth. Today the pot burner is making no impression on downstairs and gales are blowing through that room and up the stairs. I guess the next thing is to remodel that to conserve heat. I always have to have a project!

Two weeks ago Rita and I went in, I to attend the Knox County Med. Soc. meeting and Rita to enter Pen Bay. She had been ill for about two weeks and needed medical attention. She did well, stayed in for some follow-up tests and came back on Friday looking great.

I went off to Augusta and was able to put off a meeting with a member of the Public Utilities Commission re our telephone service, scheduled for Wednesday, until the following week, so took off for a visit with Emma in Epsom Manor and on to Concord with the Gardners, Friday I had luncheon with Nancy in their attractive apartment in the shadow of the Mother Church and glad to see her looking so well after her chemotherapy. Saturday Elizabeth and I went down to Kingston to see Aunt Helen, now 94, and took her to lunch at Bert's in Marshfield (?), which she thoroughly enjoyed. A stop back at Epsom Manor on Sunday and on to Rockland where I found Rita out of the hospital so we went over to the Samoset for supper - the first time she had seen the new building, she hadn't been off the island in seven years! It was a bit like being with Rip Van Winkle. Among the people who came to see her in the hospital were Zimmie Brackett, Sarah (her daughter-in-law). She called on Penny Akin and went with June and Henley Day to Maine Central Institute to see Courtney, who is enjoying his year there. Today, Sunday, she was back at the register at the Store as usual. Also saw Ivan Masser's ingenuous solar heated house in Concord.

Wednesday Kim Kenway, a staff lawyer for PUC, came over to investigate the telephone service here. We held an open meeting that night at the schoolhouse, well attended, and he explained the legal aspects of continued service from New England Telephone once our cable breaks again. NET says they can't continue to serve us once that happens. We are trying to say that they can't cease now they have served us. And so the battle shapes up. It will be a long time before anything is settled and we all hold our breaths everytime we have a wind and heavy seas such as we had last night.

I am back on generator power. The snow piled up on my solar panels and froze. Today's bright sun was not hot enough to melt the ice - it got up only to 18 today and the NW wind didn't touch the snow on the southern side of the roof. The wind indicator also stayed frozen in place so, although it blew a gale most of the day, it registered a flat calm.

The big news is that Dan and Barbara Rothstein will be parents of twins in August, making nine children between his, hers, and theirs. He had better not think about claiming the gold-headed cane when I'm away! Mort Bernstein, who beats us both in age, is spending much of the winter at the Trade Winds in Rockland.

Rickie Cundy did a great job plowing the roads last night with the new plough; then he sanded all the hills today, so it is easy getting around. I took my little red sled to the Store and stocked up on heavy items today. Someone made a huge snow person on the lawn of the Library, complete with oranges for eyes, a carrot nose, a string tie, orange rubber gloves on long branches for arms and hands - quite an engineering feat and work of art. Ears were huge pine cones.

If any of you have any good ideas about the telephone situation send them along. We need to have a plan should NET leave us flat, and we need a show of interest to help convince NET we need phone service. There's always the specter of MONEY.

The lobster catch has fallen off dramatically in the last week or so and most boats are at their moorings much of the time. Alan has been here helping Bobby, having put up the Seagull for the winter.

My Harvard chair arrived and the Mount Holyoke one is expected any day. A buffet and corner cupboard will complete the new furniture. The bookcases and desk cupboards are already over crowded!.. Phoebe and Margaret, Rothstein's cat who has adopted the house, are sleeping together on the new davenport. Tut, tut.

I guess I've covered the news for the month. Keep your letters coming.

As ever,
Alta, Phoebe & Margaret

Monhegan, Maine
March 15, 1980

Dear Friends:

Can spring be only five days away? The wind is howling, and has been for the past 36 hours, and even longer with one beautiful, calm day thrown in. The trip out this morning was the worst I have ever taken. More about that later.

First about the telephones. Kim Kenway called a while back to say that Public Utilities Commission feels New England Telephone has indeed given us service and has an obligation to continue to do so. People have been good about getting copies of their charge accounts to PUC, showing heavy phone usage, which helps. I'm now waiting to hear from PUC as to what the next move is. At Plantation Meeting on Monday a $3000 budget was voted on to use for expenses and legal advice and I was voted in as coordinator of effort, so to speak, and have the go-ahead to hire a lawyer, if necessary.

On March 2nd we held our first Republican caucus, also the first I had ever attended, and I got elected Chairman and delegate to both the District and State Convention, ho! ho! They will be held in Bangor in mid-April. Today I got a call from Reagan headquarters with offers to help in getting reservations, etc. as a less than subtle means of swinging my vote. We couldn't decide on which candidate to back at the time of the caucus and I am to go uncommitted. It is getting to be quite a horse race at this writing.

Annual Plantation Meeting was held on the 10th. I was going to run for either Planning Board member or Third Assessor, was nominated for the latter but after giving a little speech about idle funds lying around in certain accounts, such as the cemetery and Emily Tomkins memorial, declined the nomination so as not to be tied up with town business and not being able to leave the island for a long stay next winter. The telephone business is enough, plus the column and my historical commitments which must be resumed now the new room is finished well enough so that the typewriter sits on the desk and doesn't travel from dining table to couch to my bed. What a relief! Even the duplicating machine is up here so I don't have to freeze downstairs when running off these letters.

There are new faces again among town officers. The assessors are Billie Boynton, Ruthie Stanley and Rusty Court. Judy Sproul is still Collector of taxes, Jeanne Gibson (Stevie Rollins' friend) is Treasurer.

Barbara Rothstein will be the new School Board member. Dougie Boynton is Road Commissioner and a member of the Planning Board. Otherwise all the officers remain the same. Ron Stanley came out on Monday just to be here so he could continue as Fire Chief. There was much discussion about spending $700 for the disposal of plastics during the eight winter months, but it finally passed because we were afraid without it the harbor would become cluttered again. A $10,000 lowcost loan was tabled, to be acted upon if and when it becomes necessary to expend it for legal fees concerning telephone service. Meanwhile we hope to be able to solicit funds as necessary from the community or by holding cookie sales, etc. The appropriation for roads and bridges was inadvertently left out of the town report so we'll have to have a special meeting when this year's surplus is exhausted. Luckily we needed to have the roads plowed only once during this very strange winter.

David Sleeper came out last Monday to finish wiring my photovoltaic system. It should be just great. The inverter is in the line so I can type with sun energy stored in the 12 volt system, or by 110 volt current if the generator is running, without having to shift plugs. Most of the dials indicating state of the batteries, rate of charge, whether or not the inverter is on or the generator is running are all on a panel in the new laundry room, very handy. Now, just as long as I don't start the washing machine or the electric iron when the inverter is on, everything should work out fine. Unfortunately all the wall plugs in this new part are on the same line, which is the line the inverter is hooked into. We are hoping to be able to change this when David comes out to do another job.

Bob Boody has his wind generator on a ten foot tower and has not been able to run it for the two weeks it has been up because the winds have been too high! I'm really glad I gave up on the wind, the sun is working out much better, now we have the kinks out of the system. I found only one battery had been hooked up, thus the small storage capacity. The cells start charging at dawn and long before the sun comes over the hill and hits the panels, and they keep charging until sunset. Now my problem is that the generator runs so seldom its battery runs down! I'll just have to vacuum the rug more often.

Monday when David came out he brought Tux Turkel from Maine Audubon who wanted to write a story about both me and David and the solar energy system. Tux was a food nut, didn't eat red meat, bacon or things "with chemicals" or white sugar (poison!) He also didn't like classical music - and I have Robert J. on all morning with Morning

Pro Musica (why did they stop his broadcast of the news? His delivery gives me time to think over each item). Tuesday it came down in buckets, David was busy running from the cellar to the instrument panel, so Tux and I were here virtually alone. The wind came up in the afternoon and blew steadily up until about nine o'clock Thursday morning. Both Tux and David had commitments in shore and just had to stay here and chew their fingernails. Thursday, by the time the boat came in there was an almost complete calm. How does that ocean flatten out so quickly? I went in with the fellows along with a dozen others who had planned to go on Wednesday. Shermie took even more in early Thursday morning in the North Wind. We didn't get as early a start as we had hoped because someone, who had come out for the ride, didn't come back even though we waited about twenty minutes. In all the confusion I left Phoebe waiting for me at the Post Office! Rita took care of her that day and Robert Rothstein on yesterday.

That afternoon I drove to Portland, took old town reports to a printer for binding and went on to a meeting on medical care. A huge storm was predicted so I didn't stay over at the Sleepers in Falmouth, as planned, but did have a delightful time at dinner with Phil, Ginny Baker and their daughter Linda Gifford, then rolled into the Samoset by 11:30 and found a fishermen's conference which had taken all reservations for Friday night. That didn't bother me because I thought I'd be gone. The storm hit about 2 AM and continued until late afternoon with snow, freezing rain, rain and gale winds. The winds picked up again this morning but Rusty what's his name took the wheel of the Laura B. Jimmy Barstow was at the Samoset. I had spent the night at the East Wind in Tenants Harbor - very nice. There I met a Ruth Spaulding Millinger who grew up in Lexington with all the Ballards, Nichols (including Phoebe Nichols Palmer) so we had a big gabfest. Then in walked Mike Moser and Susan Webster, en route to the fishermen's conference where Mike was to speak, but there was no room for them at the Samoset.

Don Stone has bought the Remminger (Corbett-Staples) cottage and Miriam Barteaux is staying on as teacher.

The wind picked up again in the night and was blowing hard by nine o'clock when the boat was to leave. I would have preferred to stay in but was worried about Phoebe. Elise and Rickie Cundy wanted to go out as well as Billie Boynton and his two guests, but Jackie was reluctant. It was the roughest crossing I have had yet, and I've come over in some bad ones. We took green water over the bow several times.

Jackie and one of the guests were sick and I would have been in another few minutes. Rusty had to stop the boat once while someone crawled down into the engine room to secure something which was rolling around. The milk left on the foredeck tipped over and half gallons slid here, there and yon. Water poured into the aft cabin under the doors, through the roof and side panels. We were all soaked. It took us nearly two hours to come across and Rusty had to take the boat way to the westward before he could turn and make a run for the harbor. I swear I don't want to leave this place for months but have to go in again the end of the month. This typewriter is still surging up and down in front of me and I haven't lost my sealegs yet when I try to walk across the room. What makes people live on islands?

It is now Sunday morning. The wind is still blowing and the temperature is in the teens. Last night the wind hit 58 knots between times the needle stuck at 47. I have to stand there and tap it loose. And to think I was about to turn the water on in the cottage for David Sleeper last week! Ida Proper's snowdrops are up and we had a brief mud season, now we're again in the deep freeze. Lance Burton came around last week and split a lot of wood for me which Tux piled up close to the addition. I was just about out of pieces small enough to fit into the Jotel.

In all the confusion of boarding the boat on Thursday, leaving the visiting photographer and Phoebe behind, David Sleeper and Leslie Whitney got their duffel bags mixed up so I now have David's sitting by the door downstairs awaiting his return. I believe he has two installations to put in here next month.

The Lords are back on the mainland. Jeanie is again in the hospital and Victor is caring for Kole. He had been doing a lot of renovations on the Monhegan House, including a new roof, but has had to stop for now. The Tribler (Golden Gull) Cottage has a new brochure. There are two apartments for two and two or three people for $275/week spring and fall, and $300/week during the summer. They have also three rooms for $24 and $16. Two have baths and one a shared bath. No pets.

Jack Cabot is back at the Eastland, but is still unable to eat solid food. He has been treated for cancer of the throat - that ever present pipe. Martha Yandle and Richard Farrell are parents of a baby girl born less than 24 hours after we gave a stork shower. Fortunately she delayed her arrival until Martha reached the Camden Hospital. Rita says she is beautiful. I have yet to see her or hear her cry. Judy Sproul is in shore waiting for her stork. Winnie Murdock isn't due until June. I hope I can finish her blanket before then. I'm still working on Newt's sweater which I wanted to give him this winter as a grateful gesture for his having cut my wood.

My planter is full of flats seeded with herbs, lettuce, and various vegetables to transfer to the cold frame in a month or so. It really is that time even though it doesn't feel so. Enough for now and best regards from us - Phoebe is still speaking to me even though I forgot her last Thursday!

Alta

Box 87
Monhegan, ME 04852
March 30, 1980

MONHEGAN TELEPHONE SERVICE UPDATE

Dear Concerned Resident/Visitor:

After nearly two years since the telephone cable to Monhegan was repaired "for the last time" there appears to be some progress in our fight to maintain service from New England Telephone Company. As you may remember, the Coast Guard Service said they could not repair the cable should it break again, and NET said they didn't have equipment to do so. The Coast Guard no longer needs telephone communication and the cable was to have been taken up and sold for scrap in 1980. They have now decided to abandon the cable "in place".

In 1979 Continental Telephone Company, after studying the situation, notified us that they were not interested in taking on Monhegan as one of their services and a private individual who came to the island in June of that year to investigate the possibility of setting up a Monhegan phone company has not come up with a proposition.

Late in 1979 the Maine Public Utilities Commission was again contacted to see what we could do to retain service from NET. Mr. Kimball Kenway, a member of the legal staff of PUC came to the island and addressed the community. According to him it must be proven that NET has extended service to the island and has considered phone holders as customers. Once this is acknowledged by NET they will have to come up with a plan for continued service, but they can still refuse to do so on the basis of its being too costly and therefore unfair to other NET customers. He advised us to employ an attorney to represent us. Meanwhile we were told to gather as much information as possible to show that telephone service is essential to the life of the island.

In December a petition signed by forty residents was filed with the PUC and a hearing was held on January 8th. At this hearing PUC concluded that Monhegan did have a case and should be heard.

On March 28th Alan Stone of Auburn was appointed legal counsel, after a meeting with the Assessors and some interested residents. He will present our case at a meeting of the legal staffs of Maine PUC and NET. I shall be there as community representative and it is hoped other members of the community will also attend. A definite date for the meeting has not been set. NET is expected to present four alternatives to the use of the underseas cable. These will be presented to the community at a special Plantation meeting at which the community can accept any one of the four or reject all and ask NET for reconsideration.

At our annual plantation meeting I was elected "community representative to co-ordinate plantation efforts against discontinuation of island telephone service" and $3000 was appropriated for legal representation to oppose termination of telephone service.

There may be a long legal battle ahead of us which will quickly exhaust the monies appropriated. In fact, even the preliminary study for the up-coming meeting may do just that. We therefore come to you with hat in hand to receive what donations you feel able to give to help obtain satisfactory telecommunication between Monhegan and the mainland. Any donation will be appreciated, the bigger the better! Make checks payable to **Monhegan Plantation Telephone Fund** and send, along with comments and suggestions to the above address.

Thank you,
Alta Ashley, M.D.

Monhegan, Maine 04852
April 23, 1980

Dear Friends:

Here it is time again for one of these letters. I found I had left out one whole page of H's in my March mailing so if you were left out just let me know. I have a few, but rather poor, copies left.

Much has happened since my last letter. The house is just about finished, with a control here and there, a light fixture, and the screen door on the outside door to be installed. If it doesn't come soon Phoebe will have scratched her way right through it.

The flats of seeds never did sprout so I dumped them into the cold frames I haven't had time or inclination to plant and now some sprouts are appearing here and there. It is still quite cold but when the sun is out it feels nice when out of the wind. I am not the speedy girl I was last year. Nothing planted as yet, indoors or out - just can't get up any gumption.

The solar system is working well, but I think I need more storage capacity. Now the problem is that I use the generator so little I can't keep those batteries charged! Also, the gas line is so long it takes the generator about two minutes to finally get started, which is pretty bad for the deep well pump. You can't win!

The telephone situation seems to be humming along. We have a lawyer, Alan Stone of Auburn, working for us. He spent the past four years on the legal staff of the PUC, specializing on telephone problems. We now have a complaint signed by 46 people against New England Telephone saying our system is inadequate and unreliable and we feel we should have adequate and reliable service, for the health, welfare, and happiness of the residents. The meeting with NET officials is now put off until mid-May. Meanwhile I've written to our legislators, to John Chancellor and David Brinkley and will tackle Walter Cronkite tomorrow. Meanwhile letters are pouring in from people I have contacted with checks to help with expenses. People have been most generous.

NET first said they would offer four alternatives to the cable, all apparently in some form of radio telephone, which we were prepared to refuse. Now they have "reconsidered" so we aren't sure just what is in the air.

On Friday Doug and Harry took me in early on the Chris so I could get to Bangor to attend the Republican State Convention. It was a dreen tide, the slip was broken, and I had to climb down the farthest ladder, most of which was covered with green moss and **slippery!** When we got to Port Clyde I had to go under, then over Laura B's lines. Jimmy Barstow and some stranger hauled me up by both arms while I felt I was tipping back and into the Chris. Wow!

The convention was my first. I got under Ginny Baker's wing. She is an old hand at it, but comes from Cumberland County, so we were separated during the whole affair. I shook hands with David Emery and Bill Cohen, both of whom told me to write letters about the phones, and with George Bush saying he would hear from me about the phones. Muskie, of course, wasn't there. A fellow sitting beside me had a cute knitted elephant named Andrew, which I fell in love with and, against my better judgment, wrote out a check for. Emery and Cohen shook his trunk and he got his picture in the Portland, Boothbay & Maine Times papers. I pinned Bryan Hitchcock's alternate badge on him and covered his front with campaign buttons. He was the most popular personage there. The Bush people had things so well organized he had run off with all the delegates before anyone knew it. That doesn't mean too much because the Maine delegation always goes uncommitted and can change its allegiance at the last minute.

I missed getting back to the island by the North Wind on Sunday, when June and Henley came back after spending the winter in Tenants Harbor, so took the ferry to Islesboro to visit the Bates, Elizabeth's friends from Concord who are now living on the Rogers' estate. Len is the caretaker and gardener and Toni takes care of the houses. They have a cute apartment over what will be the greenhouse - a really nice set-up for them. Len raises all the vegetables and berries the family and guests will use during the summer. Quite a change from insurance adjusting.

A call just came in saying the informational meeting with NET is now to be on May 2nd., which fouls things up because I just told the census taker he could stay here while he picks up forms from 37 houses and lists the rest of the 129. I just counted them for the telephone statistics. That Monday will be my seventy-fifth birthday!!! It is hard to comprehend it, but it is true. Maybe that's the reason I get behinder and behinder as the days go on.

Town water went on this past week. Paul Horn had very little trouble. There was a big leak from a frost heave up by the tanks, so he was delayed somewhat from pumping, but there were only minor breaks in the distribution lines in contrast to last years experience. Clothes lines blossomed with washings and the sun shone brightly to dry then. Shampoos and baths were enjoyed by all those without running water all winter. Thank heaven for my well, even though the pump is a problem.

Now the sun is higher I am burning more wood in the Jotel because there is little heat from the windows. It is warm enough, though, for me to sleep in my bedroom, which I had closed off all winter.

The town is thinking of installing solar systems in the school and it is possible we could use solar power for the new telephone system, if and when we get it. David Sleeper could have some business if these plans go through.

In my last letter I forgot to tell you about Livka Yandell Farrell, but now I see I did. We are having a shower for Winnie Murdock at the Store this afternoon and it looks as though Winnie might beat us to it. Her due date was moved up about four weeks and I barely got the blanket finished in time. Doug and Harry are coming to dinner later and I should be cleaning house instead of typing. It is about time for the boat. The Laura B is off the run this week for annual maintenance and Alison didn't bring out the Rita B in yesterday's storm. Boat fare is up to $10 round trip, Sunday and double runs start June 21st through Sept. 21. No news yet from Alan and the Seagull, --later. The shower went off well. Thirty-two women were there, and the presents were beautiful. Sally had decorated a raffia bassinnette which was filled to overflowing. We sipped champagne punch then gorged ourselves on goodies prepared by Sally, Katy and Ruth. Winnie is due to go in on the 8th and w-a-i-t.

It is now mid-week. We are enjoying a nor'easter stalled in the Bay of Fundy. The census taker is picking up forms and staying here. She had to list all vacant houses (101) and occupied (45), quite a few more since the electricity survey in 1964. Every other one has to be long form! It's raining hard and she has been out all day. I have a hot chowder waiting for supper to warm her up. Doug gave me a cusk yesterday.

It is now May 7th, so I had better finish this or give up. All last week it was so rainy I put off typing until the sun shone, which it did over the weekend, but I was in shore and now it is raining again, which means the solar cells aren't charging very well, but surprisingly more than I had hoped.

On Friday there was a meeting in Augusta with officials from New England Telephone, the Public Utilities Commission, our lawyer, Billie Boynton (First Assessor) and me. NET offered as their best solution an improved mobile phone system allowing four telephones but only two simultaneous calls and only after Monhegan puts up $64,000. They had put together some figures for micro wave without first looking the area over. After three hours of harranging the meeting broke up with NET saying a study would be made. Yesterday five people were here to look the place over. It was raining little fishes and some of them went out to see the site of the geodetic marker on the highest point of land which is now deep in the woods Green Point way. They feel the water towers would not be suitable places on which to mount the "dish". We should be hearing from them in a few days. Doug & Harry Odom own Station Hill area (marker).

Since I was in shore I decided to go to Concord, stay over until Monday, when I entered by diamond jubilee year, and attend the Knox County meeting on Tuesday. I thought I'd stop to see Marge and Phil Mitchener, only to find they were in England, and then thought I could see Emma on the way down or back, but she said she was too busy to see me. Elizabeth and Bruce gave me a fun birthday party for which I was the cook - rice, chicken in mushroom soup, salad and strawberry-rhubarb pie. Elizabeth and I went to see Aunt Helen (94) and took her to dinner (but she paid!) to Plymouth, then stopped to see Tommy Ashley, who is a patient at the West Roxbury V.A. Hospital having had his second heart attack. He is doing well and should be able to go home soon. Aunt Helen, as usual, is blooming!

Things are buzzing along here. The Coopers came over on Friday, George & Ruth Colton on Tuesday. Jo Day is here and about ready to open the Trailing Yew. The Periwinkle has a new soda fountain and bright orange booths. Bob and Mary are working hard refurbishing the Inn and Victor is reroofing the Monhegan House as well as repainting and replacing many windows. Jeanie is back from the hospital and can watch Kole while Victor works. Maryann Boody has charge of the Nelson houses - Elise died a few weeks ago.

Alan has sold the plane and will be returning to Alaska and bush piloting in hopes of buying a helicopter some day to replace the Seagull. Thank goodness Jimmy has closed in the Laura B so it is comfortable, even now that he has removed the wood stove. New benches in the back and strap hanger rails, without the straps, help in keeping one's balance.

I forgot to say that the final frustration of the weekend was the postponement of the medical meeting to next Tuesday! Danny Rothstein and a Pen Bay physician and I all got caught. I had forgotten Phoebe's dishes and my medications when I stayed overnight on Friday in Gardiner, so drove back, retrieved them and called on Louise Webber, having seen Lura Titcomb and Lucile Cheney on Saturday before leaving for Concord. I also stopped at Margaret Smith's to show the suit I had bought 25 years ago, and Bill Walters recognized it! (It is very tight through the middle!).

The children are on a field trip in all this rain, fortunately not planning to camp out. There will be at least ten in school next fall. Bobbie and JoAnn Mersfelder are hoping to be able to move into Freda Stanley's house. Don Stone and family are now in the Staples cottage (Reminger). Must close, no more room.

As ever,
Alta

Thanks to all Monheganites who sent checks re the phones. Official thanks & update will be sent soon.

Monhegan, Maine 04852
August 1980

Dear Friends:

Can it be that no letters have come off this typewriter since April? Where has the time gone, and what has kept me so busy? The summer season does have its interruptions. Every time I think there will be a nice period of time to catch up on things - filing news items, writing letters, or even dusting - there is someone at the door wanting to see the solar electric generating system, asking about the cemetery or friends appearing out of the past and most welcome visitors.

All summer long I have failed to keep my journal, chiefly because I'm no longer recording wind velocities for windmill data and the book is never in the same place two days in a row. And I thought I'd be better organized once the new room was operational!

Last night was the annual meeting of Monhegan Associates, INC. Ted Cochrane did a great job presiding over the meeting. He even gave me a few minutes to summarize the telephone situation, which is still in limbo. I've been waiting for Alan Stone to okay a new questionnaire, including a thank you to all those who have contributed so generously to the fund. At least now expenses for postage, telephone calls, and stationery don't all come out of my pocket. We are waiting for the Public Utilities Commission's answer to the Summary Judgment statement concerning NET's obligation to serve and what constitutes adequate and reliable service. Bell Labs has just written to say that solar power is not feasible and would not come up to the "high standards of AT&T". It certainly works well for me and my demand is greater than that needed for a telephone service. Certainly it is more reliable than four small generators run on LP gas. Grrr--.

Steve Rollins and Jean Gibson were married at the Lighthouse on

August 2nd. Jean's father almost didn't make the wedding. He was one of the crew of the Navstar that sailed without electronic aids to Dakar and back, and returned to their home in Newagen scarcely a week before the wedding. They are working hard getting their new house finished. Steve's three boys are now living here with Steve and Jean, two will be in school here, Jeff will go to high school in shore.

Ron Stanley married a nurse early in the summer. She has two girls who will also be in school, so there will probably be twelve or thirteen children here in school this fall. Lance leaves soon for school in Albuquerque, Bob and Mary will follow when they get the Inn put up in mothballs for the winter. We will miss them a lot and will look forward to their return in the spring, when they will be doing more remodelling and repairs on the Inn and Pierce cottage.

Jack Cabot left a couple of days ago to go live with his cousin, or nephew, George in Brunswick. He was unable to eat because of a recurrence of the tumor in his throat, but hopes to get some strength back with tube feedings for a while. Everyone wishes him well.

The entire Nelson family has been here this week. On Tuesday they had an internment ceremony at the cemetery for Elise conducted by the adopted son whose name I do not know. He hasn't been here since I bought the Mooring nineteen years ago. Can it be that long?

The other morning Lance Burton's boat, the Sea Squirt, was found filled with water and then upside down in the harbor. He had all kinds of electric equipment on it plus seat cushions, etc. They found the cause of the accident a loose fiberglass panel on the port side. He thinks all his equipment was ruined, but hopes it can be dried out.

The big news this week has been the sale of the Colton cottage to Carl Hiller, who has rented the Bodanov place for the past several years. Now that both of their offspring are in California the Coltons feel they should spend more time on the west coast, rather than here.

Bill Sweetland, the carpenter who worked on my place, is now doing the finishing touches on Shermie and Ruthie Stanley's house. I'll be glad when he comes back to finish things up here and when the screen door to the garden arrives. Phoebe will have dug a channel right through the wooden door before long. She has been taught to scratch at the door rather than to bark, but that was because the outer doors on the rest of the house are aluminum and don't scratch. I have ordered a self-storing panel door, which has been back ordered about six months!

This summer I have been off island more than usual. First there was a reception for the Zorachs, followed by an annual luncheon at the Farnsworth Museum. I combined that with a trip to Emma's to see John and Rue Riddle, on from San Francisco. Then I went off to Pittsfield, Mass. to visit Margaret Sizer Dorr and four other MHC '27s - Tony Chickering Weathers, Mary Elizabeth Edmunds Hultz, Elizabeth Leavers, and Hallie Patey Long. Hallie drove out with me and nearly lost her composure several times as I got confused in traffic. I realize now driving in Maine is far different than in Massachusetts where four or five streets come together so often. We had a great time, went to Tanglewood twice - the first night for a picnic supper and the concert, and to Tosca the following night - a memorable performance. The last trip was to my grand niece Mary Alice Jackson's wedding which took place at Rochester Institute of Technology auditorium. I stopped off to see Marjorie and Paul Mitchener en route to Concord, having spent the night at John and Ruth Hosmers' on Rackliffe Island. Elizabeth and Bruce Gardner and Tommy Ashley and I went to Syracuse in my car, then to Rochester in a bus the Jacksons chartered for the twenty-five guests from the Syracuse area - a good and thoughtful arrangement. We spent the night back at the Jacksons' then back to Concord and Tuftonboro. Phoebe found a place at Palmer's kennel, after much pleading, so didn't have to take the long hot ride to Syracuse. I nearly missed the afternoon boat on Monday because of the slow and heavy traffic from Bath on. Fortunately I didn't, because my house was on the house tour for Museum Festival week, to show off the solar power set-up. Twenty-one people came.

The solar power installation has worked out well. The cells function even in dense fog, but not in heavy clouds. I've had to charge the batteries by the auxiliary generator only twice since February and have used the generator only 40 hours since June 7th, instead of 4-6 hours a night as before. That's good, too, because gas is now \$27 a cylinder, compared to \$8.50 when I first got the cottage and \$19 last year. I should be paying for the photo-voltaic cells sooner than I expected to do in savings.

Gertrude Anderson, for many years resident minister for the island and for ten years a resident of the Baptist Home in Chestnut Hill, died last June. Mary Taylor, once the hostess at the Inn diningroom, died in May in England where she had returned a few years ago.

According to Nita Minewski, the memorial show and reception for Alex were well attended and received. Alex died soon after leaving here a year ago. Unfortunately both the show and reception were held while I was away, as were most of the events of Museum Week.

Bob Boody has his wind generator up and working, but I haven't heard how well it is doing. One thing, wind machines are noisy and require a fair amount of maintenance, which solar cells do not, unless they get covered with snow. Last winter was no test, since we had only one moderate storm after the panels were mounted.

Ted Kennedy is due here, according to rumor, for some R and R at the Wyeth cottage, now that Jamie is back. Bush said he would be out, but he is off to China. Maine has certainly come to the fore with Bush and Muskie in Kennebunkport, Hodding Carter buying Norumbega in Camden (which looks like an enlarged model of Grampa Woodbury's house in Allston, built in same year). Maybe someone will realize we are still a part of the United States of America.

We have had a poor summer for the garden. First it was so cold beans just rotted in the ground and had to be replanted twice. Then the drought hit and is still with us. It has been unusually hot and sultry, but nowhere near as bad as in the rest of the country. Now the winds are coming from the NW and probably the beans and lettuce I just planted won't amount to a thing. I have a lot of gourds all mixed up with squash. Last night I cooked what appears to be a gourd, but it tasted like yellow summer squash. Italian cucumbers look and act just like zucchinis, of which I have plenty, as everyone does. The best crop has been Italian runner beans which are filling my freezer with gastronomic joys to come. No more bush beans for me another year. These are so tender they require only about 5 minutes of steaming to cook. Twelve plants of salad bowl lettuce has supplied me and my tenant, Eugenia Halbmeier, all summer, but are now running out. Steve Whitney took down two dead spruces in the front yard the other day and of necessity had to trample down most of my cultivated raspberry bushes which have supplied my breakfast table with luscious berries for two weeks, and now black and blueberries are ripening. I feel as though I were living on a farm with my postage stamp garden and berry bushes out front. Day lilies, those left by the deer, have been spectacular for over a month now. But Fall is on the way. Barberry bushes are coloring up and flocks of cedar waxwings can be seen on their way south along with skeins of cormorants.

I've just about had it with the booklet "Introduction to Monhegan" which is supposed to be handed out on the boats to trippers and new arrivals. First the Monhegan Associates footed the bill for printing, having suggested someone write such a booklet, then withdrew support because "it publicized Monhegan too much" but they then came out with the trail maps! Commercial establishments paid, but with rising costs for printing they are becoming restive, some even refusing to pay their share. I set up boxes with "a dime will help" and those have been rifled and now the printer says I owe $140 which was paid over to them from checks received in early July. The booklet really needs rewriting, but I guess I'll let someone else take over the job and its headaches.

Enough for now. I'm having trouble with my duplicator, so please excuse the quality of your copy. Thanks again for Telephone Fund contributions. We need and appreciate them very much.

As ever,
Alta

MaryAnn Boody says the wind generator supplies enough power for their lights, they rely on the regular generator for heavy power demands only, as do I. That's nice to know.

Today we encountered a doe in the woods which **chased Phoebe!** and scared her to death. It nearly came up to have its ears scratched until **brave** Phoebe, on the leash, barked. No problem with deer chasing here.

A & Phoebe

1980

Energy System Mooring Complex Monhegan, Maine
Alta Ashley, owner Photo-Voltaic System with
L.P.gas powered Kohler generator
auxilliary and back-up
power

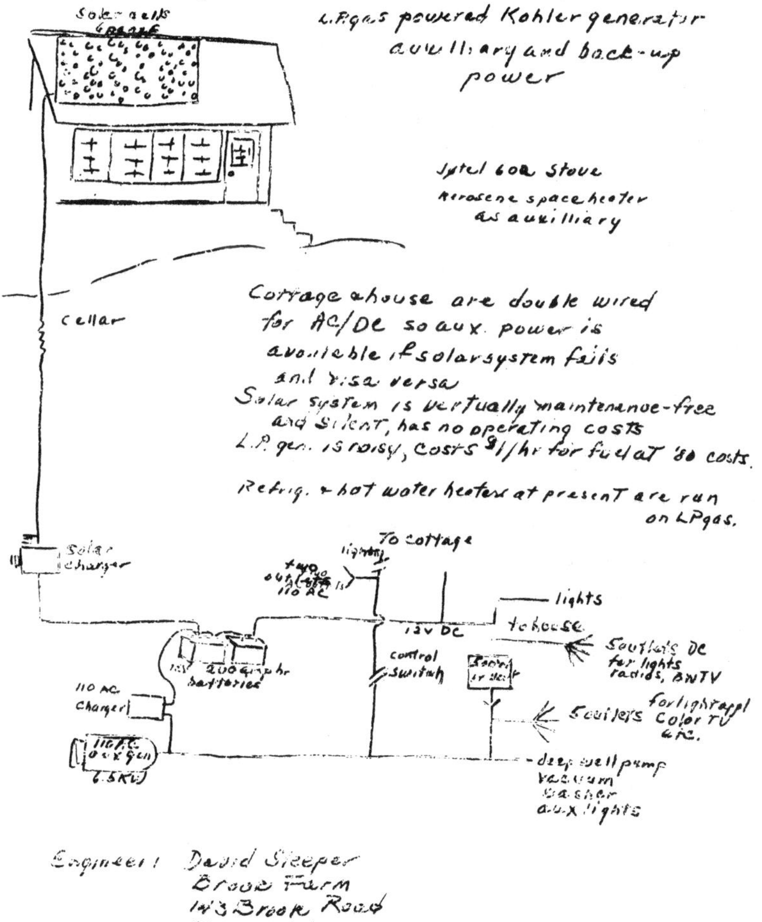

Jøtul 602 Stove
kerosene space heater
as auxilliary

Cottage & house are double wired
for AC/DC so aux. power is
available if solar system fails
and visa versa
Solar system is virtually maintenance-free
and silent, has no operating costs
L.P. gen. is noisy, costs $1/hr for fuel at '80 costs.

Refrig. & hot water heaters at present are run
on LPgas.

Engineer: David Sleeper
Brook Farm
143 Brook Road
Falmouth Me 04105

Monhegan, Maine 04852
October 20, 1980

Dear Friends:

At long last I am able to sit down to write a much over-due letter, but not by the power of the sun. For weeks I have had trouble starting my generator, but after a few minutes of laboring on its part it has come on to do the things it is supposed to do - pump the water, charge the batteries when they are low, run the washer and vacuum cleaner to keep me in some semblance of cleanliness. But for the past week to ten days the thing just won't turn over until it gets kicked here and there, two or three screws are turned and even a new starting battery has been installed. Each time either Paul Horn or Bob Boody comes down to look at it the thing starts up almost immediately for them and they say "There's nothing wrong with the gas pressure/the generator." I have also blown out the inverter.

On top of all this we have had a series of cloudy days and there have been people in the cottage thus putting a demand on both the water supply and the electric (battery storage) power. All of a sudden the photo-voltaic system is failing to store enough juice to get through an evening and without the generator to supplement the lights have dimmed to a dull orange glow.

On Wednesday I spoke to the Library League in Tenants Harbor, supposedly on the joys of living on an island. That morning everything went bad. When we finally got the generator to go we started to vacuum the soot from the space heater in the cottage. There was a hole in the bag allowing the soot to come out in great clouds all over the living room and us before we could find the switch to turn off the vacuum. It was a cold, wet trip in against a NW breeze blowing up to 30 knots. I decided to get a new suit for the lecture but could find nothing wearable bigger than a 16, which I fill out far too amply. Then I went to get a new bulb for my projector and there was none in stock which would fit. Had not there been a genial cocktail hour and delicious dinner at my hosts, the Haskells on Hart's Neck, I think I would have opened the talk in tears.

While we were setting up things for the talk, using my small Minolta projector which necessitated picking out the slides from the carousel, turning them inside out and right side up, which didn't always happen, I started to explain why the confusion plus all the other things that had gone wrong and launched into my talk without being introduced! The house was packed and we got through only one set of slides because of the time it took to load the little projector as well as the many interruptions by questions from the audience. Even so I was there talking or chatting informally for nearly three hours and even received a request to speak sometime soon in the Springfield area. Maybe I am about to embark on a new career - the lecture circuit!

Jack Cabot died early in September and his ashes are buried in the cemetery. Next summer it will seem strange indeed not to see him waiting for the boat up on the Inn front lawn. Alfred Fuller died, also, in September and Earl Field, Sr. in August. "The old order changeth..."

The Yew closed on Friday and that afternoon Jo Day took a hike to White Head, having served her last meal for the season that morning. I was stuck here at home waiting for a phone call so didn't see the crowd at the dock either then or yesterday. Our population is now down almost to winter proportions. After Tuesday I am planning to close the cottage - too many problems unsolved at the moment, too hard to heat.

All summer long the photo-voltaic system worked so beautifully I kept saying how fortunate I was to have such an enjoyable place in which to live. Now I'm not so sure. I should go back to oil lamps, buckets of water, the broom, and the use of the clothes washer at the Store, I guess. A household of gadgets is no place for an aging, slightly over-weight female. How is it men have such a knack of getting mechanical things to run even if it is only by a rap with a hammer? They just find the right place to hit, apparently. Bob Boody's wind generator is down more than it is up, so I'm glad I didn't go that way. David Sleeper is bringing out more or different batteries when he comes out next month, meanwhile I think I'll take a few days vacation in shore awaiting his arrival.

Wanted: able bodied unattached male engineer, carpenter, yardman, mechanic etc. and entertaining raconteur to share bed(?) and board in island home with aging, overplump female getting beyond her depth!!

Joan and Robert Rothstein have sold their house to her father, have bought a house in Falmouth and will be off tomorrow after a farewell party tonight at Herby Kellem's where they are now living. They are taking Margaret with them so I may have to get a cat to keep the rodent population in check. Right now she is in here happily guzzling Meow Mix, much to Phoebe's disgust. They studiously ignore each other.

Steve Rollins is working hard getting his house closed in. While John Kordak was here he gave Steve a much needed helping hand with placing the rafters. Jeff, by the way, is in eighth grade here so we have thirteen children in school. Billy Boynton is coming along with his house. He and Jackie should be able to move in fairly soon. My house still lacks a storm door on the new room, a generator switch, and new shingles on the west roof; plus there are dangling water spouts which have to be replaced. Maybe next summer...

As yet we haven't had a killing frost. I can't keep up with the broccoli; brussel sprouts are coming along; runner beans still provide a meal twice a week and the Italian cucumbers are still forming. My acorn squash amounted to nothing, though, but I have lettuce in the window greenhouse which finally was set in place a few weeks ago. Three dead spruces were taken down and their brush finally taken to the dock by my renters on Friday. It took five trips with the golf cart to get it all down there. Now I have to cut blackberry vines from around the cedar trees and get them to the dock without bleeding to death.

The lawyer, Alan Stone, and his family are due tomorrow. We have a new questionnaire to work on. Things are beginning to gel and we are in hopes of convincing NET that a satisfactory system can be put up for about $100,000. They came over from the Energy Resources Dept. a few weeks ago with an offer of a **free** wind generator, but the specifications for location were such it could have been erected only on Manana or somewhere off Gull Cove. There is to be a seminar with NET the PUC, Monhegan representatives and Dave Sleeper on the 28th concerning costs, etc. Let us hope something concrete will come of it. We have been in limbo long enough concerning the telephones.

It is close to 60 degrees right now, before noon, and the cells are charging well for the first time in days. Maybe we can watch TV tonight without having the picture squeeze down to nothing. Right now I am typing, have just vacuumed and Ginny Baker, my renter, is using the washing machine, so everything but the inverter is back in working order, a note most fitting on which to end this epistle except to say I had a nice visit from Elizabeth Gardner and Elizabeth John in September and wished I could go with them to Emma's, to a family dinner and a ride over Cape Cod to Nantucket in Arthur and Carol's plane. A great trip. Phoebe thanks you all for her many pats, I thank you for your interest in the telephone problem.

Happy Hallowe'en!
Alta Ashley

October 27, 1980

Dear Concerned People of Monhegan:

I thought I would take this opportunity to give you an update on the telephone situation. As you recall, since the cable broke 2 years ago, leaving us without service for nearly 4 months, and raising a serious question about the future of our service, we have been trying to get the Public Utilities Commission (P.U.C.) in Augusta to resolve the issue of New England Telephone Company (NET) obligation to provide Monhegan with reasonable, adequate and sufficient telephone service.

In July, 1980, we asked the P.U.C. for a legal ruling that NET has a continuing legal obligation to provide telephone service to Monhegan Community and that NET may not discontinue telephone service without first getting P.U.C. approval. We felt, on advice of our attorney, that such an order by the P.U.C. was necessary in view of NET's officially stated position that it has no such legal obligation and may terminate our service "at will." We also petitioned the P.U.C. to investigate the adequacy of our current level of service. On July 7th, 1980, a public hearing concerning these matters was held before representatives of the P.U.C. at the Monhegan Community Schoolhouse. Present at the hearing were members of the Commission, members of the P.U.C. legal and engineering staff, members of NET and its legal staff, and our attorney, Alan G. Stone. The day long hearing consisted of cross examination of witnesses and public testimony from members of the community, who spoke of the need for a reasonable, adequate and reliable telephone system on Monhegan, that NET has historically been furnishing telephone service to Monhegan and should continue to do so, and that the existing service is far from reasonable or adequate to meet the Island's needs, and is unreliable because it is dependent upon an underseas cable which could break at any moment.

We are happy to say that on September 12, 1980, the Commission issued an Order that:

> "1. NET has incurred a continuing obligation to serve the Monhegan Community on the basis of the service it has provided since 1919; and
>
> 2. NET must petition this Commission under 35 M.R.S.A. §212 prior to a discontinuance of service to Monhegan."

The Commission, however, reserved the question of what type of service is adequate, reasonable and sufficient for Monhegan, and will be holding hearings on that issue at a later date.

We have taken the position that adequate service requires the installation of a reliable system capable of providing direct dial private line telephone service consistent with that provided elsewhere. This could be furnished reasonably by means of a point to point microwave telephone system between Monhegan and the mainland designed to provide all classes and grades of exchange service.

NET has refused to install such a system voluntarily, and has taken the position that their obligation to provide adequate service to Monhegan would be satisfied by replacing the existing service with a rural VHF radio system (Motorola-Multi Access 2 channel Rural Public Telephone Service), with only 4 telephone stations available and 2 channels. According to NET, this system would permit a maximum of only two telephone conversations at any one time, requires manual outgoing calling (as opposed to direct dial), and permits privacy of conversation "to the extent that radio is private." NET acknowledges that this would provide "lesser service than is currently available."

In order to determine how you feel about this situation, and the level of need for telephone service on Monhegan, we are asking you to compete the following questionnaire and return it to me within a couple of weeks.

We wish to thank those of you who have already expressed your concern by your donations to the Monhegan Plantation Telephone Fund. In order to continue our efforts to secure reasonable, adequate and reliable telephone service to Monhegan, further donations are desperately needed.

Thank you in advance for your interest and support in this vital matter.

Sincerely yours,
Alta Ashley, M.D.
Coordinator for Telephone
Service for Monhegan
Willard J. Boynton
1st Assessor

Box 7
Monhegan, Maine 04852
December 4, 1980

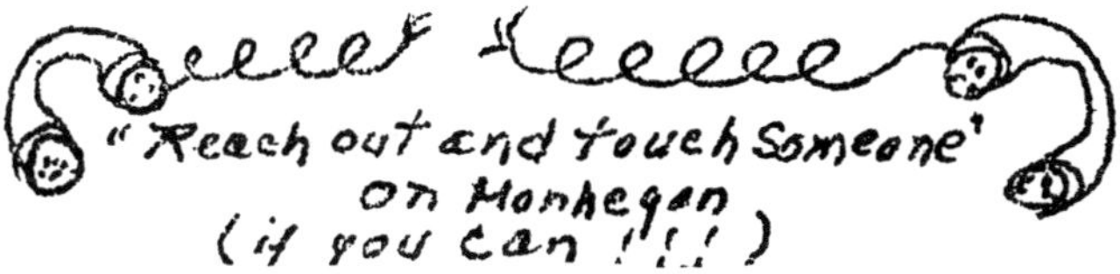

Dear Friends:

Season's Greetings from a windswept island! There are a dozen or more things I should be doing, but today is perfect "Dear Friends" letter weather with a nor'west wind churning up huge seas and icy blasts creeping in through every crack. Fortunately this new room is snug and draft free, which can't be said about the rest of the house.

Phoebe and I returned yesterday after nine days ashore to join Emma in celebrating Thanksgiving in Center Tuftonboro and then attending the Knox County Medical Society Christmas party at the Sail Loft on Tuesday. I thought surely the predicted storm would cancel the party out, but it didn't strike until after midnight so my guest was able to return to Pemaquid Point before the wind, rain and lightening, which scared Phoebe to death, blew through. Then there followed a period of relative calm so our trip out on the Laura B wasn't too bad. She had just returned from Boothbay Harbor, had her stove set up and going, and the scuppers blocked off so we had little water on the deck despite our plowing through some pretty big seas. It is almost a joy to make the trip now!

The island was pretty quiet over the holiday. Were it not for several visitors there would have been only about fifty people here. But the Boehmers had a houseful, Miriam Barteaux's two daughters were with her and other families had guests for at least over a boat. The phones went out in Friday's storm but NET men were able to repair the lines on Tuesday. According to them it was probably the last time they can make repairs. Water has seeped into the cable and is corroding the wires.

Meanwhile we wait for a decision on the part of New England Telephone as to if and when they will furnish us with an adequate and reliable system. The mobile phone unit they want to supply will be scarcely better than the VHF system some of the boats have and calls will be just as costly as dealing through Camden Marine -- a $6.00 or so hookup charge plus long distance (operator assisted) charges from Camden. David Sleeper, who put in my photo-voltaic system, is having a hearing with NET officials this morning in regards to the use of the cells to power a phone system on Monhegan. A cost analysis hearing will be held later this month or in January - the sooner the better, we say. We need the service! What town has no phone service for its town clerk, head officer, school or physician, even in the best of times? We don't feel our demand for service is unreasonable, especially since more than fifty phones have been requested through the last survey, plus additional pay phones requested by those not wanting phones in their own homes. Response to the questionnaire has been good. Thank you all who sent in donations to help the cause along. So far we haven't had to beg for more town funds because of your generosity.

I am still unable to use solar power to type with but Dave Sleeper is due tomorrow with my inverter which was not badly damaged after all when I forgot to turn it off before starting the generator. Let's hope it will work all right. Bob Boody found I needed a new water pump, fan belt, and brush for the generator. With all those replacements the thing is working like a charm. This morning I tackled the pot burner again, using my best vacuum cleaner to suck out the soot. All went well for a while, but then the soot came through and filled the air, the rug, me, and my velour house coat with grime. My hands will have to wait for a new layer of skin before they can recover. Solo, Bold, All and Spray 'n Wash will all have to be used to clean the clothes! By the way, I received no answers to my desperate help wanted add in my last letter. A couple said they might apply but were already happily employed! My cousin Helen Woodbury reminds me that after trying many years to visit, first in Farmingdale and now on Monhegan, they finally made it out for an overnight stay this summer and I omitted telling you about it! It was great having her and Bob here. We **did** the island in the twenty-four hours they were here, leaving a spot here and there so they will want to return.

The school children put on a play last month, written by Jon Bogdanove who is serving as teacher aide this winter. It was a clever spoof of island life and some of its best known characters - Bob Burton and his coffee cup, Zimmy Brackett and his crown of snowy hair, Barbara Rothstein and the twins. The kids did a great job and both audience and cast had a great time.

Joan and Robert Rothstein are living in Cape Elizabeth, not Falmouth as I reported last time. Ron Stanley left with Mary before

Thanksgiving. He has not been well for some time. Linda, his wife, took the girls in the previous week. They will be living with their father in Dresden. Bob Burton has joined Mary and Lance in Albuquerque for the winter. All our human robins but Mort Bernstein have left. Cindy Barker took off yesterday on the return of Katy Murdock from California. Cindy had been taking care of the animals during Katy's absence. The Rollinses have moved into their new house and I believe Billie and Jackie Boynton are now in theirs.

Early in November I went in for what was thought to be an eight hour test, but ended up in a three day stay at Pen Bay, due apparently to intolerance to the dye used in the myelogram. Phoebe had to stay at Dr. Sidwell's for the duration, then we took off for Massachusetts, stopping to see Marge and Paul Mitchener before going on to Concord. That Saturday Elizabeth Gardner and I went to Newburyport, no, Marblehead, to Ulriche Welsch's open house. Uli was recently in Peru and came back with a full show of her experiences. She is back with the Boston Globe, teaching and free-lancing - a great photographer. The following day we joined the Peters (my niece Nancy) for brunch at the St. Botolph Club, complete with a pitcher of Bloody Marys but didn't realize Don Stone was having an open house at the Copley Gallery, so missed that. Many Monhegan people made it, though.

Huey Kurz has just been here and given an estimate of the cost of putting in a new chimney for the cottage so I can install a wood or coal stove. I'll have to eat hamburg and spaghetti for months to foot the bill! By the way, the cottage is unencumbered until the last weekend of August, should you know of someone wanting a cottage for one or two months, preferably the latter.

It is now Saturday. Dave Sleeper came over yesterday - the Laura B came out despite the gale warnings. So now I have my inverter back and once more am typing by solar power - a very satisfying activity. Dave is here to look over the building which might be used to house telephone switching equipment and to discuss solar systems with two or three interested persons on the island. It would be funny if little Monhegan led the way toward photo-voltaic cell energy systems in the country. Once the cost of the panels and batteries can be brought down it will be the method of choice, in my opinion - no noise or odor pollution, no fuel expense once the system is established. Now to devise a small wind generator, such as an electric fan, as a back-up to replace petrochemical generators. Then we really could tell the Arabs to fly their kites. But we have to convince Mobil, Exxon, Gulf, et al to get on the solarwagon.

The Christmas play will be on the 19th and the Christmas Community party will be held on Monday the 22nd. I hope someone will be going in on Tuesday. I hate to wait until Wednesday to leave for Christmas. And so on that note may you have a beautiful Christmas and a rewarding 1981.

With holiday greetings from
Alta & Phoebe

December 20, 1980

To those who rediscover this box:

May you have as much delight in the contents of this box as I had when it was found a few days ago.

In May of 1961 I became the owner of the "Mooring," having bought it from the estates of Father Sutton and Ruth Parry who, I believe, bought the cottage from the Moores. It took nearly a year of negotiations to obtain the deed.

In 1969 I built an addition to the cottage which would allow me to live on the island year round and in 1972 the cottage was winterized by Jimmie Barstow and his crew. Hot fires required during the winter took their toll of the chimney which was considered unsafe by 1980.

Hubert (Huey) Kurz, a mason recently from Rochester New York, now residing on Monhegan, removed the old chimney and rebuilt a new one, including a brick facing which will absorb and radiate heat from an airtight stove. This stove, which will have visible flames, will help warm the cottage occupants in the winter as well as during the spring and fall and on cool nights and rainy days of summer.

Into the box I have placed a 1980 brochure from the Island Inn, a copy of the Monhegan Museum Brochure and one of "Introduction to Monhegan" as well as a newspaper account from the Rockland Courier-Gazette, Dec. 16, 1980 written by me and a grocery ad from the same paper. The Moore's letter and copy of Science and Health have been replaced.

Best wishes to the finder by the depositor, Alta Ashley.

Dec. 20, 1980

This is a copy of the letter found in the chimney of the "Mooring" when it was demolished in December and rebuilt. The original chimney had to be replaced because of leaks in the masonry which had allowed some scorching of surrounding beams. The fireplace will be replaced by an airtight stove to conserve heat and allow for more efficient use of wood, in these days of energy economy and skyrocketing fuel costs.

Monhegan, Maine
September 1, 1926

This house was started July 29, 1926, designed and built by and for Delphine Nelson Moore and George H. Moore of Boston, Mass.

The builders are Maynard A. Orne of Monhegan, an honest and capable workman.

Harold A. Osgood of Monhegan, apprentice.

Solomon J. Higgins of Wiscasset, who claims to have laid the foundations of Solomon's Temple and to have built "never a fireplace that smoked."

Archibald Brown of Rockland, "Fighting Irish," who dug the foundation and carried brick and mortar for the aforesaid Higgins.

We deposit with this record a copy of "Science and Health" of the special edition published for free distribution to the soldiers of the world war, a current number of the Christian Science Sentinel and Quarterly and a current folder of the Island Inn.

We deposit also our hopes and good wishes for many a happy hour for the present and future owners of this house, before the fireside.

"Except the Lord build the house, they labor in vain who built it." Psa 127 - 1.

Signed Delphine Nelson Moore
George H. Moore

Holiday Greetings from Monhegan! 1980

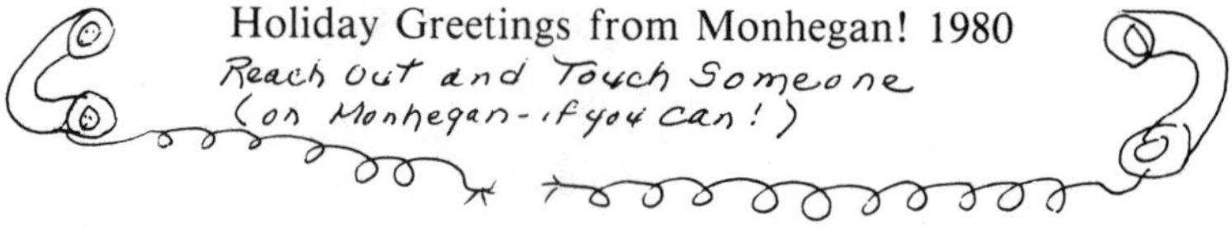

Dear Friends:

What a year to tell you about! First of all, I finally was able to move into the new room just before Christmas and had a post Christmas dinner for Rita, Miriam (our teacher), Doug and Harry, delayed a week because of weather too rugged for the Chris. From then on I practically lived in the new room, it was so comfortable and easy to heat. The window wall on the south makes fires in the Jotel unnecessary on sunny days.

David Sleeper came out and set up photo-voltaic panels which were in operation by February 1st. I now use the generator about 5 to 10 hours a week, depending on how clean I want to be. I have to use it for the vaccum, dish and clothes washers, and the deep well pump. Right now I can't tell how much I will save in gas because for several weeks I had a leak in the line and consumed three times as much fuel as I should have! I never seem to win. But is great to sit here and type by solar power instead of petrochemical (Arabian) fuel.

Early in the year I stopped of at Continental Telephone offices in Damariscotta to find out what decisions they had come to as to furnishing Monhegan with a telephone service and found that we had been turned down by the national office in Virginia. So back to the Public Utilities Commission to see what we could do through New England Telephone once more. The year has been filled with meetings, letter writing, etc. and we are still in limbo. The cable has been broken twice already this fall and the repair man says there is no hope of being able to do any more repairs. So we may have to depend on NET's "Improved Mobile Telephone" with only two lines and four instruments (and who will get those?) if and when the cable parts once more. The latest survey shows a demand for 45 private phones, six business and at least two public phones as well as several extensions, which is twice the demand of two years ago.

In March I was elected as delegate to the state Republican Convention in Bangor and took my first dip in the sea of politics. It was fun, but a bit bewildering to be among so many thousands after seeing less than a hundred people here on the island for months. I fell in love with a stuffed elephant named Andrew, and brought him home, much to Phoebe's disgust, but not before he had shaken his trunk with Bush, Emery, and Olympia Snowe. He was the hit of the convention and got his picture in several Maine papers. It was fun to meet so many people known only by name before. Once is enough, however.

All summer long people trooped by to see the photo-voltaic system in response to an article in Solar Age, a national magazine, which dealt with Sandy Dickson's and my systems - hers as that needed for minimal demand, mine for the greater demand of two houses (this and the cottage) and more light demand gadgets. The house was included in the House Tour put on by the Museum during their fund raising activities. I've certainly had a lot of exposure between the publicity for the electric system and that concerning the telephones. I hope the latter will soon come to an end and we'll have a decent phone system, one in which the town clerk, first assessor (mayor), and physician have their own phones. Is that too much to ask for? The Store and Post Office have phones, such as they are, now.

Our population took a leap forward on one end of the scale with the arrival of Livka Farrell, Kristina Murdock, Eric Sproul and the Rothstein twins Rebecca and Rachel (Dan and Barbara's). Linda and Huey Kurz have become permanent residents. Huey will be putting in a new chimney for me in the cottage so I can install a wood fire (airtight) there. Work should be started late this month or early in the new year. Lexi Krause has given up his merchant marine job and will be lobstering next year. Jackie and Billie Boynton have moved into their new house; Steve, Jeanne, and the three Rollins boys are now in their new home. But on the negative side of the ledger is the departure of Robert, Joan and Josh Rothstein for Cape Elizabeth along with Angela and Margaret. I miss Margaret and her rat-catching abilities but so far have had no new invasions by Rhoda and Ronald Rodent since they broke into the cottage in October and got fat on Decon. A new poison seems to be working, thank goodness. The Cundys will be moving to Boothbay Harbor, Liniken Bay area, at the end of lobstering.

Several old friends of the island passed away. Jack Cabot, Earl Field, Sr. and Alfred Fuller died all in September. Eleanora Stevens died on peaks Island in October or early November. Gertrude Anderson, for many years our resident minister and Mary Taylor, long the hostess in the Inn dining room, both died earlier in the year. Enough of gloom.

June and Henley Day are now in their trailer in Tenant's Harbor for the winter, Faryl Henderson and Fred Wiley are back in Kennebunkport, Don Stone and Linda have moved back to Ogunquit for the winter. Harry Bone is here for another leave from his ship and hopes to get his foundation finished before being called back.

With the price of oil soaring and its quality diminishing many people have turned to coal stoves which require attention only once or twice a day instead of constantly, as do wood stoves. I'm still trying to use my pot burner but am having a terrible time with soot build up, which clogs the air intake, thus increasing the soot. Bob Bradley brought out 24 tons of coal last month to fill the many new coal bins people have built to receive it. I'd like to go to coal, but need a steady unattended source of heat when I go in shore for a few days. I had hoped to take off for several weeks between Trap Day and Town Meeting day in March but now feel I had better stick close to this house and the new chimney in the cottage.

Alan Mersfelder sold the Sea Gull and went to Alaska to bush pilot again. He and Ellen are back in the states but came out here for only a boat in October. No more plane rides for us so we are grateful for the increased comfort of the Laura B.

In August I had a delightful time at a mini-reunion with five other Mount Holyoke gals in Pittsfield. We took in a concert and an opera, "Tosca", at Tanglewood. Then there was a trip to Baldwinsville and Rochester for the wedding of my niece Mary Alice Jackson and Earl Websterlund. Jeanne Gibson and Steve Rollins were married at the Lighthouse on August 2, with a reception on the Hudson cottage lawn. We need at least one such celebration a year to bring the islanders and visitors together, always such fun.

Bob Boody put up a wind mill. When it works it works well, but he has to take it down because of high winds quite often, so I am more than satisfied I went the solar and not the wind route.

Let us hope that your holidays will be enjoyable and that 1981 will have much good fortune in store for you. And so, "Reach out-----"

MERRY CHRISTMAS AND HAPPY NEW YEAR!

Sincerely,
Alta Ashley and Phoebe Starling
Monhegan, Maine 1980

Box 87
Monhegan, Maine
January 1, 1981

HAPPY NEW YEAR! To prospective Monhegan telephone subscribers.

It is with pleasure that I greet you today with some hope that real progress concerning our telephone service will be made in the next few weeks. A hearing will be held with New England Telephone Company sometime later this month in which there will be a cross examination of NET's cost analysis of six proposals for services. Four of these proposals involve radio mobile systems which require antennae and 12 volt power provided by the customer. The fifth is a microwave system, and the sixth would involve replacement of the present cable. No direct dial charges would be available with the first four proposals. A monthly rate of $13.75, plus $50.00 monthly or an installation charge of $2,700 to $4,000 per instrument would be required for these systems. Rates for microwave or cable systems would be set by Public Utilities Commission, which would be in line with charges for other neighboring services.

New England Telephone Company feels that revenues generated by calls to and from the island do not justify the microwave or cable systems and the number of subscribers would also be too small. We, therefore, are asking you to make application for phone service directly to NET as soon as possible so that NET will have first-hand knowledge of the demand for service.

For those requesting residence service please indicate 1, 2, or 4 party service requested and mail your request to:

Peter J. German, Manager
New England Telephone
212 Main Street
Waterville, Maine 04901 Tel. 872-9911

For business and coin telephones:

Harry W. Miller, Manager
77 Exchange Street
Bangor, Maine 04401 Tel. 947-9911

It would be helpful, also, if you could report the volume of telephone service you have made in the past year or so. If you have any questions I will be glad to answer them to the best of my ability. Thank you for your cooperation in this important matter.

Requests for service at this time are not binding.

Sincerely yours,
Alta Ashley, M.D.
Coordinator for Telephone
Service on Monhegan

Box 87
Monhegan, Maine 04852
January 3, 1981

HAPPY NEW YEAR !
1981

Dear Friends:

What a year this is promising to be, at least for early January. We are in a deep freeze for the third time since Christmas Eve when the mercury fell to -15 degrees here and much colder inland. Tonight the forecast is for even colder temperatures. Right now I have an Aladdin heater out in the generator shed hoping to thaw out the generator so I can put on the pump before the water lines freeze. There is another heater in the cottage striving to keep the mortar in the new chimney from freezing. Huey nearly finished when he had to stop because five "birders" arrived for the Christmas bird count on Monday, an ideal day in which to lay brick through the roof. Now he'll have to wait for a January thaw, I guess.

The "birders" weathered below zero temperatures in the cottage with the two aladdins keeping the temperature just above freezing. Snow and sleet cut down on the number of birds coming out of shelters but they did get forty species without seeing a blue jay, starling or grackle, all of which were at the feeder after they left.

Today the temperature was -6 degrees this morning outside and 50 degrees inside. Needless to say I have all three fires going. There is so much vapor that it is flowing over the island in great clouds. Between it and the amount of snow on my solar panels, as well as the low angle of the sun, my batteries are pretty low and I'll have to stop typing unless I can get the generator going to charge up the batteries. We had about six inches of snow which has been whipped into huge drifts by the wind which gusted over 45 knots (needle stuck there) last night. Doug, Harry, Charlie MacDonald (here to help with setting), and Miriam Barteaux all braved the storm and came here to dinner last night. I'm glad I was the hostess and didn't have to go out. D., H., and C. had come over New Year's Eve with a bottle of Charlie's homemade champagne from mead he makes from his own honey. I got out the champagne glasses and we toasted the New Year about 8:30, after which they went home and I crawled in bed with a book, fell asleep, and never knew when the ball slid down the pole at Times Square. Really lively bunch, we!

Phoebe and I left for Christmas the Monday before in the Chris, not daring to wait until Tuesday, because of the weather, so we missed the community Christmas party, got to Elizabeth's in Concord just minutes after she had left for a party. I found some cross-word puzzles and spent the evening doing them until Elizabeth came home. I had arrived a day early, of course. Wednesday Robert and I went down to see Aunt Helen because of some problems which had arisen. After speaking to several people **and** Aunt Helen we hope things have smoothed out and that she can stay at Evanswood, which she thoroughly enjoys.

It is now Sunday Jan. 4th. What more can go wrong? Yesterday the solar cells refused to charge the batteries because of a coating of snow and ice, and the generator refused to start because of the cold. After six hours heat from an aladdin the thing churned over and I was able to charge the batteries. But last night the mercury went to -10 and hasn't gotten up beyond -6 all day and everything in the old cellar - pump, pressure tanks and water pipes are now frozen solid. Fortunately I had a few bottles and one pail of water so have survived today. They were left over from the freeze and broken pipes encountered when I returned with five "birders" last Monday and found water pipes hanging like festoons in the cellar. It was mild that day so the pipes had thawed and were in working order after Bill Sweetland came around and resoldered them. Not only are the pipes, pump and pressure tanks now frozen, the lock on the garden door is frozen solid so I have to go down stairs and let Phoebe in and out. Let's hope the space heater downstairs doesn't catch fire before I can get the lock thawed.

Going back to Christmas - Tommy came out to Concord and then Cindy, Scott and the two boys came up for a late Christmas dinner. The boys brought all their toys, including computerized tanks, machine guns, games etc. so we had dead bodies strewn over the floor, Phoebe was run over by the tank and the games beeped all day. It was great to see every one but I decided to take refuge elsewhere and drove to Emma's Friday - she had been alone for Christmas and was **very** glad to see me and Phoebe. We had a quiet time of it at the Farm and I left early Sunday in a snow storm, stopping in Falmouth to see the Sleepers and Bakers before going on to Rockland, all in the snow. Monday was mild and the ride out in the Laura B. was very pleasant. I had missed the Chris by only an hour or two on Saturday (connections, I mean.).

The five "birders"stayed in the cottage. The first night it wasn't bad, but the next day it was only 32 degrees in there and below zero outside. I see I've already told you about them.

When Huey took down the chimney he found a box placed there by the original mason, Samuel J. Higgins of Wiscasset. The cottage was built in 1926 by Maynard Orne as builder. In the box was a letter telling of the owners and designers - George and Delphine Nelson Moore of Boston. Besides the letter they put in a brochure of the Island Inn which gave weekly rates as $21 to 24/wk and fare on the Gov. Douglas from Thomaston or Boothbay as $1.75 round trip! They also put a copy of Science and Health of the Christian Science Sentinal. I replaced the letter and the Science and Health and Island Inn brochure and added a 1980 Inn brochure, a copy of "Introduction to Monhegan", a brochure of the Museum and a copy of the Dec. 16th Courier-Gazette article on Christmas on Monhegan, all of which I had written, plus a letter telling about what had happened to the property since the Moores left. I wonder who, if anyone, will find it.

A telephone hearing is to take place later this month in which costs will be discussed with NET and consultants. I've spent two days getting out letters to prospective phone applicants as well as sending thank you notes to the seventy-five or so people who have donated to our telephone fund. Actually there wasn't much else to do - too cold to ventrure out. It is now -12 degrees a little after 10 p.m., with quite a bit of wind.

Trap Day was put off until Monday because of the weather and colds among many of the fishermen. Then Friday we had a good snow storm which buried everything and for the past two days the vapor has been so thick it has coated everything with rime, including my solar cells. At times it was as thick as fog and shut out the sun most of the two days. I keep running the generator to deep it from congealing. I wish some of that heat could get to the cellar and thaw out the pipes and now the pressure tanks even are frozen. That means a new arrangement for plumbing for cellar heating, or just not staying here in the winter, if we are going to have such weather. No one has tried moving many traps to the dock as yet so Trap Day won't probably be tomorrow either. I haven't stirred from the house in two days. It was so cold this morning even by my gas line froze, shutting off all the pilots including the refrigerator's. When will it end?

Trap Day 1981

This year not only was the weather bad but several of the fishermen of Monhegan were **under** the weather on New Year's Day so that, unlike 1980, no traps went under the water on that day when legal lobstering can begin in the waters around Monhegan. After a series of "lobster meetings", when all concerned gather to discuss the whys and wherefores of setting or postponing the opening of the season, it was decided at last night's meeting that there would be another meeting this morning at 7. Much chatter about the weather and health of the fisher folk went on over CB radios during the evening.

Early this morning (Jan. 6th) Don Cundy heard a favorable weather report for today and was up and loading his truck with traps, then down to the dock to load his boat, with the help of daughter Kelani, son Rickie and Elise Burnham. It wasn't long after the Prudence A. with Don at the helm left the harbor for his favorite setting grounds. It was almost an hour later before anyone else began loading. Once again the old tradition of waiting for all boats to be loaded before anyone left her mooring had been broken.

This year there will be fourteen boats in the fleet plus Victor Lord's dory from which he sets only a few traps. There will be three husband and wife teams plus three other women who will be fishing this year.

Today was just about perfect for the opening of the season. There was an almost complete calm, the temperature rose from mid-teens to the low thirties and a bright sun shone most of the day. High tide was around noon, so that most of the loading was done on a coming tide, with low tide arriving after dark. But what will happen tomorrow and the next day even the weatherman said could not be predicted because of a complex weather system coming our way. At least some traps will have slipped under the water. Let us hope the seas will not be too violent in the next few days.

Once more the waters around Monhegan are dotted with bright pot buoys and the hum of boat engines fills the air. The 1981 lobstering season on Monhegan has begun, not to end until June 25th.

Above is the story I have sent to the papers. I have just enough room to tell you that Debbie Bates and Dougie Boynton were married here by Winnie, on their return from vacation during which Kappy, their dog, died after suffering a stroke. Today there are nine snowbuntings and three horned larks in Rita's yard. My door can now be opened but the pump is still frozen. I am typing once more with solar power - snow has melted, only to make way for a new fall tonight, and I'm catching water off the roof.

Alta

Monhegan, Maine
Feb. 1981

Dear Friends:

Although it is only the 28th of January I am going to start writing this letter because I am having trouble with my duplicator again - and now what is new? - and may have to to take this in to be run off when I go in next week, if I do.

Ever since the "birders"left the end of December I have been having one problem or another with the plumbing. Bill Sweetland got the pipes together while my company was here, but another cold snap really did a job on the pipes down in the old cellar. He or Billie Boynton would come by and mend a break and leave before I could start the pump to see if there were any more breaks. All told there were **four** and yesterday they were all repaired. Gleefully I begun flushing toilets and found that all sewage was coming up and **out** of the downstairs john. No one seemed to know where the "snake" was so I decided to give up and go in shore today, unless some way could be found to free the drain.

The last evening and into the morning I poured scalding water down the john as soon as enough water had drained out to allow for more. Finally I put some draino down, much against instructions on the can, and whoosh! the water flowed away in a great eddy. Now the clothes washer is humming merrily with a three week accumulation of dirty clothes. The dishwasher will be put to work tomorrow **if** the generator behaves itself. It began to do funny things in the second hour of "The Bunkers" so I switched over to my solar batteries and was able to finish out the movie with them. David Sleeper found they were charging only 70 hours (amp) and not holding their charge during the night. So I'm due for a new set. They are guaranteed for five years and supposed to last for ten. Maybe by spring all my woes will have been dispelled, I can't take much more and keep my sense of humor.

The lobster catch has fallen off earlier than usual this year but the smack paid $4.15/lb last week, which made up for the poor catch. So far our fleet hasn't been iced in as the boats in shore have been. Jimmie Barstow has had to be cut or towed out of the ice in Port Clyde two or three times so far. The poor Laura B. is all scarred from the ice and looks as though she hasnit been painted in years. The cabin is nice and cozy though with the wood stove and the blocking of the scuppers from the cabin. Only a little water comes in when a wave breaks over the bow.

Several others have had frost problems. Barbara and Paul Horn came back from six weeks ashore and found everything frozen stiff. Bob and Mary Ann Boody have had frozen pipes and drains, and Doug and Harry have had their problems in the old Post Office and are now in shore to get showers and have laundry done. The Store is short of water because of the lack of rain or snow and several cisterns are dry. What a winter!

It is now the day after Ground Hog Day.As has been the rule for several years, we had a big storm, so big, the Laura B didn't come out yesterday. Now I am waiting to go in with a group of engineers from NET to go to the dentist. I broke another tooth last week and now it is thumping. What next?

The storm carried high winds and loads of rain which just about washed away the ice, but then more formed last night and tonight we are supposed to go back into the deep freeze.

Feb.5. I'll try to finish this today and get it off to you by the beginning of the week. I've just come in from the generator shed where the Kohler is shivering in the 10 degree cold but warming up with an aladdin heater. One glad note - the solar cells are churning out 6 amps even though only two are cleared of snow: two are half cleared and one has a small coating of transparent ice through which ultraviolet rays must be penetrating.

Tuesday five engineers and assorted personnel from NET came out to look things over. They viewed the fuel storage building, then took sightings from the Lighthouse tower, went over to Hardwood Hill and the hill beyond to see the height and type of trees which might interfere with the line of sight to the Rockland area. I elected to go in with them, thinking they would beat the Laura B in. She had come out early, however, to make up for the lost trip on Monday. I had an appointment with the dentist in Rockland for 2 p.m. and we finally left here at 12:30 to go to Tenants Harbor. My car was at Port Clyde and Ivan nowhere to be found to change plans. We reached Tenants Harbor at two and started whizzing down to Port Clyde in one of the telephone company cars and passed Ivan on the road bringing my car up. I changed cars and flew off to Rockland, then had to wait a half hour! At first Bruce thought he could save a half of my tooth, then found I had split it right up through the root, so out it came. Novocain and all, I set off for the Sail Loft and a Knox County Medical meeting only to find it had been postponed! Back to the Samoset, some bourbon and aspirin and sleep.

Eleanor Dexter wrote last week that Bob, her son, has undergone a colostomy for cancer. He probably is back home in Georgetown, MA by now, I hope. Eleanor says he loves to get cards from those who know him. But the news that has really saddened everyone is that which came yesterday. Glenn Layman called Winnie to say that after two weeks of illness Elsie died yesterday in Florida. How wonderful it was that they found each other and had those few years together. It will seem so empty not to see them both working in the garden which has always responded so well to their care.

Jeanie Lord (I reverted and called her Rollins in my column, I see) is still in shore trying to get regulated on a new medical regime. Ron Stanley does not have an aneurysm but is in kidney failure and goes to Portland for dialysis, which they hope soon can be carried out at home. His wife Linda is a nurse.

Skating has been good this year. they have used both the Ice Pond and the Meadow. The joys of living in Maine. Ivan had gone uptown Tuesday morning and on his return he saw my car still at the dock. He called Winnie who said I had gone in to Tenants Harbor; that is how he knew where I was. The telephone man took Ivan back home so I could go on. Thank you, NET. And speaking of NET, the next thing on the program is for our testimony and the presentation of new figures by NET at a hearing expected to be held here in March. Just keep praying that the cable holds until we can have a better system - not that radio phone which NET and Ma Bell think would be just dandy for us. Wouldn't it be nice if some of that biggest profit in history could come this way?

I am still seething about being called Ms. by UPI. When the

reporter asked my marital status I said it had nothing to do with the story and she should address me as she would a male physician. The inverter keeps cutting out so I'm skipping letters etc. Enough for now.

Best wishes to all, and thanks for your letters. REACH OUT AND TOUCH SOMEONE - ON MONHEGAN, if you can! This is a pretty disjointed letter. Better luck and better news next time.

Alta

Alta Ashley, M.D.
Monhegan, Maine 04852
February 9, 1981

Hon. William S. Cohen
United States Senate
Washington, D.C. 20510

Dear Mr. Cohen:

At long last I can report to you that New England Telephone sent out a team of engineers and consultants last week to study the feasibility of erecting a microwave telephone system for Monhegan Island. They spent several hours here looking at the Lighthouse site, the oil storage building which might house equipment, and took sightings both from here and from Dodge Mountain and Benner Hill in Rockland to determine if indeed there was a direct sight between here and those points on the mainland. I believe their findings and cost estimates will be presented at the next hearing scheduled for March, at which time we will also present our data.

As far as I can determine there has been a request for fifty-five phones plus an indeterminate number of coin telephones for the hotels, Store, and possibly the Post Office and Schoolhouse (Town Office phone). All those who might wish service were asked to contact the business offices directly, so I don't know how many such requests have been made.

I for one will be glad when this matter can be settled equitably for all concerned. Let us hope it will be before next winter so I can take a care-free vacation and escape the cold for a few weeks.

One of NET's arguments against giving us a full service facility is that it is unfair to the rest of the customers because of the expense. Yet they say that diminished service anywhere diminishes that to the entire body of customers. Then doesn't lack of adequate service to Monhegan diminish service to the person unable to complete his call because of busy lines or breaks in the cable? Communication is a two-way street, is it not?

Thank you for your continued interest and action in our telecommunication problems. Andrew sends his best regards with a wave of his truck.

Sincerely yours,

Monhegan Corr.
Alta Ashley

Shakespeare Visits Monhegan

Feb. 14. Those Monheganites who were up and had not taken off in their boats for the fishing grounds must have been somewhat startled to see William Shakespeare or one of his contemporaries walking up the road to the Schoolhouse yesterday morning about nine o'clock. Waiting at the school were all the Monhegan School children, as well as an assortment of dogs. For two weeks the children had prepared for this visitor. Sally Murdock, who was active in an amateur group of players prior to her coming to the island as resident and co-owner of the Monhegan Store, had supplemented the sixteenth and seventeenth century studies presented by Miriam Barteaux, our schoolteacher, with talks on the history of the theater and readings from the Merchant of Venice, Romeo and Juliet, and other Shakespearean plays. And so, as Shakespeare, otherwise known to modern day audiences as Duncan Inches, approached the school dressed in doublet and tights, his balding head fringed with shoulder length hair, gleaming in the sun, they knew immediately who the visitor was and greeted him enthusiastically.

For two-and-a half hours the children, ranging from kindergarten to eighth grade, sat enthralled as this accomplished actor took them dramatically through the life of The Bard.

That night islanders gathered at the Schoolhouse to witness Inches' presentation of "A Little Touch of William in the Night," written by Ian John Katee. This is an interesting interweaving of excerpts from Shakespeare's plays, songs, and sonnets, and based on the seven ages of man. Inches was the sole performer, although he did come down off the stage to involve some of the children in the act. He took awe-struck Heather Boody's hand at one point to express his love, another time he seemingly attacked Jeff Rollins with his foil, whereupon Jeff fled to the back of the hall, probably all in make-believe. There were many props which helped to transport the audience back into Elizabethan times. Inches took two parts on occasion, leaping from one side of the stage to the other or even off the stage entirely. He danced, sang in a rich baritone voice, whipped his foil about with abandon and thoroughly enjoyed himself as he brought pleasure to his audience.

The performance was fittingly brought to a close by the line from Hamlet, "The rest is silence," but the enthusiastic applause from the large gathering of islanders belied this statement.

Following the play several families stayed for box (lunch) suppers, giving them an opportunity to meet and talk with Inches and his props manager Pam Scott, as well as with Sally and Bud Murdock.

Duncan Inches is a Massachusetts native and son of actors. His father was a Shakespearean scholar and instilled his love for The Bard in his son. " At twelve I played Prospero and at fifteen Richard III," he says. He is a graduate of the University of Arizona, which has an excellent drama school, and has taken graduate courses in theater at the University of New Hampshire. "But," he says, "I learn mainly from other actors." He was inspired to play the life of Shakespeare by Massey's "Abe Lincoln in Illinois" and has been working on his production for the past five or six years. He is now the artistic director for the Commonwealth Shakespeare Company with headquarters in Watertown, Mass. The company has no theater, but travels about on invitation, particularly to high schools. This is the first time that Inches has played to a group so disparate in age.

One feature of the performance which must not go without comment was the beautiful flute playing which introduced both acts of the play, so ably done by Barbara Johnson, a Monhegan resident.

The islanders will long remember this night and all hope there will be more entertainment of this caliber. We are indebted to the Parent, Teacher, Friends Club of the Monhegan School for sponsoring the event, and for Sally Murdock for finding this interesting performer.

Monhegan, Maine 04852
March 1981

Dear Friends:

If I don't get this out by the 18th I'll have to pay 18 cents a letter, so here goes. First of all I must tell you about the trip I made to Peak's Island, which I forgot to mention in my last letter. It was the middle of January, actually the last really cold spell we had. Since I had to be in for Telephone Hearings in Augusta the following Monday and Tuesday, Phoebe and I took the opportunity to visit Eugenia Halbmeier on Peak's. We stopped for a delightful evening at the Hosmers' on Rackcliff Island then drove off to Portland and the ferry.

The Casco Bay Lines have nothing to offer over the Thomaston-Monhegan Line except a warm (horribly run down) waiting room in Portland. I had to pay $4.00 to park my car on the wharf, and what a scary time I had trying to back it off in two inches of snow and ice! The architecture of Peak's leaves much to be desired, too. Eugenia's house inside is very attractive and comfortable, even though the temperature hovered around zero all weekend. Two black cats kept Phoebe at bay most of the time, though she became brave when she saw them tail-on. On Sunday morning we took a walk along the shore and attended church services and met the last Trefethen, who said she knew nothing about the family, so that was that. It was a pleasant visit, but I'm still glad to be on Monhegan.

Bob Campbell of Acton, MA has brought the Balmy Days and will be moving to Boothbay in April. He hopes to have the boat ready for Memorial Day weekend, which will be May 23rd this year, and plans to run regular trips from June 11th through September 20th. He didn't tell me what his fares will be. His Boothbay phone will be 207-633-2284.

On Valentine's Day Duncan Inches from Boston was here in his role as Shakespeare, complete with doublet, tights and fencing sword. He spent the morning at school, introducing the kids to the Bard. That evening he played to a packed house of adults, reliving Shakespeare's life through his writings, ending, as expected, with "The rest is silence." We need more of such things.

The following week I gave a talk on women in medicine at the Maine chapter of Medical History Society in Augusta, at the invitation of Richie Kahn from Pen Bay. Phoebe and I left on Friday and were entertained by the Kahns at their farm in Union. That evening they were hosts to a covered dish supper put on by the members of the Rockland Unitarian-Universalist Church, a most interesting group of people of all ages. Pat Kahn's parents, the Gambles now of Tenants Harbor, were there - a delightful couple. He grew up in Belfast, Ireland, and still has a soft brogue which was put to full use when we embarked on a songfest. It was the first time I had sung in months and found my range was more like three notes than my once two octave span. There were two things I wanted to attend the following weekend so I decided to spend the intervening week at Emma's in Tuftonboro. I stopped for lunch at Lura Titcomb's before my talk and caught up with news of my old pals of Augusta days. It was great to see Clara and Ted Nilson at the meeting, and Lorna and John Richardson, now of Searsport, but once neighbors and confreres of Newton-Wellesly Hospital days.

The week at Emma's was quiet, chiefly because of weather which gave us quite a snowstorm in mid-week, despite temperatures in the forties. We did go over to Moultonboro to see Frances Stevens who has assembled a great deal of information about the Trefethern/Treffren family and I was able to fill in a few holes in my material, but could only guess at a possible connection between them and the Woodburys, who settled in Salem, MA and moved on up the coast with the Treffethens to Kittery who then went on to Monhegan and Peak's, joining with the Bracketts. As Papa used to say "Scratch a New Englander deeply enough, and you have found a relative."

We drove down to Kingston and took Aunt Helen out to dinner at her favorite restaurant, Bert's, opposite the reconstructed Plymouth Plantation. It was a beautiful day after the storm which had filled the parking lot with sand, seaweed and stones. Ninety-five year old Aunt Helen was thrilled to see the breakers rolling in much like at Wakiki. We drove down and back in one day, a surprisingly easy trip right through Boston on I-93. (I wonder if I'll ever stop off and see the new Quincy Market and Fanueil Hall reconstruction. I haven't been down in Boston for over ten years, my native land!)

On Sunday I left for Portland with the intention of taking Dave and Jeanne Sleeper to a reception for MPBN contributors. We drove up to the Ramada Inn and found no one there and no knowledge of where we should be so we returned to the Sleepers to cook a frozen chicken by micro-wave. Two days later I remembered we should have been at the Roma Cafe, it was at the Ramada Inn that the reception in Bangor had been held the week before!!

Monday I went back to Augusta, had lunch at Skip and Jackie

Thurlows' just two doors away from Ashwood, where Mother and I lived. Adele Nichols and I went to Louise Webber's to dinner and brought her back to Harlow Gallery in Hallowell to a most interesting meeting of Kennebec Valley Art Association. The speaker was the Unitarian minister from Augusta whose talk centered around a motet of Pachelbel, quite familiar as the introductory music for a program I can't now recall. The Knox County Med. Soc. meeting on Tuesday brought this ten days of activities to a close. After all this I am more convinced that a couple of months or so in shore during the winter should be on my agenda from now on. Does anyone know of a house-sitting job, preferably in the Augusta area where I can have access to the archives, for next winter from Christmas to Town Meeting?

That brings me to Plantation Meeting which was held last Monday. So many people were away that there were only thirty-five voters present. Rickie Boehmer acted as Moderator and had her job cut out for her, for it was a testy group, to say the least. Katy Murdock is Third Assessor, Danny Bates is Road Commissioner and School Board member, Jackie Boegel is Tax Collector and Jeanne Rollins, Plantation Clerk. Victor Lord is Fire Chief and Planning Board Member, Mort Berstein also Planning Board member and Steve Rollins Constable and Harbor Master. A budget of nearly $80,000 was passed, including major repairs to the Ice Pond walk and $5,000 for the telephone service effort which will be taken from surplus and the Farrelly account. Vehicles will be taxed $20, whether they run or not. It was decided to think further about the pros and cons of becoming a town. School, garbage and plastic disposal and roads accounted for the major part of the budget. We need to have a few palaces built to absorb the increased assessments necessary.

Word has come of the passing of Lauline Paranov, a summer visitor for many years, though confined to her home for most of the time by illness. Jean Lord had been in shore but is expected back shortly. Victor and Kole went in to get her this week. More houses on the market - the Sprouls and the Hornes hope to sell. Bill Sweetland is building a solar workshop where he expects to run his tools by solar power and a new kind of energy efficient inverter. Right now this substitute inverter isn't working, so I am typing by generator power while the dishwasher is running. The dishes are done so I'll sign off with a tail wag from Phoebe.

As ever,
Alta

TELEPHONE UPDATE

We are nearing the end of our effort toward, not only retaining, but augmenting telephone service to Monhegan from New England Telephone Company. It has been established from previous testimony that NET has served and is obliged to continue to serve the island with a reasonable, adequate, and reliable system, but what constitutes a "reasonable, adequate, and reliable" system has yet to be determined. This is expected to be done following a final hearing to be held on the island, probably in early June.

In order to strengthen our testimony it is necessary for us to know the level of requested service. All those requesting should apply directly to NET by writing as soon as possible to:

Peter J. Gorman, Manager
NET
212 Maine Street
Waterville, ME 04901

for residence service

Mrs. Joyce Stevens, Manager NET
77 Exchange Street
Bangor, Maine 04401

for business service or coin telephones

To date several have done so, but many have not. Would you let me know, if you have not already done so, what type of service you have applied for. It is difficult for me to obtain it from the company.

These applications are not binding and there is no charge for them. Service charges, if and when we obtain expanded services, are set by the PUC and must be reasonable. Since no pole line will be a mile long there probably will not be any pole charges.

If you have not already sent in your questionnaire would you do so as early as possible? Our case will stand or fall on the outcome of the coming hearing. Unfortunately we do not have extra copies of the questionnaire. What is needed to be known is (1) whether or not you feel telephone service is necessary (2) adequate, reasonable and reliable (3) if a four-manual two-line radio phone proposed by NET as a substitute for the present system would be reasonable, reliable and adequate; (4) what type of system would be r, r & a (5) what type of service you would like, if any, in your home; and any comments you wish to make.

Thank you for your cooperation, both past and future. Let us hope that a satisfactory resolution of the problem will be arrived at before the present cable ceases to function.

Alta Ashley, M.D.
Coordinator for
Telephone Service for
Monhegan

Monhegan, Maine 04852
March 28, 1981

Monhegan, Maine 04852
April 1981

Dear Friends:

This is a rainy, dreary afternoon with possible thunderstorms predicted, much like the first Good Friday, certainly not one to put anyone in the mood of Easter. I started this thinking I could type by solar power again, because new batteries were installed last week. But now I find there is trouble in the inverter switch which makes that cut in and out, mostly out, so the old generator is churning away. David Sleeper is due tomorrow, if the winds don't pick up too much again, and he promises to have something by which to repair the inverter. Really, is my life style provoking all these problems. Do Pheobe and I deserve all this?

We left the island last Monday with a string of appointments and errands - getting this and that repaired and tested. First was an eye exam required to renew a driver's license, then a road test, which I had hoped I could have on Tuesday, but the appointment was set in Augusta and couldn't be changed, so the day was spent shopping. Then Wednesday I had a session with the dentist, but not before I took by car in to have snow tires removed. I made a fatal mistake of going over to the Buick place to wait for the tire mounting and got talked into buying a '81 Skylark. They suggested I not pick up the new car until after my road test on the following day.

Thursday morning I got the old car packed and ready to go for my test. While I was signing out of the hotel someone backed into my car and bashed in the right front door! I had to have Phoebe in the car because we were to be on our way after the test, but the examiner wouldn't allow her to stay, the registry office wouldn't keep her and by that time I was in tears. Fortunately, the Army Reserve office opened its arms to her. With all that I forgot where the light switch was or how to put on the turning signals for the first couple of minutes - how do you dim your lights or signal a turn without the motor running, as I was commanded to do? Well, I passed, picked up the new car and set off for Pemaquid for a visit with the people over there, some of whom I had met at the Republican convention just a year ago, and some I had known in Augusta. It was a wonderful visit with a dinner party at the Bradley Inn close by. The next morning we were up bright and early to watch the blast off of Columbia on a black and white TV set on the dining room table while we munched our delicious French toast. Well, as you know, the blast-off didn't come about until two days later. Actually it gave us more time to talk, which was nice.

On my way to Center Tuftonboro that morning I stopped off to see Bobbie and Ruth Ives who are running a boat shop, giving instruction and employment to disadvantaged boys. Ken Brookes, the minister at Augusta's South Parish when I lived there, assists the Ives in this project. He happened to have the day off so I missed him. Ruth's hand, which was so badly burned by boiling oil last fall, is healing well and she is able to care for the house and family once more. After an overnight stay with Emma we were again on the move to attend the wedding and reception celebrating the marriage of my niece Susan Peters and Hank Kezer in Melrose. At the wedding there were but four of us on Susan's side of the chapel, Nancy and Tote Peters, Varney Mead (grandfather) and I. Varney looked great, all brown from working in his garden in Wellesley Hills where he and Gertrude moved fifty-five or so years ago. Susan looked lovely, as did Sally who was one of her sister's attendants. During the dinner the orchestra played "Alice Blue Gown" while I sang and acted out the chorus! Will I ever outgrown being a ham? I don't know what happened to the cake or bouquet. I guess the cake cutting came while most of us were gathereed like flies around the bar in the other room.

I spent the night in Concord with Elizabeth and we watched the delayed launching of Columbia the following morning. Then off to church (Concord Tricongregational). The service was fine, the music superb. Among the new members received was David Nethercut (I see he is Richard D.), who turned out to be a cousin to Jane Varnum of the Trefethen mansion! While Elizabeth was busy with a blood bank program I went off to visit Marjorie and Paul Mitchener in Andover.

The Massachusetts countryside was beautiful. Magnolias, dogwood and forsythia were all in full bloom as well as some azaleas. Tulips and daffodils were out and crocuses were still in bloom here and there. Titmice and goldfinches were busy at the Mitcheners' feeders. Marj and Paul, after a twenty member family dinner on Easter, will be off for Egypt and then a week on Corfu. Monday I was back at Emma's, then off for Rockland and the Samoset on Tuesday expecting to go to the island on Wednesday. That afternoon we had a terrible rain storm. I was thoroughly drenched as I went from place to place getting new lenses, teeth, and what have you. The following day the wind was worse and the boat didn't run. PUC engineers were due to visit the island on Thursday but had to cancel because there were too many people returning to the island for the small boat to handle - the Laura B is off the run getting refurbished for summer.

It was still pretty rough and windy on Thursday but we did take off, leaving Dyan Fitzpatrick, Barbara and Paul Horn on the dock. Dyan stayed in but the others finally got out with Bryan. Paul was anxious to get out because he wanted to turn on the town water, which finally went on Friday afternoon.

The PUC engineers are due this week. They will gather data for a cost estimate for an expanded telephone system to replace the present system which is dependent on a decrepid cable which will not be replaced once it ruptures for good. There are three more hearings on the docket, all to be conducted in the Hearing Room of the PUC building in Augusta. The first is scheduled to start at 9:00 a.m. on Monday, June 8th, to "continue day-to-day until cross-examination is complete". My testimony is 16 pages long. As many people as possible are urged to attend all meetings to testify as to the inadequacy of the system proposed to replace the present system as offered by New England Telephone, namely the Rural Radio system with two lines and four instruments. A hearing on June 26th (Friday) will permit presentation and cross-examination of New England's rebuttal case and will also be continued on as many days as required. The third hearing will be on Friday July 7th for the presentation and cross-examination of the surrebuttal cases of the PUC and Monhegan Plantation. After that we await the decision on the case. I for one will be glad to have this thing, which has almost completely monopolized my energies for three years now, end. Every time I planned to do something it seemed that I had to defer plans because of telephone business. Billie Boynton, after going around with a pocketful of coins, has been granted a credit card for the Plantation, which should be a help in keeping track of expenses. Coin phones aren't the most convenient of instruments! Our case stands or falls on the results of the upcoming hearings so any of you who are truly interested are urged to attend at least the June 8th hearing. I'm sorry it couldn't be held here.

Easter was a beautiful day. I awoke at 4, dressed by moonlight and set off for White Head, gathering Miriam Barteaux on the way. The sun rose as a red ball from the ocean just after we reached the headland. A group of about a dozen were assembled on Burnt Head. It was so clear we could even see the light on Matinicus Rock flashing twenty miles away. A full moon was close to setting in the west as the sun rose, a beautiful sight. At six a group gathered at Pemaquid Point for a service so I waved my blaze-orange hat in hopes they could pick it up by telescope.

There was an egg hunt later in the morning for the smaller children. Annie Court and friends colored five dozen eggs and Sally Murdock contributed candy eggs, all of which were hidden on the Monhegan House lawn. Phoebe tried to get in the act but I stopped her before she had found any hidden treasures. That night I had a combination birthday-Easter dinner for Rita whose birthday is the 21st. Charlie MacDonald, Rita, Miriam, Harry and Doug were here for ham, scalloped potato, avocado salad, ice cream and cake. I tried a dream whip frosting which was a disaster and ran all over the place. But the cake, with the help of Duncan Hines, was the best ever.

A couple of weeks ago both Carl Wincapaw and Tigger Hitchcock picked up balloons set free from the Junior High School in Wiscasset. Carl found two floating behind Manana and Tigger found his at the foot of their garden. They were set off as a pen-pal stunt in a writing project but unfortunately the date of their release wasn't included in their messages so we don't know how long it took for the balloons to float over here. The Bartows sent a clipping telling of John Gummere's being made President of Phoenix Mutual Life Insurance Company of Hartford. Congratulations, John.

Word came over the weekend that Mary Stanley is seriously ill in the Bath Memorial Hospital. At last report she was doing a bit better than expected. Ronnie, meanwhile, is at Maine Medical Center because there is no one to take care of him. Linda and the girls have gone again.

Don and Linda Stone are back. Don has a photo-voltaic system at present mounted on the ground. He is going to have one put in his studio later. Bill Sweetland is building a shop and will have photo-voltaic power for that. Other systems are due to be installed soon here on the island. So, per capita, Monhegan leads the world in photo-voltaic power! Right now the inverter is working while I am typing. My big inverter should be back on line soon. David was out here this weekend and put in fluorescent lights for the bedrooms and bathrooms, which ought to cut down on the power draw. They use so little for so much light.

Chris Nicholson is back so Doug and Harry have moved into the Underhill house again. But their pump doesn't work so they are without water while the rest of the island is luxuriating in hot baths, etc. Chris came up from Florida with Bob and Mary Helen Purinton who left in the Chris on a very stormy day. The wind has blown so long and so hard all winter and spring it is hard for the men to go out for lobsters. Everyone will be glad to see the Laura B back with her new coat and rebuilt cabin and head.

Every day I get farther and farther behind. The garden hasn't been turned over or peas planted. Some seedling are sprouting in the window greenhouse, but temperatures of 30 or below, as we had this morning, make gardening as well as cleaning the frigid cottage, things to be put off. My magnolia has one ragged blossom and a few spring beauties have bloomed. The daphne bush is in bloom but no forsythia as yet and no daffodils. Spring is passing us by!

I forgot to mention the delightful dinner and evening spent with Ruth and John Hosmer on Rackcliff Island last week. They had just returned from a trip to the south where they saw John's sister Libby who was one of my campers at Camp Low these many years ago.

The school kids put on a clever skit the other night, "The Case of the Missing Parts of Speech" and did themselves proud. A towhee is at the feeder, a mockingbird on the fence and cardinals are calling. Time to close.

As ever, Alta

REPORT on TELEPHONE HEARINGS HELD in AUGUSTA, MAINE JUNE 8-9, 1981

On Monday and Tuesday, June 8th and 9th, hearings were held concerning the proposed discontinuation of telephone service to Monhegan Island by New England Telephone Company (NET) at the hearing room of the Maine Public Utilities Commission (PUC) in Augusta.

The first witness to testify was Guy Twombly, Chief Engineer for PUC. He stated that since NET has provided telephone service to Monhegan for many years and the Monhegan lies within the franchise of at least one NET exchange, telephone service must be continued by NET. Furthermore, according to stated company policy, demands for service should be met in a reasonable manner.

Cost figures for a 48-line analogue (ordinary dial) system, proposed by five consultants, were in the neighborhood of $50,000 less than those the company had estimated for substituting a micro-wave system to replace the present deteriorating cable now serving the present five-line system. The 48-line system would satisfy present demand and would not, in his opinion, constitute an unreasonable expense to the company.

Twombly's figures were substantiated by five consultants, with the addition of costs for switching equipment inadvertently omitted, amounting to approximately $7,000, still well below NET's figures for the five-line system.

Parker Vincent, Electronics Engineer, who installed for NET the present telephone system on Mt. Washington, said that this micro-wave system has been extremely satisfactory in contrast to the mobile radio-phone system, such as that NET proposes to install on Monhegan to replace the present system once the cable breaks. He stated that there have been no problems with this new system since its installation a year or so ago. He also stated that if NET would "depart from the book" to accommodate special conditions involved in service to Monhegan that substantial savings could be made - e.g. no heating or air conditioning for the service building would be necessary except when someone is working inside, special types of hook-up etc. could be used. A micro-wave system would allow for private conversation, would be more reliable and would allow for more simultaneous incoming and outgoing calls than the radio phone proposed, which would be affected adversely by atmospheric conditions, its channels could be crowded and any of the hundreds of owners of "scanners" could listen in to any conversation. Furthermore, power for these phones would be at the expense and discretion of those housing the instruments, certainly a far from reliable situation.

David Sleeper and Peter Talmadge, solar and wind power experts, respectively, gave figures on wind/solar/mechanical power generation systems which would use one, all, any combination of systems to supply power for the 48-line analogue system proposed by PUC engineers. Wind power would require two wind generators mounted on 100' towers at least 50' apart erected on Lighthouse Hill. Solar cells to furnish adequate power could be ground-mounted on platforms occupying approximately 640 sq. feet, and would require virtually no maintenance. Both systems would require battery banks for power storage and should have auxiliary power available from mechanical (LP gas generator) for emergency power.

Testimony by William Flynt, Utility Engineer for PUC, and of Clint Coleman, Chief Engineer for North Wind Power, were filed without cross-examination. Flynt's testimony dealt with costs for microwave towers and equipment buildings; Coleman's substantiated the testimony of Peter Talmadge concerning wind power's suitability for Monhegan.

The chief areas of controversy in these hearings, as in previous ones, were what constitutes telephone "service", "exchange service", reasonable costs, and NET's legal responsibility to serve Monhegan with how much.

Public testimony from three islanders, Jim Barstow, and Gil Hall of Pen Bay Pilots Association was entertained. Billie Boynton as 1st Assessor spoke of the difficulty in carrying out town business under the present system, let alone the proposed substitute. Dan Rothstein, M.D. stressed the need for more, not less, accessible phone to save precious minutes in cases of emergency, as well as the need for privacy and the difficulty in keeping in touch with children in off-island schools. Sherm Stanley spoke in addition of business needs for adequate phone service. Jim Barstow, Captain of the Laura B., told of the problems encountered during the four month absence of service in 1978. Gil Hall said the Pen Bay Pilots serve as "eyes and ears" for the Coast Guard and other law enforcement services in cases of smuggling and other infractions of marine law and that privacy and availability of service is essential.

My testimony was also filed with only a few added comments.

Our case is somewhat weakened because only eighteen persons have applied directly to NET for service, although according to the survey returns forty-nine phones requested. I urge you to write **IMMEDIATELY** to:

Peter Gorman, Manager NET
212 Main St.
Waterville, ME 04901
for residence service

Mrs. Joyce Stevens, Manager NET
77 Exchange St.
Bangor, ME 04401
for business or coin services

if you have not already done so. These letters should be in hand before NET's rebuttal scheduled for June 25th. These requests are not binding and monthly charges are set in accordance with standards applicable to any similar exchange service.

It now remains for the Commission to determine what is reasonable, reliable, and adequate service for Monhegan. A surrebuttal by Monhegan is scheduled for July 7th in Augusta.

Our by-word is:

REACH OUT AND TOUCH SOMEONE - ON MONHEGAN, IF YOU CAN!

Alta Ashley
Coordinator for Telephone
Service on Monhegan

Monhegan
June 15, 1981

June 14, 1981

Dear Friends:

Here it is the middle of June and time for another of these letters, which coincides well with the latest telephone report. For those who won't be receiving one, the Public Utilities Commission presented their cost analysis of a full service micro-wave telephone system for Monhegan, backed up by testimonies from five experts in various fields, as well as testimony from islanders concerning the need for such service and the total inadequacy and unreasonableness of the proposed radio-phone system to replace the present system once the cable breaks for the last time. The hearings were in Augusta last Monday and Tuesday. A dozen islanders and persons concerned (Jim Barstow and Gil Hall) attended, and five were heard. I was prepared to be grilled all day, got a haircut and put on a new suit and found that in the matter of saving time my testimony was filed as printed, with only a few pertinent comments from me. We now await the decision of the Commission after possible rebuttal by NET and surrebuttal by us. PRAY!

Summer is upon us. Lupine and rugosa roses are out. Snow in summer is at its prime, lemon lilies and some of my hybrid lilies are out. Mary Burton has her gardens full of blooms and hanging baskets of fuschia and lobelia decorate the Inn porches. Bob has completed remodelling the two downstairs suits so one doesn't have to cross the first to get to the second. Victor has finished reroofing the Monhegan House and has done a lot of shingling of the sides. The Inn opened yesterday, the Monhegan House and Tribler the first of the month and the Yew the middle of May. The Balmy Days began regular daily trips on Thursday and Jimmy will have two trips and Sunday trips beginning next Saturday.

Steve and Jeanne Rollins' dog wagon is doing land-office business each noon until the boats leave in the afternoon. They offer franks and Italian sandwiches, with all the fixin's.

The Curtises were here a couple of weeks ago to get the cottage ready for Mort and Mae Cass who should be arriving any day now. Anna and John were here to clear out the Jessie Dunbar cottage for Winnie, John and Kristina Murdock to move into. The Cundy House has been vacated and the family is now living in East Boothbay on Liniken Bay. Patsy is still at the Cundy Cottage. I believe Elsie's cottage will be for sale. Her niece who inherited it will be here next week with Glenn. She has had double hip replacements and a broken leg and will have to be brought up from the dock in my cart.

The Sprouls have their place up for sale, and also the Horns. Ralph Hinchman wants to sell his place and the two Mersfelder places down at Green Point are also on the market. I don't know what is happening about the Loew house and studio but the Shumaker place is also for sale. All are still carrying inflated price tags and aren't moving.

Mary Stanley is back. She had a rough time this spring with pneumonia when Ron was in the hospital, but is now recovered. You probably know by now that Ron died in May after being weeks on dialysis.

The Barkers are in their cottage for the week. Barbie graduated from Dartmouth so many of the family came east for the ceremonies. Sandy's new husband, Bud Davis, didn't come on but will come with Sandy in the fall.

Jackie and Julie Hudson are here for a few days and then are off for England. Marjorie Mitchener wrote telling of Nan McKeen Stanley's death in April, the first of our old gang to leave us. We'll miss her at any Goodaway Club reunion.

My garden has been awfully slow, it seems, but is beginning to look pretty lush. I've harvested radishes and lettuce but nothing else except for a few sprigs of dill and some chives. One or two stalks of asparagus have appeared, but not enough for a meal. Thank goodness for the three pounds I bought in Center Tuftonboro, which I am eating at about five stalks a meal so it will last a long time. I still have a few of last year's carrots, otherwise all those vegetables are now gone. The strawberry bed had to be all dug up. It took so long for me to put in new plants most of them died in the spell of dry weather we had in May. Blueberry bushes took a beating, and the magnolia was so badly frost bitten it looks again like a French sycamore.

I have new batteries now, so the solar system is working well, but the relay hasn't been installed so I have to type when I'm washing clothes or dishes, unless I want to run the generator just for this foolish typewriter. I wish I could find one with a 12V-DC motor. Then I could take power off the batteries without having to invert the current. I still delight in making use of sun power, even though it is a bit limited at the moment. Dave Sleeper should be out soon, then maybe things will all be back to running order.

The school children outdid themselves in their presentation of one act of Treasure island, and my pictures of the production came out very well. I certainly am glad I invested in a Minolta camera for my press photographs. At $5 per print from the paper I should have the thing paid for fairly soon. I only wish I could get prints into all three papers, instead of just the Courier.

Last fall I met Richard Saltonstall at dinner and found he was interested in obtaining material to publish about Maine, so I trotted my opus to him and he in turn sent it to David Outerbridge saying "You have a story there". Well, in May Richard died suddenly and David sent back the manuscript. Part of it has been sent to another publisher in Woolwich, but no word yet has been received. Meanwhile I've been having a great time rereading it and wondering how I ever wrote so much (and forgotten so much!) Here's hoping.

I am now back on the inverter, having finished the wash. It keeps going off in the middle of words and ideas, so bear with me. One time, many years ago I watched a doughnut machine. One poor little doughnut didn't get turned over. He kept spinning around, getting browner and browner, finally really black on one side but try as he might he couldn't get into the right position to be turned over. I often feel that that doughnut was me (I). It often seems as though I am continually spinning around and not getting in the right place or doing the right thing, try as I may, especially lately. All sorts of things happened this spring, including having my manuscript returned. Then Ed Dunn of Cincinnati days finally planned to come for a visit, only to turn a trip to Monhegan off for a trip to China. But really, I don't blame him. I'd do the same thing, I'm sure, had I the chance. The latest, but probably the softest, blow of all was that Doug appeared last night with an empty pail which had contained fish chowder planned for lunch when Jean Hayes and her family arrives Saturday and which I had stored in the Odom's freezer because of lack of space in mine. Harry served it to Doug and Mort Bernstein last night!

The island is beautiful. Lupine is everywhere, lilacs were more luxuriant than ever, rugosa roses and peonies are filling the air with perfume. Fishermen are busy getting in their gear. The dock looks more like Trap Day than late June, so many traps are piled up waiting to be trucked home. Enough for now. Reach out annd touch someone on Monhegan!

As ever,
Alta

Emma is a proud grandmother. John and Rue's Suzie has been accepted at Harvard summer school program for gifted students. She will be a third year H.S. student this fall. They are on from S.F. to enroll her. Wish they could come over.

Monhegan, Maine 04852
July 1981

Dear Friends:

Summer is in full swing and it is time for another letter, even though many of you are on the island and receiving news first hand. The solar system is back in good working order. I now can type without having to put in a wash to justify the use of the generator. The trouble with the inverter was finally found and corrected, right now I can't remember what it was, and the new batteries are doing well. We have had marvelous weather most of July but even in thick fog the cells work well. Only during a deluge or when a black cloud covers the sky do they stop churning out power. Great!

As many of you know, shortly before I mailed out the last letter I received a call saying that Emma, my cousin, had died in her sleep the night after John and Rue had started back for San Francisco following a happy two week visit at the Farm in Center Tuftonboro. She had seemed well during the visit and was planning to come over here in a few days. She had organized everything so well that the services held four days later on July 2nd went off as she would have wanted. John and Rue returned almost on the next plane after landing in San Francisco. Many of the Riddle family were at the interment service in the Tuftonboro cemetery in the morning and many friends gathered in the garden of the Farm in the afternoon. Emma had made many friends in the community with her work at the Library, the Historical Society, the Arts and Crafts Guild and just being a good neighbor. It was wonderful to see them all. I shall miss the many contacts I made through Emma there in Tuftonboro and Wolfeboro. After all these years it is hard to realize we won't be sharing glad and sad times. But life goes on. Maybe Suzie can come to Monhegan after her Harvard course ends, before she starts back for San Francisco.

Last week I went in to visit the Hosmers on Rackcliff Island in order to see John's sister Elizabeth. After rummaging around I found several pictures of our Girl Scout Camp Low. Libby and I were there with our Life Savers' badges, pigtails, head bands, middies and serge bloomers - the best camp styles of the twenties. We had a great time laughing over them. Ruth and John had arranged a gathering to which the Charlie Seibs came. I was glad to meet them because several years ago Charlie crowned me Queen of Country Correspondents, and we had never met, even though he is a summer resident of the region.

The memorial plaque for Emily Tompkins (Farrelly) was finally unveiled a couple of weeks ago. Fortunately I had asked Caroline Steiner if she remembered Emily and found that it was because of a friendship with Caroline at Smith College that Emily had come to the island. So Caroline spoke at the dedication. It was with money left the town by Emily that the firehouse and fire equipment were financed. Later there will be a memorial service for Ron Stanley who did so much for the fire department. By the way, Mary has been well enough to open her house to a few long time tenant friends. Fan Kaplan has been here all month.

Dick and Edith Cooper expected to spend their vacation at Mary's, but an accident a month earlier in which he hit his head ruled otherwise. He and Edith appeared at my door the morning I was going in to see the Hosmers so I took him to Pen Bay. From there he went to Lewiston and then to Maine Medical in Portland, was successfully operated upon and should be home now. Not knowing about the accident I thought he might have a brain tumor. Thank heaven for the fortunate outcome.

This Saturday a man rowed from Boothbay Harbor to Monhegan in a new type of boat - an aluminum hybrid affair adapted to sailing, rowing or canoeing, called the Saroca. It was so calm and the rower (oarsman, more properly) so skillful and the boat so well designed the man, Bruce Lewis of Portland, arrived two hours ahead of his schedule. I was so involved in the event I forgot all about calling Elizabeth on her birthday a few days earlier! No, that day!!

The Alex Minewski show at the Museum has gone very well, which pleases Nita very much. I went to see her studio yesterday and was most impressed with her work which is in many media and styles. Right now she is involved with scratch board. Last year she did beautifully things with stitchery. We are promising each other that come the New Year we'll get back to our collaboration on the Monhegan stories. Meanwhile I am having to hunt for the notebook which held many of them, including the write-up on Alice Stoddard's show in 1972 which has all the pictures Alan Faller took. I think I took it to Belfast for Richard Saltonstall to look at and since his sudden death this spring we haven't been able to find it.

Last night the Laura B made a special trip, bringing out lumber for more photo-voltaic installations, some passengers who missed the afternoon boat, and Doug Odom's new (to him) truck, a '71 International. The old jeep went back on the return trip. Just after all vehicles were landed there was a gorgeous sunset, one of the best of the summer. The sky was covered with rose, orange and gold-edged clouds against which were silhouetted hundreds of gulls in their evening ballet. Saturday night we were treated to a beautiful show of Northern Lights. How nice it is to live in a smogless world.

The Elsie Rowe Layman house has been sold. Before he left Glenn gave away things to many of Elsie's friends. I received the organ, a beautiful piece of furniture with interesting stencilling. But there are no bellows! I'm hoping we'll find a **small** patchable hole. Meanwhile

I'm getting a lot of finger exercises in, using it as a practice piano. Maybe when it is repaired I'll really be able to play the wedding march, a lot better than I did last fall for the Gifford-Baker wedding.

I wonder how many alarm clocks will go off at 4:30 tomorrow for the royal wedding. I think I'll be extravagant and turn on my color TV. If you have it, use it, I say.

The telephone situation is still, or again, in limbo. We have submitted our brief, answered NET's brief, but I haven't received a copy of theirs. We are now waiting for the examiner's recommendations, after which both sides have a chance to make objections or accept them. But nothing can be done until the PUC makes its decision, sometime in September. If NET is to go ahead let us hope they can do it before winter sets in. Meanwhile the system grows steadily worse and the confusion with my phone and the Barkers' is still with us. Their cottage is going to be occupied now until into the Fall. The difficulty is that this is an extension so has no identifying ring.

Needed repairs to the schoolhouse are to begin in August, so the Selver-Brooks classes won't be able to be carried on there. There was talk of building a school but with winter families moving off the need for more space has eased. The Sprouls are moving to Liniken Bay, Barbara and Paul Horn want to move to New Hampshire where she can continue her college courses. The Cundys have gone, but the for sale sign is still on the house.

A new edition of Introduction to Monhegan is in press and should be out in a few days, rather late in the season but I wasn't convinced that there was a demand for the booklets until the present supply was exhausted, or nearly so, a few weeks ago. Costs have gone up awfully. I wish I could think of a good way to collect money. The tin boxes have helped some, but I've realized only $200 from them in the past two years. it doesn't seem fair for the boats and hotels to have to carry the weight of the costs, but I guess that is the way it has to be. I'm hoping the community and the Monhegan Associates will help. After all, the booklet seems to help in keeping the roads and trails clean, the wild flowers protected and having people better shod for climbing. But there was a bad accident at the base of White Head the other day. The girl had to be taken off from the rocks in Doug Way's boat. He took her right in shore where she was stitched up and back on the island in a few hours.

David Sleeper is spending this week here preparing mountings for at least three new photo-voltaic installations. This is an ideal place for such - good sunshine, no commercial power to compete with and all the expense and difficulties of servicing private generators, let alone the noise they make. The silence of photo-voltaics is **wonderful!** I can't say enough for the system. Thank Goodness I didn't go the wind-power route. Bob Boody is none too happy with his, but glad he is wired for 12 volts so he can store energy from his generator, cutting generator time at least in half. Mine is down to 15-20 hours a **month.** That will be less if we can rig up a system for running the pump and washing machine on 12 volts or by inverter.

The Winter Works shop is having a good year with more members. But contrary to the story in the Courier-Gazette (mine rewritten by their staff) membership or acceptance of consignment articles is not open to summer residents. It is strictly a **winter population** endeavor. The work turned out is outstandingly beautiful and of fine quality - no pot-holders.

Elizabeth Gardner, Priscilla, Libby and Amy Jackson are due here next week so the Mooring will be a busy place. Due to shifts in rentals I had a vacant week which they will fill, something I've tried to do other years, but times have not been right. The garden is full of peas and cabbages. I planted a flat of 18 plants, thinking they were broccoli seedlings, but they turned out to be cabbages. I guess I'll have to go in for making sauerkraut and giving Morse's a bit of competition. But there are no carrots even though I have planted three times!

The other day I received cards from the Audubon Society asking for information on nesting birds. How does one know? I'm going to have help from Rita White and Billy and Jackie Boegel Boynton for this task. We do have nesting mocking birds this year, and some cukoos. Some people have seen bluebirds, but I'm not sure they stayed to nest. We didn't get the gypsy moth infestation that hit about everywhere else and the grasshoppers haven't been as bad as they were last year.

The day Emma died I received word from the Salem NH Historical Society that both the Asa and Isaac Woodbury residences are still in existence, although both have been converted to condominia. But no further news from the lawyer as to what, if any, inheritance the many heirs should expect. I'm afraid we are due for back taxes, not gold. At least one of the properties went up for a sheriff sale back in the days of the Depression. But take heart, Linden Farm still exists.(Great-Great-Grandfather Asa Woodbury's home).

More than enough for now except that because of rising costs of postage, paper, etc. I am going to have to have a subscription of $3/yr from non-family recipients of this letter come September. I now prepare 100 copies.

Best wishes for a wonderful rest of summer from Phoebe and me.

Alta Ashley

Monhegan

Alta Ashley, Corr.

THE CABLE HAS BROKEN for GOOD

Sept. 8, '81 - Yesterday there was a tap on the door and a man in green uniform said "Your cable is dead, no repairs are possible any more". Thurs., we on Monhegan learned that the day, which had been dreaded for more than three years, had arrived. The underseas cable, on which our telephone service had depended since 1959, had ruptured through the wave action of Hurricane Emily, which has brought us spectacular surf over the weekend. On Friday the aging cable ruptured, admitting salt water which quickly corroded what was left of the wires causing the phones to roar and squeal, the bells to jangle day and night. Now all is silence.

Ever since the cable snapped in great storm of February of 1978 Monhegan has tried to obtain an adequate and reliable telephone service, both through small private companies and from New England Telephone Company, which has extended service to the island since 1919. Through a series of hearing before the Public Utilities Commission Monhegan has tried to prove that NET has a continued obligation to serve the island. Cost studies have shown that a micro-wave full service system can be built, taking advantage of all existing facilities, at a reasonable cost and satisfying the demand for 40-50 phones on the island. It was hoped that such system could be established before this dreaded day arrived.

We now await the decision of the PUC as to what system is reasonable, reliable and adequate. Meanwhile we must rely on other means of communication. Let us hope the day will not be too long forthcoming when satisfactory telecommunication is re-established. Meanwhile

"REACH OUT AND TOUCH SOMEONE" on Monhegan, if you can!

September 13, 1981

Dear Ray:

Your lead editorial in the Courier-Gazette of Saturday disappointing and distressing. It is disappointing because it shows how little the real problem of telephone service is understood, even by our neighbors, and distressing because we are having to fight this battle without local support.

You say "Everyone would pay". What about the ferry service to the Penobscot Bay islands? Who pays for them? Monhegan has no state ferry. The round trip fare for most has risen to $12 on the Laura B. and $14 on the Balmy Days. We cannot commandeer either boat for emergency transportation, as can Vinalhaven and other islands. We ask for no state aid for our roads and bridges.

The telephone company has leased the Coast Guard cable for at first $12 and now $24 a year to bring telephone service to the island. We have benefitted, yes, but who pocketed the profits? According to NET they have expended $29,000 for service to the island since 1959, yet their projected revenue stands at $30,000 a **year**. Who has subsidized whom during these 22 years?

When telephone service is interrupted everyone suffers. Not only is it difficult for the islanders to conduct their business, those on the mainland dealing with the islanders are inconvenienced. The longer the phones are out the greater the loss to all, including the telephone company.

Since phone service was established early in the 20th century we have had to pay for all calls, including local calls. Long distance calls are all billed at operator assisted rates. The proposed rural telephone system proposed will still be expensive as well as inadequate (two lines and four instruments to take the place of an inadequate system of five lines and twelve instruments).

Monhegan is tired of subsidizing the phone system of the rest of the state. We feel our contribution to the economy of Knox and Lincoln Counties more than justifies our existence. Day trippers by the hundreds come to spend a few hours enjoying the beauty of this small island. Summer people often spend a day or more along the coast en route to the island. Monhegan lobsters and fin fish supply a significant amount of the seafood sold in the state. We are not an isolated community of dreamers sucking on a blade of grass. Summer and

Winter the People of the island are hard at work carrying on the business of the day.

And so we feel you should be able to "Reach out and Touch Someone" on Monhegan as easily as you do people in other parts of the state and nation.

Telephone Service to Monhegan - What Does the Future Hold?

By now the world is pretty well aware that Monhegan Island, Maine and her six hundred or more inhabitants have been entirely without telephone service since the aging underseas cable succumbed to Hurricane Emily's fury a week ago (Sept. 4, 1981)

Ever since the end of WW I. Monhegan has been served by New England Telephone Company over lines leased from the Coast Guard Service through their cable laid down in 1919. In 1959 a new cable was laid which allowed for an expanded service consisting of five two-party lines and two off premise extensions, all coin operated phones. The economy of the island has grown around the presence of telephone service, limited as it has been. Life of islanders and visitors alike has been geared to availability of the phones.

Now the phones are dead due to a break in the cable about three miles from either Monhegan or Burnt Island, necessitating the services of a cable repair ship. Neither the Coast Guard Service nor New England Telephone is willing to assume repairs. the Coast Guard says they no longer need the cable, in fact, they may have already turned ownership over to the General Services Administration. NET says they only lease and do not own the cable, so have no responsibility for its upkeep.

Meanwhile, Monhegan must rely on communication through Ship-to-Shore services over VHF radio, using either Camden or Boston Marine operators. This system is both inconvenient and cumbersome. It is a "push-to-talk" system which allows for single voice transmission and a second wait at each end of a transmission, a system not understood by conventional telephone users. It is limited to ship-to-shore use, and each call is limited in time. It is also not private; conversations can be picked up by anyone having a "scanner" or is standing by on the channel being used. Waiting periods of half hour or more may be necessary before one can take a turn on the channel. There is a hook-up charge for each minute and all calls are operator assisted. Thus, each call is quite expensive.

Can you imagine putting in a grocery order to several distributors, discussing a private matter with a relative or a physician for all to hear, or making or changing plane reservations under such conditions? One hotel owner has said he has missed eight or ten reservations because the phones have not been working this week.

In 1978 Monhegan was without phone service from the Great Storm of February 5-6 to May 22. Ever since then the community has been trying to get a reasonable, reliable and adequate service installed, through appeals to and before the Maine Puplic Utilities Commission. The last hearing was on June 7th in which consultants for the island presented cost studies for a direct dial full service system which could be installed, using present facilities, for a figure at least half that estimated by New England Telephone, and which would satisfy the demand for the 40-50 telephones requested by the islanders. NET said the cost was unfair to the other subscribers in Maine. Yet the company has expended approximately $29,000 in services to Monhegan since 1959 while it estimated income from the island is about $30,000 **per year!** Monhegan people feel they should not be subsidizing the rest of the state any longer, and that they should join the rest of the world in adequate, reasonable and reliable services.

New England Telephone has offered a push-to-talk emergency radio telephone "to tide Monhegan over" until something better can be done. This telephone can be used only for true emergencies - ordering food and fuel is not considered in that category - must be attended and available 24 hours a day, the recipient must supply 110 AC current probably 24 hours a day. The company magnanimously says there will be no installation or service charges. So far no one on the island has volunteered to house the instrument. No one has twenty-four hour power except that to run the town pumps until town water is turned off in late fall.

Before any decision is made we want conditions of accepting the radio phone to be in writing and to make sure that acceptance of this service will in no way jeopardize our chances of getting the micro-wave 48-line service we feel is the only reliable, reasonable and adequate system for the island. We await the report of the court examiner concerning the hearing of June 7th which should be made public in a meeting at the Public Utilities Commission building in Augusta this coming Wednesday or Thursday.

Meanwhile - "REACH OUT AND TOUCH SOMEONE" - on Monhegan, if you can!

Alta Ashley, M.D.
Cordinator for Telephone
Service for Monhegan

September 11,1981

Tel. repair man just reported
the cable has gone completely!
No repairs possible.

Box 87, Monhegan, ME
04852-0087
September 1981

Dear Friends:

The organ works! Ray Stineford came over last week to see if he could diagnose the problem but neither of us had time to take the organ apart until this morning. The only trouble was that the patch which covers an air outlet at the base of the bellows had come off and landed on the top of the bellows, invisible until the back of the organ was removed. A bit of scotch tape and a screw to put together the one stop that was thought to be broken and abacadabra, it sounds beautiful. Ray played it a bit before he was satisfied that all was okay. I could have listened to him the rest of the morning; my lack of proficiency leaves much to be desired, even with the finger exercises I did earlier.

Things are winding down fast. The Plantation Gallery and Periwinkle will close for the season this afternoon; the Museum is open for only two hours daily and will close on the fifteenth. The Balmy Days will probably stop running that day, too and the Laura B goes on her fall schedule of six trips a week beginning the 21st. By then the Inn and the Monhegan House will be closed. Only the Yew will be open and that only through Columbus Day weekend. Where has the summer gone?

We have had no rain since August 5th when we had three inches. The barometer has been 30.20 or above nearly all that time. There has been fog, heavy dew and a bit of drizzle so the rain barrels have some water in them, but both split up the sides when I put in plastic bags to hold the water against the tiny leaks in the side. They look like big toads squatting under the down spouts.

My sweet peas did beautifully until last week's NW gale which tore them off the side of the house. The morning glories, planted much too late, however, are beautiful. I had a dozen huge blossoms out this morning and scores of buds to develop. The ten-headed sunflower has succumbed to the ravages of the drought and deer hunger, but not before being photographed by many visitors on their way to or fro the Lighthouse.

Last week the Cooks, my August tenants, went in shore and said they would bring out a flat of twenty pounds of blueberries. Dave couldn't find any fresh blueberries so bought thirty pounds, frozen! I spent all weekend scooping the half-frozen berries into every container I could find and then hunting for freezer space to put them. Doug and Harry's freezer took the bulk, others went to neighbors as gifts. I made pies and blueberry muffins and am about to make another pie to take to Natalie and Charlie Johnson as a hostess gift for dinner at the Wik Wak this evening. If we can find a place to keep those berries frozen they will be welcome in pies and muffins all winter. My hands and fingernails are just now losing their cyanotic hue!

Last night the Inn held its final buffet supper and I went as guest of the Nesbitts, back for their usual fall vacation in the Mooring. As usual, everything was delicious. It was my third buffet this summer. It is about the only place one sees long-term cottagers, except for those who go to the boat and wait for mail.

Elizabeth Gardner, Priscilla, Libby and Amy Jackson - my nieces - had a good four days here, even to getting caught in the tremendous thunderstorm which dumped on us three inches of rain and dozens of bolts of lightening all one afternoon. We went to the Inn for one meal and a lobster dinner, so I had little cooking to do. It was great to have them here.

The new brochures finally arrived in mid-August, just as I found a huge cache of the 1979 edition, after crying for booklets all summer. A notation that donations are welcomed has helped. Maybe enough will be collected to half-finance the next printing, if no changes have to be made.

In mid-August I went over to Pemaquid to visit the Aldens and attend a Republican Fish Fry in Bristol. It was a lot of fun. I stayed an extra day so had a chance to swim at Pemaquid Beach with Ruth Brookes of Augusta days and took in a brunch and sale at Willing Workers' Hall in New Harbor. Just before that I went over to a Mount Holyoke picnic in So. Bristol. Sherm took me over by Phalarope when he went in for a fishing party and picked me up again after the picnic for the trip back. It was a gorgeous day and the mirage of Monhegan and Manana on the eastern horizon was fantastic. The only thing to mar the day was that on my return I found that Glen Krause had died that noon. He had worked on a painting all morning, expecting to finish it after lunch. MPBN was having its fund-raising campaign just then so we raised $292 in Glen's name. A day will be dedicated to him and music which he liked will be played. It is hoped that a print of one of his paintings can be used on the cover of Airplay for July '82 and the day chosen will be July 30, the anniversary of the day he died.

A man named Hackett came to see me the other day with a scheme called "Wind-heat" whereby he expects to create enough heat by stirring water by wind power to heat a small house. He feels my place would be ideal, but I'm getting tired of pioneering. I don't think the batteries for the solar system are working right yet, or possibly my gauges are off. David Sleeper is trying to get his truck loaded with solar gear for the new systems over on the Laura B. But at first the tides were wrong, then Emily came along and made the seas too rough, so he hasn't been around to check things. Emily gave us a wonderful spectacle with tremendous surf both at Lobster Cove and the headlands. It was weird with bright sunshine and no wind.

The gypsy moths didn't shun us entirely. They finally came out and did quite a job on the English elms down at Newt Searls'. I hope they won't now succumb to Dutch elm disease which they have resisted up to now.

From all reports Dick Cooper is doing well since his sub-dural hemorrhage and clot was removed. Glad it wasn't a malignant tumor.

No further word on the telephones. Meanwhile they are completely out of order as of Friday noon. The Coast Guard just finished stringing a new cable to power the light and many are convinced some harm was done to the cable. I doubt that because their phones aren't dead, they are too noisy for conversation, so there is power still. Peter Gorman from the Waterville office had just finished talking, without trouble, with his office when we began having trouble. Let's hope something will be done to speed things up now, and **no substitute radio phones!**

School started last Tuesday with both Terjanians teaching. They are staying at Max Rosenthall's until they can move into the Mooring in mid-October. I wish some other plan could be made because I want to close up and **go** when and if the spirit moves come cold weather. No more struggling with frozen pipes and drains.

My notebook with several stories came back, also my opus returned with another pink slip. Guess I won't be known as an author!

Best to all from Phoebe and me,
Alta Ashley

Box 87, Monhegan, Maine
04852
Oct. 1981

Dear Friends:

Two days without mail to sift through! All because Columbus supposedly discovered America and we have a Monday holiday therefore. Winter had arrived in the northeast far too soon. Today the temperature didn't get above 40 degrees, but the wind subsided a bit so it wasn't as cold as the days have been since my return to the island in mid-week. Jane and Bill Curtis had their annual fall picnic for us old timers, but it was even too cold to hold it on their cozy porch so we ate indoors. Those present were Zimmie Brackett, Julie and Jackie Hudson, Phoebe and I. We are the oldest old-timers left, even anti-dating Jo Day as to arrival on the island. Rita White couldn't make it because this was her afternoon at the Store. What a gabfest we had!

Phoebe and I left the island two weeks ago. Jimmie had the boat come out that Sunday to take back a wedding party and a group of Audubon people so we hooked a ride and went off to visit Elizabeth in Concord, leaving the apartment to John and Mary Yrrizary and Mary's 82 year old mother, Geneva Mathiasen. They left the apartment so clean I feel I should have them come spring and fall to do the cleaning!

While we were in Concord I went over to Carlisle to see the "solar house" and then had lunch at A.E. Banfield's, who writes for the Carlisle Gazette and calls himself a card-carrying columnist. It was a delightful afternoon. But the house, which is supposed to be the first home with photo-voltaic energy - mine is really the first one engineered although Sandy's was put in a few hours earlier. Well, I can't see that the Carlisle house does any more than do the electric company out a bill. The whole front roof is covered with silicon cells which churn out power and turn the electric meter backwards, but there is no attempt to store energy so that at night power comes from the "grid," Boston Edison or what have you.

We drove down to Kingston and took Aunt Helen out to Bert's for lunch. She ordered a strawberry daiquiri, then saw my martini and decided to drink it, which she did non-stop. Due to some mix-up we didn't get our order and soon she was beginning to sway in her chair, a bit glassy-eyed. I got some food from the salad bar and we passed that crisis without further trouble. After a huge scrod dinner she ordered a chocolate ice-cream cake which was big enough to sate a horse so I helped her eat a part of it. Yum, yum. As usual I got lost getting back to Kingston, trying to avoid Plymouth, but the countryside was beautiful and we enjoyed wandering around the byways.

On that Thursday we set off for Dixville Notch, the same hamlet that reports it presidential votes a minute after midnight. It is also where The Balsams is located and which constitutes the entire town except for the Tillitson Rubber Company right next door. These are the same Tilletsons who lived on the Sprague Estate in Wellesley and whose children were once patients of mine! Small World Department, again. I went over and found that Neil Tilletson owned the Balsams and that the sons run the factory and had a chat with Rick, who brought his wife over to the cocktail party we had that night at the hotel. The Maine Medical Association meetings were being held, thus my visit to the Balsams. En route I took the Kangamagus Highway from Lincoln to North Conway, passed the Presidential range, whose craggy tops were snow covered, gleaming over the most brilliant fall colors I have ever seen. Jane and Bill Varnum, you should see our White Mountains in the fall. They really are spectacular, even though not as high as your Rockies.

Unfortunately it rained most of the time we were at the Balsams so I did no swimming or sunbathing and got in only two walks. It was bitter cold, too. But as soon as we climbed over the pass into Errol, NH the sun came out so I decided to drive over to Otisfield in the Sebago area to call on the Cooks, my erstwhile tenants, and ended up spending the night there. They have a big blueberry farm (shades of the thirty pounds of frozen berries David brought back last month).

Monday I stopped in Auburn and had a session with Alan Stone, the lawyer for the Monhegan Telephone service, and got the low-down on that situation. More later. That night I spent with the Darlingtons in Augusta. We all went to the KVAA meeting in Hallowell that evening to hear a report of a trip to Chinese Museums. Brint and Alice Darlington had been on a People to People Mission to China a year ago and were most interested in the slides shown, which were excellent.

When I got to Thomaston and picked up my mail I found an answer to a letter I had written to Alexander Stewart who writes the column "Fun in Knowing" for the Courier Gazette and we had lunch together at the Helms in Rockport - most delightful. Phoebe was a guest of Dr. Sidwell's, having a cyst removed from one of her legs. A session at the dentist's and dinner at the Sail Loft with Knox Co. Med.

Society completed the activities ashore. Can I settle down to just plain island living after that?

The island is still active with people. Jimmy brought the Laura B out again today (Sunday). About fifteen people came over filling up the Yew, the only place open besides the Tribler, so Zimmie had to take in a couple. My place is vacant for one night only. The Bakers will arrive tomorrow with a couple of 1/4 cord boxes of wood and a chain saw. After them the new teachers move in. They are an attractive couple. After complaining all summer about breaking down and renting the cottage to them I have now come around to thinking I can even go away and let them have my apartment and close up the cottage in the really cold weather. Now that there is no telephone service I am planning to fix up downstairs into a studio apartment with a hot-plate for short term guests, or even for me while the teachers are upstairs, if necessary.

And now about the phones. No sooner had I finished my last letter than the cable went out for good. After three years of fighting for a better service, or even some kind of service before the cable would break or be taken up, the inevitable happened. The hearings examiner finally came out with his report last week. In essence he states that NET must give us at least as much service as we have had - namely, five lines. **But** he thinks five public stalls scattered over the island would be the way to go because he states that probably most of the calls are made by **day visitors.** How ridiculous can one get? Also, because of our "isolation" we should come up with a capital contribution of $54,500 if we want microwave full service.

Our answer to that is that we have had 13 phones on five lines and should not have less, that we have subsidized the system for 22 years in that it has cost the company $29 thousand dollars, during that time while the estimated income has been $30 thousand dollars **yearly.** I forgot to say that the capital contribution would be only if a 24 or 48 line system was installed, it would cost the company a substantial amount to put in just the 5 pay-stations, as proposed. Right now we have a "fixed-mobile" station at the Store whose line we share with forty other customers in Knox County. This is definitely a "temporary and emergency" service and is supposed not to jeopardize our chances of getting the full service system we are asking for. Given the costs of any system the micro-wave system does not seem unreasonable. A box to receive donations for the privilege of using the phone will be placed by it. I tried to use the phone twice but was unable to complete my calls so I don't know how it really works. We just hope that we won't have to be limited to this phone for long. It, of course, is not available when the Store is closed and is supposed to be used only for emergencies, and for the Store to place orders for food and fuel. A dozen times a day I think of going to my phone to call, then realize the grim truth. Telephoning is the way of life these days, we all know now.

Monday

Alan Stone, the lawyer for the Plantation, is due out here the end of this week and the Commissioners' recommendations are to be made public sometime next week. Meanwhile we are running out of funds to fight this case, besides coming up with capital expenditure.

David Sleeper is practically a resident. He has been working on several new photo-voltaic installations here. Don Stone, who has about the most sophisticated system, is having an open house this afternoon to show it off. I'll tell you about it later.

Don and Prue Cundy are here closing up the house which has not been sold yet. They are both enjoying living on Liniken Bay. Craig Sproul is getting his gear together and invested in bait Saturday when the first load arrived, so I guess they have given up thoughts of selling out and going to Southport, as they had hoped. Lexi Krause has bought Mary Stanley's house and will move in when Mary closes up her business for this year. Doug and Harry have nearly finished repairing and preparing traps and will be able to take off in a few weeks. The Horns will leave when Paul gets through with shutting off the Town water come November 1st. They haven't sold their house as yet, either. It is nice to know that the Shumaker house has been sold and is being fixed up once more.

Joan Harlow was here over the weekend and says that her mother, Ruth, will definitely be back next summer. I for one hope so, anyway. No news from Miriam Barteaux. I thought by now she would be here with the two dogs.

My three good crops this year were runner beans, beets, and tomatoes. Zucchinis came in just about right, carrots and cauliflower were flops and the slugs did a job on the cabbages. We haven't had a frost yet so some things are still growing.

Tomorrow I'm having a birthday party for Harry and hope that Rita can join us. I forgot to pick up a good card when I was in last week. This was supply day, so to speak, for the island. Jimmy brought out cord wood, two quarter cord boxes of red oak came over from Phil Baker's tree farm - actually from New Hampshire which came in too large chunks for S.D. Warren pulp mill. Phil chops and splits it into stove size wood and brings out the half cord each year. He also began sawing up the cord wood I have had in the yard since last February. The teachers have ordered **five** cord to carry them through the winter in the mooring. You may not be able to see the buildings once all this wood gets piled up around them.

Wednesday

The second load of bait and salt arrived yesterday on the IO and I got about two pounds of nice herring which I cleaned and froze, except for four which I had for breakfast this morning, and weren't they good.

The birthday party went off well. It was a glorious night, clear and cold and a full moon shining. There was frost in the Meadow this morning but as yet my garden has been spared. We are having beautiful Indian summer weather.

Last week I received the sad news that Peggy Marks Barton's husband, Bernard, died early in the summer, after failing health for several months. He was buried in the family lot along with his great aunt Clara Barton in Oxford, Mass. Peggy was among the first of my new friends on Monhegan when I returned to the island twenty years ago. She also was one of my earliest tenants. I do hope she can return, but it is a long haul from Laguna Hills where she now lives.

Danny Bates has gone to Arizona. Amy is still here with the two children and looking for a place to live,. Sandy and Lisa Dickson are off on a five-week trip to the far west after attending an energy conference this weekend in Bar Harbor. Dave Sleeper is off for the conference, too. I think all his installations here are now finished, but there will probably be more. Even the phone system may go P/V for its principal source of power.

Courtney Day and Zoe Zanidakis are both at Maine Central Institute. Jeff Rollins is also there, I think. Timmy Boehmer is at Proctor, Doug Rothstein at Brewster and his sister Margaret Taylor is at Stoneleigh-Burnham. Lance Burton is at an academy in Albuquerque and his tuition payments are in doubt because Mary and Bob spend a few months there. Kathy Wincapaw is at Morse High in Bath and lives with her grandmother Mable. The new teachers, David and Lori

Terjanian, are working as a team and seem to be enjoying their work and the kids seem to be quite happy with their teaching.

The Coast Guard carried out a mock disaster drill, supposedly a hotel fire and explosion with twenty-five casualties. It was a miserable day with a cold NE gale and drizzle. The helicopters couldn't participate because of the wind, although they would have, had it been a real disaster. I got pictures yesterday from the C.G. photographer which make the thing look very real. It was held to see how the C.G., the islanders, and in shore emergency units could be coordinated and brought out several problems which can easily be surmounted. The lack of phone service played a role in some of the difficulties and the bad weather made it unwise to transfer island "casualties" to the mainland and back, so only C.G. personnel were included in that phase of the drill. It did show that the islander can get together and perform effectively as need be. The "fire" was under control within an hour and pronounced "out" shortly after.

For those of you who are really interested in the telephone situation, I'll send another letter as soon as I know where we stand. It may be that we will have to accept either the capital contribution or take the case to the Supreme Court and spend more money with the possibility of losing in the end.

Uncle Henry's has been rebuilt after the fire of 1979 and looks fine. There will be an addition put on the uphill side next spring. Peter Boehmer and Vernon Burton have done a great job, as usual.

Back to P/V installations - Don Stone has only four panels to my six, but is on a 24 volt system, which is more efficient than the 12 volt system that I have. One thing, it doesn't require double wiring because the same wire will safely carry both 24 and 110 power. I could rip out all that surface wiring put in to take the 12 volt system! Maybe I will. But it also would take a different inverter which is costly. Time will tell.

I can't think of any more island news so Phoebe and I are now signing off. I forgot to say that Phoebe had a tumor on her left hind leg removed last week and has been fine about not tearing at the stitches. She is getting gray, a bit stiff in the joints, sleeps most of the day but still loves to bound off for a walk in the woods, but no more deer chasing except to protect me. We are both getting to be little old ladies! This is the last letter unless I hear from you that you are still interested in receiving more.

As ever,
Alta Ashley

Box 87
Monhegan, Maine 04852
December 1, 1981

Dear Fellow Heir(s) of Asa Woodbury:

Many of you may have been mystified to receive a great sheaf of information on the heirs of Asa and Isaac Woodbury. This may all have stemmed from the effort Emma Woodbury Riddle and I made to find out about the possible existence of Linden Farm, the old brick house belonging to Asa. We stopped in a real estate office close by Rockingham Park and left our names. Emma had wanted to have the house declared a national monument because she thought it was a good example of old New England architecture. Unfortunately, the day I received any information was the very day I also learned that Emma had died.

Our cousin, Robert L. Woodbury, is retiring from the Massachusetts Land Bank the end of this year and he says he will be willing to represent us all at the hearing. He has someone who is looking into the situation. It seems foolish for all of us to consult lawyers, since there may be very little for any of us except possibly the payment of back taxes! My lawyer thought having Bob represent all of us would be the wise thing to do.

While I was in Massachusetts over the holidays I was able to contact Ernest Hall in Andover and he agreed to the plan. So, if you want Bob to handle things for you would you contact him or me?

Bob's address is 52 Holmes St., Needham, MA 02192.

My address is Alta Ashley, M.D., Box 87, Monhegan, Maine 04852, as of now our phones have not been restored since the cable broke in September.

It was fun to see how many relatives we all have and to find some second and first-cousin-once-removed right here in Maine. Perhaps we should have a reunion out here on Monhegan come summer! I wish I could say I would have a phone by then but we are still waiting for the Public Utilities Commission to make up its mind as to what kind of a system will be "reasonable, adequate, and reliable". They have been deliberating since July!!!! But they should worry, they all **have** phones.

January 5th will be coming up quite soon, so I urge you to contact Bob or me as soon as possible. You might find out about any wills which would affect the situation, those which are not included in the information at hand.

Best wishes for a wonderful Christmas season and a great 1982.

Sincerely yours, and hopes for more than a postage stamp for all!
Alta Ashley

Monhegan Christmas Bird Count

Dec. 31, 1981. The fourth consecutive annual Christmas Bird Count on Monhegan Island was carried out on Tuesday, December 29th in rain, drizzle and gusty winds. Those participating were Peter D. Vickery, compiler from Richmond, Maine; Davis W. Finch and Cloe Mifsud, of Epping, N.H.; and Jeanne Rollins of Monhegan Island.

The four participants observed 45 species, which is the second highest species total to date for the island. A few never before sighted species brought the four-year species total to 72.

The most interesting sightings, because of either the number of members or their rarity were: 1 Arctic loon, 1 Barrow's Golden-eye, 16 Bohemian waxwings, 171 Canada Geese, 1 merlin, and 1 snowy owl. Thousands of eider ducks were seen off-shore and two skeins of Canada geese flew by the day before the count. Birds usually seen in large numbers, such as white throat and tree sparrows, starlings, redwing blackbirds, and blue jays, were seen singly or not at all. All in all, however, it was a satisfactory endeavor.

1981 SEASON'S GREETINGS FROM MONHEGAN-CROSSROADS OF THE WORLD

Dear Friends:

As the years roll by I become more and more disorganized, make lists then leave them on the desk. So, here I am with only one red master unless I can borrow one from the school. So you'll probably have to accept one red and one purple page!

Two things seem to characterize this past year - the island telephone situation and visits from the family. Seven of my nearest relatives managed a trip to the island. My nieces, Elizabeth Gardner, Priscilla, Libby and Amy Jackson, and nephews Tommy Ashley and David Gardner made the trip as old hands at island visiting, while my Niece Nancy Peters came over for the first time after nearly twenty years of planning! It was good to see all of them and particularly to have Nancy here so she could see that my living on Monhegan is not such a crazy idea. My cousin Emma Riddle had planned to come out in mid-summer but, after a long period of failing health, she died practically on the eve of her intended visit. We were practically like sisters and discovered the wonders of Monhegan as small summer visitors more than seventy years ago. It is hard to realize she still isn't in her beloved house on the knoll in Center Tuftonboro, N.H.

The island telephone situation is still in limbo. The Public Utilities Commission finally decided that New England Telephone Company has and must continue to serve us and to do so "adequately, reliably and reasonably", but they didn't define their terms. We had hearings during the spring and expected a decision from the commission shortly after July 4th, but they are still deliberating! Meanwhile, the cable went out on September 4th and cannot be repaired. In October an emergency "fixed mobile phone" was put up in the Store which is shared by 40 other customers in shore and can be heard by anyone tuned in or with a scanner or VHF radio. Yesterday two men came out to see if a different kind of radio phone could be set up and found the signal to Portland is excellent. But calls to Rockland and Port Clyde would be long distance, ho, ho. There would be more than one phone, but only one line, so here we go again. Meanwhile we await the PUC before the full service system can be begun. Construction is estimated to take at least a year to eighteen months, so we will have at least one summer without more than minimal service. This fight has taken so long we have about exhausted our resources. So any of you who are concerned with telephone service to the island can help by contributing to the Monhegan Plantation Telephone Fund and send your checks to me. I am now confident we'll have a full service system when the final installation is made. The publicity the Boston Globe, Christian Science Monitor, TV stations and many Maine papers have given this situation has helped a lot. But I hope things will come to a head before the first of the year when I am off to East Falmouth, Mass. for a couple of months to house sit for my niece Nancy Peters. More about this later.

It is a few days later. Sorry, but I have no more red masters, thus the color change. Now, we have received good news. New England Telephone has a plan for installing a full-service micro-wave telephone system, subject to approval by and acceptance by PUC and Monhegan residents. It will mean a capital expenditure, but this will be far less than we would have to spend were we to set up our own company. We now have over 40 applications for phones.

Phoebe says she has had a hard time keeping track of buried treasure hidden in various places we visited during the year. There were many trips, as usual, to Emma Riddle's during the winter and spring. We renewed acquaintances made at the Republican Convention in Bangor in 1980 with three trips to Pemaquid, including a Fish Fry to meet candidates for election in the fall. We found that Dick and Mary Alden, our hosts there, lived close to the Hansons, Brookes, and Madeline Bates, all Augusta friends, and Mike Kearney whom I probably once met in Center Tuftonboro at Kay and Micky Muerlings'. A speech about women in medicine brought me in touch with Irene and Fred Gamble of Tenants Harbor where we were guests before and after Thanksgiving. From their house on Mosquito Cove we saw Monhegan Light flashing on the horizon. We also stayed at Richie and Patty Kahns', Patty is the daughter of the Gambles; the Drownes' in Tuftonboro and the Pettees' in Wolfeboro and, of course, at Elizabeth Gardner's in Concord. Then there was a delightful luncheon with Alexander Stewart in Rockport. He writes a great column "Fun in Knowing". Photo-voltaics brought me in contact with A.E. Benfield of Carlisle who says he is a "card carrying" columnist!

Phoebe has just reminded me she also buried bones at the Sleepers' in Falmouth, the Giffords' in Readfield, the Mitcheners' in Andover and the Hosmers' on Rackliffe Island where I saw John's sister Elizabeth, a fellow camper at Camp Low and now an ordained Episcopal priest. We also visited our tenants the Cooks in Otisfield, the Kenways in Gardiner, Louise Webber and Lura Titcomb in Augusta and Skip and Jackie Thurlow in Hayford Heights. Thanksgiving was spent in Baldwinsville with the Jacksons (niece Priscilla) which gave me a chance to visit the Jahns in their lovely home in East Syracuse.

All these contacts, and many I haven't mentioned, made me realize I needed more stimulation during the winter than Monhegan affords. So, I have arranged to "house sit" for my niece Nancy Peters in her summer home on Cape Cod. **My address will be Eel River Road, East Falmouth, MA 02536.** Think of having a phone! After the big storm of last weekend I wonder just how much winter I'll miss. The Cape was clobbered, while we had only sleet, rain and WIND. It was a wild one and probably would have taken out our cable had it been repaired last fall. At least I won't have to struggle with frozen water, oil, and drain pipes as I did in the Deep Freeze of last winter. The teachers who, against my better judgment are now in the cottage, will have to contend with them. Let us hope that bales of hay around the foundation, two oil stoves, two wood stoves and two emergency electric heaters will fend off the frost.

My opus made three more trips to publishers and returned with pink slips. How does one get published? New version of **Introduction to Monhegan** finally came out in August, in time for all the numbers to be wrong!

We lost several old Monhegan friends this year, among them Glen Krause, Jo Townsend and Elsie Rowe. Glenn generously gave me Elsie's parlor organ which Ray Stenford got into working order; it is a joy.

And now "REACH OUT AND TOUCH SOMEONE ON MONHEGAN" for you may be able to do so before long. The best of all to you for 1982!

Alta Ashley
and Phoebe Starling

1981 SEASON'S GREETINGS FROM MONHEGAN 1982
CROSSROADS OF THE WORLD!
GOOD NEWS!
REACH OUT AND TOUCH SOMEONE ON MONHEGAN!!

Dear Friends:

The good news is that, if present plans which New England Telephone Company has in the offing are accepted by both the PUC and Monheganites, we will have a full service direct dial system on the island in a year to eighteen months, once construction is started. It will mean a moderate initial charge for each telephone service, payable over a three year period, and a small mileage charge over the $9.00 plus monthly basic rate. But it will be far less than we would ever have to pay had we tried to install a system ourselves, and far better service as well. All homeowners will be receiving further information shortly, as soon as we receive it from our lawyer, Alan Stone, and can get envelopes addressed and mailed.

This has been a long fight and an exhausting one, both for our funds and for my equanimity. But if NET's terms are accepted by the PUC and us it will have been well worth the struggle. In the meantime we will probably have a limited interim system until the permanent system is operational. Men were here last week from Summit Communications in Portland and found the signal from here to their tower is excellent. NET has contracted with them to put in a radio phone system in addition to the "fixed mobile" phone installed in the Store after the cable broke in the storm we had just before Labor Day. We share the mobile phone with forty other customers. I believe the radio phone will be for our use only, although there may be only one line, but two or more phones. We will still be "biting the bullet" but won't feel as out of touch with the world as we do now. The publicity we received from the Boston Globe, Christian Science Monitor, New York Times, and several Maine papers helped our cause, I'm sure. Now we wait PUC's decisions, which we had hoped to have early last July!

This year was not filled entirely with telephone problems, although at times it seemed so. Often I had to change plans because of a scheduled hearing, but there was time for many contacts, both here and on the mainland. In mid-summer my nieces Elizabeth Gardner and Priscilla Libby and Amy Jackson came out for a few days during the week the cottage was vacant between tenants. In September my niece Nancy Peters made her first visit to the island after twenty years of trying. I hope she left feelng my living here isn't such a crazy idea after all. In October nephews Tommy Ashley and David Gardner came out for a couple of days and helped with splitting wood my tenant Phil Baker had conveniently sawed up during his stay the week before. My cousin Emma Riddle has planned to come out in July, but after a long period of failing health she died June 30th, practically on the eve of her intended visit, after a two week visit from the San Francisco contingent of her family which she had enjoyed greatly. We were almost like sisters and discovered the wonders of Monhegan as small summer visitors more than seventy years ago. It is hard to realize that she isn't still in her beloved house on the knoll at Mackerel Corner in Center Tuftonboro, New Hampshire.

A few weeks before Emma died we stopped in Salem Depot, N.H. to see if any of the family farm houses were still in existance, particularly the one belonging to our great-great-grandfather, Asa Woodbury, known as Linden Farm. Emma thought it should be nominated as an historic site because of its architecture - pink brick with end chimneys. The day she died I received word from the historical society that both houses were still standing. Great-grandfather Isaac Woodbury's house serves now as the Jockey Club and Tack Room for Rockingham Park Race Track. It didn't burn in the recent fire. Shortly after that I received a call from a lawyer asking for the whereabouts of the heirs of Asa and Isaac Woodbury and just a month ago all 32 of us received a hugh document concerning a small tract of land which we apparently still have an interest in. We don't know whether we will inherit a fortune or be liable for back taxes. Robert Woodbury, who will retire from the Massachusetts Land Court, has offered to represent us when the case comes up in January.

Phoebe says she has had a hard time keeping track of buried treasures hidden in the various places we visited during the year. There were many trips, as usual, to Emma's during the winter and spring. We renewed acquaintances made at the Republican Convention in Bangor the year before with three trips to Pemaquid, including a Fish Fry to meet candidates for the coming election this fall. We found that Dick and Mary Alden, our hosts, lived close to the Hansons, Brookes, and Madeline Bates, all old Augusta friends, and Mike Kearney and I probably first met at Micky and Kay Muerling's in Center Tuftonboro! A speech about women in medicine brought me in touch with Irene and Fred Gamble of Tenant's Harbor, where we were guests before and after Thanksgiving. Their house overlooks Mosquito Cove and on to Monhegan, where we could see the light flashing at night. We also stayed at Richie and Pattie Kahns' in Union, daughter of the Gambles, at the Drownes' in Tuftonboro and the Perrers' in Wolfeboro and, of course, at Elizabeth Gardner's in Concord. Then there was a delightful luncheon with Alexander Stewart in Rockport. He writes a great column "Fun in Knowing". Photo/voltaics brought me in contact with A.E. Benfield of Carlistle, also a columnist - he says he is a "card-carrying" one.

Alta Ashley & Phoebe

East Falmouth, MA

Dear Friends:

Here I am in Falmouth, using the old manual typewriter which lives to skip spaces, just for fun, and is quickly wearing out my fingers. Also the M-R key, apostrophe and quote keys are all in different places from where they are on the electric typewriter. Now I have readily available electric power I don't need the sun, which is shining brightly this morning, something it has done very rarely on Monhegan since mid-October.

Monday when I left Monhegan the wind was blowing a gale, but it was a southerly, so we just coasted in on the crests of the waves - not a bad trip at all, and the Prudence A kept a surprisingly even keel. When we landed at Port Clyde, however, it was sleeting and it was touch and go as to the condition of the roads. I drove Mort Bernstein to Rockland, did three errands and left for Concord by two o'clock in heavy rain. The roads were clear of ice all the way, but there was a lot of slush and water on the roads which landed on my windshield every time a truck went by. Even stopping at Kittery for happy hour supplies it took only four hours to reach Elizabeth's. The rain was so hard at times I couldn't see the lane marks in the road, so it was a bit hairy, especially when confronted with that maze of lanes near the Lowell Connector. This time I chose the correct one and didn't have to make two or three clover leaves to find Route 3, which is often the case.

New Year's Day we had quite a storm. The winds blew pretty steadily over 30 knots, peaking often over 50. Some of the men had started moving their traps to the dock the days before in hopes the storm might pass us by or be over by Friday. Naturally there was no thought of setting during the storm, which lasted well into Saturday. The boat came in that day, since Friday was a mail holiday, and had a hard time unloading and loading because traps were piled up everywhere. At a "lobster meeting" Saturday afternoon the fishermen couldn't come to an agreement about setting on Sunday. The older men said "no", that it was bad luck and that a storm was due again on Monday. The younger ones were restless and said "yes".

Sunday was dark and dismal but there was a flat calm. By noon Shermie Stanley had loaded his boat and set the first trap. Others saw his boat working the setting grounds so began to follow suit. Doug and Harry heard that lobstering had begun about mid-afternoon and were able to get only two runs in before dark. Charlie MacDonald was on the island to help them. He makes wine and champagne as a hobby and had brought some up to celebrate Trap Day. So, although it was far from a traditional opening day of lobstering he brought up a bottle to the house Sunday night and the four of us toasted the new season. The champangne was delicious.

I guess the old flavor of Trap Day is gone for good. There were no mug-ups on the dock, no baskets of food passed to the people working on the boats, no feeling of good-natured competition and cooperation, which has characterized Monhegan's lobstering here-to-fore. We older folks trust all has not been lost with the new me-ism which is so characteristic of the younger generations.

There are twelve boats in the fleet this year. Dougie Boynton may join it later. He is still working for Betsy Wyeth on Allen's Island and won't be done until spring.

The Laura B will be off the run for quite a while yet. She is having extensive repairs done to her hull, including a new ice shield. When she returns she should be a quite handsome boat.

Three Audubon people came over on the 28th for the Monhegan Christmas bird count. The day they arrived was lovely so they went about the island scouting for good places to observe birds. Of course, the next day it drizzled, then blew and by night it was quite cold. They identified only 45 or 6 species but found some "first observations" including Bohemian waxwings, a snowy owl and someone's golden eye. I left my 1981 record on the island so can't be sure of all my statistics. My memory is becoming a **sieve**.

There should be some very good news coming about the telephones. I can't say for sure because we are still waiting for the PUC to accept terms which NET and Monhegan have agreed upon. We were to come up with 40 requests for phones before NET would agree to anything. By Monday we had 48 requests and at least four sure requests hadn't come in. It was beyond my fondest dreams. I thought we'd have to wheedle several people into accepting a phone. What a relief! I had hoped everything would be signed and sealed before I left the island. My being here will possibly slow things up because I am the petitioner, and action has to go through me, apparently. At least I can be reached by phone here, which is a help. I guess I'd better order some of Charlie's brew for the victory celebration.

The need for phones was brought out dramatically New Year's Eve. Because there was no chance of Trap Day being on New Year's some people began celebrating during the evening. One helper who, unbeknownst to anyone was on medication, went around from house to house, began acting strangely, then went beserk and broke into Billie Boynton's down at Deadman's Cove. Jackie got out of the house, in her nightgown and ran for help to David and Katy's across the road from me. David got Stevie Rollins, the constable, and the three men got the boy under control. He was taken in to Pen Bay Med. Center by Coast Guard, after spending the night at Stevie's. It meant a lot of running around by several people, more than would have been had there had been phones. We got two or three applications for service after that episode. Now we are virtually cut off from receiving calls it is easier to convince people of the need for phones - a blessing in disguise?

I brought a carton of my writings in hopes of getting down to whipping something into shape. Someone is bringing over some office equipment for me to try out this afternoon. I may get this run off. I forgot all supplies - paper, envelopes, stamps (just bought $40!!) I intended to stay here at least until town meeting March 1st. So HAPPY NEW YEAR from Cape Cod.

As ever,
Alta Ashley

*Katy Murdock and David Boegel have announced their engagement and plan to be married in June.
Phoebe loves the sand and oak leaves.

Monhegan Votes on Telephone Proposal

Feb.8. A special Plantation meeting was held tonight at the School house in order for the towns people to act on a proposed settlement for telephone service with New England Telephone Company. The meeting was opened by Plantation Clerk, Jeanne Rollins. Ruth Stanley was elected Moderator by the twenty-two residents present. After an introductory speech by Willard Boynton, First Assessor, the meeting was turned over to Alta Ashley, M.D., who has acted as Coordinator for Telephone Service for Monhegan for the past four years. The lengthy proposal was gone over step by step, with opportunity given for discussion and further explanation of the meaning of each item. The meeting cleared away several uncertainties concerning the proposed service and the residents voted unanimously to accept it thereby allowing New England Telephone Company to plan for a full-service micro-wave 80 channel direct dial system for Monhegan out of the Rockland exchange. Each customer will be charged $300 for each primary telephone installed pursuant to applications for service received on or before December 1, 1982. (So far close to sixty applications for service have been received by the company). This will be a one-time charge, with $120 payable initially and the balance at a rate of $15 per month over and above regular monthly charges. There will also be a $2 per month mileage charge.

Monhegan is asked, also, to supply a building for housing necessary equipment and a suitable site for erection of the micro-wave dish antenna. NET promises, insofar as possible, to complete construction and to supply the service by May 31, 1983.

This proposal is subject to approval by Public Utilities Commission. There will also be public telephones (coin or coinless) installed on the island at locations selected by the company, to meet public need.

An interim system has also been proposed to tide the island over until the full system is available. This will consist of a two-channel service provided by Summit Mobile Radio Company but the charges will be met by New England Telephone Company. One channel will be dedicated to Monhegan use and the other will be available on a restricted basis. Calls will be limited to five minutes maximum. This system is dependent upon FCC approval, but it is hoped it will be in operation before the summer of 1982. The one-channel mobile radio phone now in operation at Monhegan Store will remain in service until the full system is in operation.

It is hoped that all parties to the agreement will act promptly in favor so that work can be started in short order. Monhegan has still to hear from Monhegan Associates, Inc. which owns the building proposed for the switching equipment and the land on which an antenna tower, if needed, will have to be erected. All who have lived though these months since the cable went out before Labor Day hope and trust work will be started soon and that the full service system will be in operation by May 1983 or before.

At the special Plantation Meeting held on February 8th the stipulation for telephone service for the island was gone over step by step with ample opportunity for discussion, following which a vote was taken. The twenty-two residents present voted unanimously for the proposed plan which, in brief, is as follows.

1. Exchange Service

NET will provide full exchange service to Monhegan based on single party lines, and having Rockland, rather than Tenant's Harbor, as the exchange with its wider calling area. There would be, however, a monthly mileage charge of $2.00 added to the regular basic monthly charge.

There will be a $300.00 one-time construction charge per main telephone per location to all initial subscribers. This is payable $120.00 down and $15.00 per month until paid up. There will also be the usual installation charges but further construction charges, as for poles, etc.

will not be made for applications received before Dec. '82.

This full service will be available, insofar as possible by May 31, 1983 - barring catastrophe, war, and what have you. It is dependent on NET's obtaining the oil storage building near the light, and a suitable area for mounting necessary micro-wave antenna(e). We now are waiting to hear from Monhegan Associates, Inc. trustees as to the terms for use of the building and whatever land, if any, is needed for a tower.

2. Interim services A and B

System A. This will be a two-channel telecommunications service to be provided by Summit Mobile Radio Company of Portland. One channel will be dedicated for Monhegan use, the second for use on a restricted basis. It will be coinless public telephone service, requiring charge card, collect or third number charges and all calls will be disconnected after five minutes, if not before, automatically, Two full duplex stations required for this service will be provided by NET, who will also pay the monthly charge for service (approx. $35.00 per station including local call charges). Long distance calls will be operator assisted. This system is conditional on FCC approval. The PUC has already approved it and has notified the FCC of their stand.

System B. This is the system already in use. It is a one-channel manual mobile service through the Rockland mobile operator and is shared by about forty other subscribers. This service, intended primarily for islander use, will remain in operation until the full service system is in operation.

3. Public Telephone Service.

New England Telephone Company will provide one-party coin or coinless telephones at locations selected by the company and sufficient in number to meet public need, as part of its provision of full exchange service.

4. Additional Agreements.

These involve the provision of the oil storage building at a reasonable cost, for the housing of switching equipment and power facilities, as well as necessary space for mounting the antenna tower, should use of the light house tower prove unsuitable esthetically or practically. Any alterations in the oil storage building will be made in keeping with the character of the building. should such alterations be necessary.

All this must be approved by the PUC before work can be started. So far we have 55 firm and three promised applications for phones. If you wish copies of the full stipulation let me know. Also any questions.

36 Eel River Road
East Falmouth, MA 02546
Feb: 1982

Dear Friends:

Here I am still in Massachusetts, but over the weekend I have to return to Monhegan for a special Plantation Meeting concerning the phone situation, which I hope will wind things up and NET can get down to engineering the system. I'll hold this letter open until my return.

It has been very pleasant here, even though the weather was much colder and snowier than normal. I was plowed in for a couple of days, but kind neighbors shovelled me out and took me to town shopping, introducing me to stores, a bank, and their personnel. We have escaped the awful ice storms and heavy snows the rest of the state has enjoyed (?). Ice on the river became over a foot deep so no longer did the ducks swim by or the tide go in and out. But for the past few days, helped by two days of fifty degree temperatures, the ice is breaking up and both the tides and the ducks have returned.

When I came down here I thought there would be days when I would be moved to work on my opus, as well as have a nice relaxed time without having to rush things to the P.O. Well, it hasn't turned out to be so. I found I have to earn 100 hours of credit for extended medical education by January '83 so I've been leaving the house soon after seven to attend meetings at either the Falmouth Hospital or the Cape Cod Hospital in Hyannis. So far I have five "brownie points" under my belt. Next week's jaunt to Monhegan will keep me from getting two more. I guess I'll have to enroll one of those very expensive courses they hold in posh resorts for a week to get enough hours in by the deadline. Las Vegas, here I come!

I've also joined the choir of the Falmouth Congregational Church, whose director is a great musician and quite a clown during rehearsals. I started out as a soprano but dropped down to th alto section last night. My five note range does better there. I sat between the tenor and soprano soloists, with an almost inaudible alto beside me, last night, and got thoroughly lost most of the time as we flew through six anthems and a Schubert Mass which is to be done the end of March, so I may stay here until then.

Nancy (owner of this cottage) has been down on two weekends and is coming down again today. Elizabeth and Priscilla are coming down for the holiday weekend, I've been out for a dinner and an evening of bridge and hope for more, so time hasn't hung heavily at all.

It is now Valentiene's Day. Elizabeth, Pricilla and I have exchanged cards and presents and have had a good laugh. They will leave for Concord this noon and I will have to turn to getting this letter out and writing about the special Plantation Meeting I had to go to the island to attend. The trip over and back left a lot to be desired. The wind blew a gale both times and Phoebe showed the whites of her eyes most of the time. Poor thing, she is so stiff it is hard for her to keep her balance, even in smooth going. We rode up and down great seas, but the Prudence A is a good dry boat. The Laura B is still being worked on and probably won't be back on the run before May.

The trip to the island was a hurried one and was taken up entirely with telephone service business and going through a huge bag of second class mail, including my Maine Income Tax form which I almost put into the fire when I got bored with going over all that mess. But it was pleasant to stop, both coming and going, with Fred and Irene Gamble in Tenants Harbor. We had had a snowstorm on Tuesday and by Wednesday afternoon the roads were still icy in spots, so I stayed over Wednesday night rather than try to go on to Concord that afternoon. The boat was late in getting in because the going was slow due to the gale wind which came up that morning, and because we had to wait an hour for the repair man who had come to fix the one island phone. For some time no one could hear the person calling from Monhegan, or only faintly. After doing a few errands in Rockland I drove back to the Gambles. That evening we went to Laurel and Don Weavers' to dinner. Don Esley, a local artist, was also there and a lively conversation kept us chatting until way after eleven. By the time we got back to Tenant's Harbor it was -2°.

This winter has been hard on the island because of the many storms and the cold. There have been many days when only one or two, or even no, boats have left their moorings. Water pipes and drains have frozen. The day I left water was having a hard time flowing out of the sink. Shades of last winter! I was glad I had decided to return to Falmouth. I didn't pick up any more news. Doug came by on Tuesday evening with 34 crab legs which I finally finished picking out yesterday. So good!

All roads were clear by the time I drove to Concord on Thursday. My niece, Priscilla Jackson arrived that day from Syracuse so she, Elizabeth Gardner and I set off on Friday for the Cape, stopping to pick up Aunt Helen and take her to lunch at her favorite restaurant, Bert's in Plymouth, where she as usual downed her Manhatten in about two minutes, sucking vigorously at its icy remains. (I wonder if I'll be able to do that at 96.) She had a great outing, but was glad to get back to Evanswood when the girls took her home. They joined me later here in Falmouth that afternoon. Had Nancy been here all three of my nieces would have been together.

On the way down I stopped in Wiscasset at the alternate energy place and found that there is a small wind generator one can mount on a roof which will put out possibly enough backup energy to preclude the need of the thermal generator for battery charging, and for a nominal cost. It sounds great, but I have to recover from the cost of my new copy machine before I think of wind generators. No excuse now for bad copy.

It is nice to hear from so many of you. I think I have had more mail since leaving the island than ever. One pack of forwarded mail arrived empty, so if your letter was in it take heart and send it directly here to 36 Eel River Road. The East Falmouth Post Office is automated and without imagination. They sent back my bank statement which was addressed to Est. River Road!

Phoebe is still reluctant to move but once out of doors enjoys a good walk. I wish I knew what is wrong and what to do . The vet charged $119 to tell me he didn't know either! We'll be here until about the middle or even end of March.

Best wishes to all,
Alta

36 Eel River Road
East Falmouth, MA 02536
March 5, 1982

Dear Friends:

Yes, I am still here in Cape Cod and loathe to pull up stakes and return to snow and ice of northern New England. Last night we had about an inch of snow, which was all gone in the 47 degrees of this morning. But more is predicted for tomorrow. This winter seems endless! Go I must next week. With the promise of better weather the rest of the Peters family is eager to enjoy their cottage, which is so comfortable. The choir wants me to stay until their concert on the 28th, but that is far too late.

I'm planning to leave here around the 14th, stop over with Elizabeth in Concord and take in the flower show before setting off down east.

No news has come from the island in nearly two weeks. I didn't go back to attend town meeting because Billie Boynton assured me there was nothing on the warrant except budget items. I haven't heard a word, even though I telephoned to the Store that day asking for news. Also, there have been no more developments in regards to the telephone situation, as far as I know. The PUC commissioners **all** backed our appeal for a speedy ruling on the "interim" radio phone which NET wants to install before summer to augment the mobile phone at the Store.

You probably will be glad to know that Phoebe is about back to her old self and beginning to jump up on people once more. The fact that she hasn't been able to do so was the only redeeming feature of her disability. For weeks she hasn't been able to jump onto my bed. But the other night we had a thunderstorm and I soon realized I wasn't the only one in my bed. A fairly good sized brown dog lay quivering beside me. We have been walking a mile or so a day, have discovered many beds of arbutus, of course all on other's property and many right in the middle of woods roads ready to be stepped on. I feel like going out in the light of the moon with trowel and sack and stealing a few plants to plant over on this lot, or even sneak some off to Monhegan, but I'm afraid I'd be committing a crime under Massachusetts law. Mayflowers have been on a protected list ever since I was in primary school.

One of my fellow altos and I are going to see "On Golden Pond". the show has been here for three weeks but the theater is open only at night and I haven't wanted to go alone. Last night at practice I found she was looking for a companion, so ...

Doris and Charlie DeCarlo and I met at Cape Cod Hospital one day I was over there for brownie points and had a very pleasant lunch and gabfest at Mildred's. They are both very busy with their art activities. Thank heaven I'm not in Hyannis, even in winter it seems overcrowded. Wednesday was a beautiful day so I drove over to Orleans to see Katherine and Ashley Day from Concord, now in the Cape Cod Retirement Center in Orleans. They have a lovely and spacious apartment, but I am not ready for such institutional living - **yet**. While there I called on Bill DeWilde, president of the Cape Cod National Bank, and had a nice chat with him. On the wall were two of Charlie DeCarlo's paintings. I drove the northern route through all those lovely villages. Also last week I took Aunt Helen out to Bert's again. She had had her hair done and looked fine, also her cough was better so she was able to go out with no problem. She didn't want to go back to the home in mid-afternoon so we drove to Marshfield trying to get a glimpse of the ocean. I guess I took all the wrong turns, got completely turned around and when we finally reached the water we couldn't see over the sea wall! I tried to find the place in Duxbury I worked in as a maid during the Depression but couldn't recognize anything.

The sun is out and thermometer is almost up to 50 degrees - it was 14 yesterday morning and the river had frozen over again. I guess Phoebe and I had better go for another walk. Let us hope I'll have some island news on the mail which is due any moment, now.

Well, I stooped to tempatation and dug up two tiny plants and dug them into a bit of mossy, sandy soil on this lot with a prayer. The mail brought two packets of mail from Monhegan and **no** island news except that my generator has given up the ghost and that all the pipes froze in below zero weather with winds topping 55 last week. Also, I got a bill for gas which has been used while I have been away and was to be paid by the tenants, and acknowledgment of my contribution to the student fund for Harvard School of Public Health with my name mis-spelt! There were two inquiries about rentals, one with no return postage all this was compensated for by my income tax return which gives me a healthy refund in both the state and federal taxes. Wow!

Sorry not to have any more news of the island. When I tried to contact Bob Boody the other day about my generator I found that the only change in town officers, apparently, was that Doug Odom displaced Mort Bernstein on the planning board. Does that mean that Doug will finally get permission to build his house, which he has been hoping to do ever since selling the Store in 1979?

I hate pulling up stakes here, leaving the choir and **bare** ground with crocuses poking up through the pine needles. It is now Monday. Nancy came down Friday, leaving snow in Boston and escaping that which fell yesterday. We had only rain. It poured when Mary Alice Kinney and I went to "On Golden Pond", which we just loved. What beautiful photography! the one thing which puzzled me was that the power boats seldom made noise, and Jane Fonda dived into the water with her watch on!

It's good I wrote this letter now because I won't be able to use the copier without a generator, also we won't have water without the pump. I wonder how long it will take to either repair or replace the generator. If only Bob Boody would call. This telephone situation is really frustrating. It looks now as if we may have to wait for the annual meeting of Monhegan Associates in August before any definite decision concerning the oil storage building and land for a tower can be made!! Will we ever have phones? FCC and PUC decisions are expected any day now. I'll hold this out a day or two in hopes.

March 8. Today's mail brought news from Rep. Curtis that PUC has agreed to stipulation presented by NET and agreed to by Monhegan, with no substantial change. So now we wait for MMA, INC. for the micro-wave system and FCC for the "interim" radio phone.

Phoebe is wagging her tail in a "hello" salute to all. She is just about back to her old self.

As ever,
Alta

Box 87
Monhegan, Maine
04852
April 3, 1982

Dear Friends:

By now I thought I would be able to write this letter and put it through my new copy machine. But somehow, the shipment of paper and "toner" needed for photocopying hasn't arrived. So back to the spirit duplicator, the master of which I am too lazy to correct.

Phoebe and I finally arrived on the island on March 22nd, with stopovers in Concord and Andover. While in Concord we, or I and Tommy took in the Flower Show, which was a great disappointment to me. Gone are the great halls filled with gardens, acacia and dogwood trees, spectacular waterfalls and lovely woodland trails bordered by wild flowers and miniature plants. Commonwealth Pier is a long black tunnel lighted by floodlights beamed on certain displays. After touring the Flower Show we went off to Quincy Market, the first time I had been there since its rehabilitation, looked at the seals in the Aquarium pool - Andre isn't there this year, but down in Mystic, Conn.

There were buses from all over New England and some from New York state. I saw four people I knew from Maine but missed the Gambles with whom I often stay in Tenant's Harbor.

Elizabeth took me to an exhibition of Regional Culture at the Concord Antiquarian Museum. They had a special showing of artifacts from Concord and Wethersfield, Conn. It was most interesting and well worth a visit. It will remain in Concord (MA) until June 5th, then can be seen at the Conn. Historical Society, ?where, from July 5th to September 26th and at Wethersfield Historical Society October 18, 1982 to January 14, 1983. The exhibit is called TWO TOWNS.

One afternoon I went to see the nursing home, Sudbury Pines, where Ellie Kelly works. It is a very nice place and very well planned. The ambulatory and with-it residents are in one wing and the incapacitated in a separate area. Those in private rooms can have their own furniture. There is a very good activities program for those able to participate. Ellie does a great job in keeping people active, both in mind and body. While I was there many of the residents were attending a concert. As we went by the room I heard "There is a Balm in Gilead" and asked the singer-guitarist if he would want to do a duet. So we sang a couple of verses with me taking the obligato. What fun! I never would have dared had I not been singing so recently with the Falmouth choir. He may come out here and give a concert this summer, if there is any interest in it.

En route to Maine I stopped off to visit with Marjorie Mitchener in Andover. Paul was away so Marj and I had a real gab-fest. We have known each other for three-quarters of a century! She was the little girl across the street in Allston.

The Gambles took me in on Sunday night, then Phoebe and I were off to Monhegan on the Penobscot Gulf, a converted oil tanker which used to service the island. It is so full of electronic equipment it looks like a spy ship. We were able to watch the Columbia lift-off on a television set in the cabin. She is a slow old boat, very narrow and not too steady. Most passengers these two weeks have been sick. We've had wind, wind, wind. More to come this weekend.

This letter was supposed to be full of hope and joy: "The spring's at the morn ---." We came over here on the 22nd in order to be ready to meet with a team from NET who were coming over the following day to look over possible sites for the permanent full service telephone system and to find two locations for the "interim" system to be set up in May to serve us until the other system is operational, promised for May or June '83. Four men showed up and came to the house for coffee and returned after the meeting for a bit of cheer before they set off for Tenant's Harbor in their chartered boat. I happily showered, washed dishes, feeling quite secure because Bob Boody had said the generator would probably run enough to pump water intermittently. Then, on Wednesday morning I found most of the water etc. on the floor of the downstairs lavatory. My heretofore dauntless spirits just about hit bottom and have been sagging badly ever since.

Bryan Hitchcock and Peter Murdock came up finally on Friday the 26th , dug frozen ground off the top of the septic tank and dipped it out. (We don't have the luxury of a honey wagon.) But the water wouldn't go out of my house, the cottage drain was okay. So I was advised to go ashore until warmer weather. Then Bob Boody came along and said he thought the drain was plugged, not frozen. He bored a hole in a visible drain pipe in which to insert a snake, found the obstruction, which wasn't ice, and finally freed it with a charge from the hose, but now nothing would go **out** of the septic tank. That was Thursday. He was to return with the back-hoe, but it was disabled. Then came last night's heavy rains, (it is now Sunday). This morning he appeared with a pick ax, dug away at the absorption field and found it full of water. So now that has to be attended to. What next!

Add to all this, my hot water heater smoked up yesterday and filled the place with soot. Even Phoebe is a dingy gray and my hair keeps getting darker. By the time I leave tomorrow for my usual trip for medical society dinner I'll be a brunette, for the first time in my life!

The new generator is here but the conversion kit for LP gas didn't come with it so we are still limping along with the injured old 6 KW Kohler. Before this is through I should know how to install plumbing, septic systems, gas heaters, and wire up generators or corral a POSSLQ (person of opposite sex sharing living quarters) who knows how.

The Lewises and Rothsteins returned Friday from Florida, all well tanned but a bit green about the gills after their two hour trip on the Penobscot Gulf, aptly dubbed "The Sneaker" because of its shape. The poor Lewises found their water line frozen, but I'll gladly change places. It is easier to get water **in** than **out**, when there are problems.

Most islanders left on Friday for the Fishermen's Festival in Boothbay Harbor. Some waited until Saturday to leave after the high winds let go about noon. Then the winds came up again and rain came in torrents. Harry T. and Rita braved the storm and came over here to dinner. I just needed to get my mind off drains etc.

Fishing is slow and the price of lobsters at the boat is nearly $4.00. The other night, however, Doug and Harry decided lobster was cheaper than meat here, so we had a lobster feed, my first since early January before I left for the Cape. They have moved out of Chris' house and are back at the Underhill house until they can move into the fishhouse and turn the other place over to the Schweiers.

Paul Horn is expected any day now to work on the town water supply. He and Stevie Rollins should have things ready by the 15th. Let's hope it warms up and stays warm by then. No frozen pipes.

I just found all of page two was typed with the shield in place! Now for the telephone news. NET has given up the idea of using the light tower on which to mount the micro-wave antenna and probably also use of the oil storage building. There have been so many requests for phones that the building will not be large enough to house the necessary equipment. They still think they need twenty-four hour generator power and three generators. The men looked at the Lighthouse and water tower areas to see where they could mount the free-standing radio tower. All sites considered are owned by Monhegan Associates Inc., so it is up to that organization to lease or sell the land necessary. Nothing can be done until FCC approval has been obtained and that can't be done until the land is available. We are all hoping for a quick decision on the part of MAI. The "interim" system will be far from ideal. There will be two radio phones beamed to Portland, supplied by Summit Communications and paid for (basic charges) by NET. Calls will be billed on the basis of operator-assisted calls. Incoming calls will be picked up by an answering service which will be contacted by holder of the phone four times daily and messages posted on a nearby bulletin board, as plans are now. One of the phones will be "dedicated" to Monhegan, the other will be available only from 7 or 8 PM to 7 AM., outgoing calls on both lines will be charged calls, no cash calls will be possible. Local calls will be to Portland, not Rockland. Sharpen your pencils, get in a supply of stamps and postcards for this summer.

The Postal Service is planning to build a new post office with twice the space they have now, "but no increase in service." One area under consideration is the lot Craig Sproul now owns on the road between Newt's and what used to be Elsie's cottage. That would be nice and convenient. We all hope the lobby will be large enough to house waiting patrons in inclement weather.

Linda and Don Stone have been all winter expecting to be able to move into their new home in shore. That is now about finished, just in time for the summer season and Don's various classes here! Katy and David are moving shortly to the apartment above Don Cundy's shop. Their wedding is scheduled for early June, the 5th or 6th, I believe. The school board is busy again reviewing applicants for the position. Lori and David Terjanian are not coming back another year. One year with all the woes of the Mooring was too much, I guess.

A few bluebirds have been spotted by Rita, the Boyntons and Victor Lord. Victor is replacing the porch at the Monhegan House. He hopes to open in late May or early June. Jo Day is due May 1st and will open the Yew on the fifteenth. Bob and Mary Burton are due back any time now and Lexi Krause is getting his house ready for summer tenants. The season is about upon us. The Laura B is having an extensive overhaul and won't be ready for the run until May. They say we won't recognize her when she hoves into sight. Harry T. Bone is here for a few weeks between boat assignments and is fixing up his yard and garden. Mort Bernstein is in Boothbay Harbor. He has had some heart problems but is now able to return to the island.

The winter on the Cape must have made me soft. I can hardly wait for good weather. No female should live here without a POSSLQ, one gifted in all sorts of handyman activities. Maybe when I return in a week things will be working well and life will be pleasant and fun. In May I'll be off to South Hadley to join the class of '27 in our 55th reunion. How the time has flown since we received our blue ribboned diplomas.

Phoebe is blossoming on a diet of raw liver and extra Vit B complex and intermittent cortisone.

As ever,
Alta

Monhegan, Maine
04852
May 1, 1982

Dear Prospective Monhegan Telephone Customer:

We have reached our final stage in our struggle for adequate, reliable and reasonable telephone service for Monhegan. A stipulation of Agreement was accepted unanimously by Monhegan residents on February 8th at a special Plantation meeting, signed by New England Telephone representatives and finally accepted by Public Utilities Commissioners on April 6, 1982.

The Stipulation of Agreement, in brief, is as follows:

1. New England Telephone will install a full service direct dial exchange service by micro-wave from Rockland, to be completed on or before June 1, 1983.

2. A one-time construction charge of $300 per main telephone per location will be made to those subscribers who request service before December 1982. Payment terms will be $120 down and $15 per month, once the system is operational, until the balance of $180 has been paid. Monthly service charges will be similar to those for subscribers in the Rockland exchange area, plus a $2.00 per month mileage fee. **No private property construction charges will be imposed for construction undertaken pursuant to applications received on or before December 1, 1982.** Long distant calls will be billed as direct dial calls, according to tariffs as published. All services, both residential and business, will be single party.

3. Because not all persons will have access to private phones the company will set up public phones in up to five locations of their choosing. These phones will be non-coin and all calls will be operator assisted and paid for through credit cards, charges to a third number or reversed charges, using dial but dialing "0", rather than "1" to signal the operator. Maintenance and service to coin operated phones would make costs for all phones on the island prohibitive. Then too, the company is phasing out such phones over its entire system.

4. An interim telephone system will be installed by Summit Communications at the expense of New England Telephone, who will also cover monthly service charges. These phones will be dial radio phones beamed to Portland. They also will be non-coin, requiring credit card, reverse charges or charges to a third number, local calling area will be Portland, and can be used for out-going calls. There will be two such phones on separate lines. The phone at the Island Inn will be "dedicated" to Monhegan 24 hours a day. The phone at the Trailing Yew will be available for out-going calls only from 7 p.m. to 8 a.m.

Incoming calls will be stored by an answering service and picked up by Monhegan on the Inn phone at certain times in the day. Calls then will be posted on a convenient bulletin board for return. Truly emergency calls will be sent through directly to the phone at the Yew.

The number for in-coming calls is **(207) 784-9161** after mid-May 1982.

This service will continue to serve the island until the full exchange micro-wave system is operational in June 1983.

The service for the summer of 1982 will be limited, indeed, but it will be for only one summer. By the summer of 1983 we should have a far better system than ever before. We ask you to bite the bullet for the time being, with equanimity and hopes for the future.

Even though you may have already made a non-binding application for service New England Telephone has requested formal applications be made now, as suggested in the enclosed form.

Thank you for your patience and cooperation.

Alta Ashley
Coordinator for
Telephone Service for Monhegan

(This is not my copier! Can't make it work.)

P.O. Box 87
Monhegan, Maine 04852
May 1982

Dear Friends:

Although I have dated this May I am starting it on April 28th because I want to be sure it gets copied and mailed before I leave for a day or two next week, so all you who may still be worrying about my various problems can rest assured that things are just about up to snuff with the arrival of toner for my copier which I have been looking for since the day I left Falmouth.

Bob Boody laid a new leach field after he found the one which has been in service since 1979 consisted of one short pipe and a round area about three feet across for a leach field. There are now four ten foot pipes laid across the front yard, which should do good service as long as I will ever want it to last. The new generator has been set up but has no battery until Bob returns with one from the mainland this weekend. The old generator is limping along and is supplying necessary power so far. Maybe when I turn on the copier it may conk out, so I'd better wait to reproduce this letter.

Paul Horn finally came by and cleaned out the burner for the hot water tank, so that is no problem now. I spent a day washing all the windows up on this floor and wiping soot off all horizontal surfaces. My hair is back to its natural very blond color and Phoebe is again golden and white with black accents where they should be.

After Bob Boody laid the new leach field he smoothed out the front lawn. My garden is pretty well ruined because he kept trying to find the grass, not realizing I had a flower garden there. But the lawn now has a nice gentle slope, without a hollow in the middle and should look much better than before. We had to move two of the cedar trees, but the young maple tree didn't have to be moved. Raspberry bushes are now in a pile by the barn ready for replanting as soon as I have courage to spend the time in this raw, windy weather. The two good days we had last week were spent in spading up the vegetable garden and planting peas, beets, and lettuce. Will warm weather ever come? The wind has blown hard all month and it feels like early March instead of late April. Crocuses and snowdrops are about the only flowers that have bloomed so far. Daffodils have budded but none has blossomed. Lilac buds are swelling and peonies are sending red spikes up through the ground. Even day lilies are hardly greening up, but lawns are beginning to look good. The one out back is about ready for mowing.

Ruth Lothrop gave a shower Sunday for Katy Murdock,(a very pretty party), her granddaughter, whose wedding to David Boegel will take place on June 6th at Patty Miller's cottage, followed by a reception at the Lothrops' Ice Pond Cottage.

About the only human robins to return so far are the Lewises who finally got their water pipes thawed out. Doug and Harry have been working on theirs which I think by now are feeding water to the cottage. Town water finally went on Sunday. Steve Rollins and Paul Horn had to replace more pipes than ever before because of winter damage. They got a late start because of our April blizzard which dumped a foot or more of snow which formed into huge drifts. All that has gone now, even in the woods. Phoebe and I took our first walk through Red Ribbon trail yesterday, encountered no snow but many trees down and across the trail in several places. No sign of lady slippers as yet.

The blizzard hit the day I had two dental appointments and had planned to go to our country medical society dinner at the Sail Loft. By four o'clock visibility was about nil, several cars were reported off the roads so I made my way back to the Gambles' in Tenants Harbor, ran off their driveway over a small stone wall, unable to see where I was headed.

Fortunately the wall was low and Fred Gamble was able to drive the car back onto the driveway without damage to garden or car. I stayed in shore all that week to meet with Peter Gorman of NET and attend a photo-voltaic conference at the Samoset. The Gambles housed Phoebe and me all week, instead of the expected two days. It was very enjoyable being with them and sharing in their Easter festivities, including a "present hunt" and an Easter dinner along with the Kahns (Patty is the Gambles' daughter) and the Esleys. He is a local artist and she is in the administration of Pen Bay Medical Center. It was a most welcome relief from the trials and tribulations still awaiting me on the island - now solved.

Lynn Hollis, who has been our special ed teacher for the gifted, accepted a position in Molokai and left on Monday after a big send-off party at the Rollins' on Saturday. I didn't know of the party so didn't attend. The remodeling of the Hausser cottage is coming along. Bricks, building blocks, mortar and masons arrived on the boat. Phoebe and I must go down that way soon. More P/V systems have gone in. I believe there are twelve either already installed or ready as soon as David Sleeper can get out here with his Freeman Clark. It has been so windy for the past month he hasn't come, after being turned back about three weeks ago by a sudden storm. He comes up from Portland or Falmouth Harbor.

Two men from NET in Boston came out today to see what can be done about the emergency phone at the Store. They found some mis-matched equipment and installed a whole new phone so now **we can be heard!** Jack Kearns, micro-wave specialist for NET, came out, too, apparently just to get the lay of the land. He talked to several of us, looked at my solar system, which was churning out good amperage in the bright sunshine. He thinks power for the telephones will be a combination of diesel power and photo-voltaic. They won't try wind, after "picking up pieces of wind generator all over Block Island." Survey crews should be out shortly and construction crews will follow early in the fall, as soon as all easements have been cleared. I spend hours on reproducing tax maps, locating buildings and making lists of places where telephone service has been requested. Formal requests are now in order, accompanied by a down payment of $120. I'll enclose forms for those of you who may need them, or send along soon under separate cover.

Back to island news. The children are off island, expected back about six p.m. on Friday. The Trejanians took them to Philadelphia where they have been visiting historic and interesting sites in that area, including a Phillies game, and a trip to Longwood gardens, Brandywine and Gettysburg battlefields. Even young Sammy Rothstein went along.

Jason Sproul fell out of a tree on Tuesday and became impaled on the piece of jagged wood which entered under his chin. He was taken in by Shermie who made the trip in 35 minutes, fog and all. An ambulance whisked him to Pen Bay where Dr. Morse removed the stick lodged dangerously near the carotid artery, but with no severe bleeding. Latest reports of his condition are good. He probably couldn't have been treated much sooner had be been in a large city. He is back and doing well.

Last night we had a glorious sunset which lasted more than an hour. The sun set just north of Mt. Washington, which loomed up soon after the sun disappeared. Sky and sea were a combination of pink, orange and emerald and sapphire. Beautiful! I guess that is why I like it here, with all the problems which seem to arise. It's nice to have them come all at once and get overwith quickly!

Balmy Days starts weekend service May 29-31, June 5,6 and full service June 10 - Sept. 19, then the weekend of Sept. 25, 26.

"Interim" phone number will be 784-9161 after May 15 or thereabouts.

Enough for now. I plan to be off island quite a bit in May, picking up necessary CME points - as many as my pocket book can stand.

Best to all,
Alta

May 30, 1982
Box 87
Monhegan, Maine 04852

Dear Friends:

I hope that this is the last letter to come off the spirit duplicator. After struggling with the Sharp copier which was transported from Falmouth and refused to work, I took it to Rockland, during which trip it was put on its side making necessary a whole clean-up job, to find that the developing fluid as well as the toner had been removed. So, no copies came through, no matter what I did. I used a whole bottle of toner at $50 a bottle, trying to make copies! In Rockland I found a mimeograph which will allow unlimited copies, and a second hand 3M thermal copier (I can't think of the name right now) for things needing only a few copies. It won't be like the Sharp, but I can't cope with transporting that twice a year, if I decide to spend winters off island. So - next time the mimeo.

The island is crawling with people this weekend. Ninety Audubon people came over yesterday on the Laura B, plus many cottagers over to open up for the season. There were so many passengers and freight the Laura B had to make two trips, despite pea soup fog. Yes, she is now on the run looking very spiffy in her new coat of paint, glass cabin windows and a flange (I don't remember the proper term) around her water line to divert spray. Her cement balast has been removed and time was too short for attaching a weight to her keel so she rolls quite a bit. This will probably be done at least by next weekend, before she starts her full schedule on the 19th.

The Balmy Days came in yesterday for a special weekend trip, but with few aboard because of the weather. She is coming over again today to celebrate her 50th anniversary on the run, as well as the birthday of her skipper, Bob Campbell, with an open house on deck this noon. Right now it is raining and thick o' fog but it is supposed to clear. Poor Bob and Diane, they haven't had very good luck in their off season trips (last year and this).

Zimmy is back behind the counter at the Spa. For a while he was thought to have to undergo chest surgery, but the trouble cleared up with antibiotic therapy. He looks great. Doc Rothstein is back after his coronary a few weeks ago. He is still confined to the house, however. Gil and Ruth Lothrop came over on Friday, both recovered from their hospital stays. Ed Hubert is now on the sick list but is said to be doing well following his heart attack. Teco and Mort Bernstein are still struggling with their shingles, the popular island disease these days. Vernon Burton is expected home in a few days following surgery at Pen Bay and should be back to his usual activities soon. I guess that covers the sick list for now.

The main topic of conversation is the telephone situation. On Thursday Summit Communications of Portland put in the two radio phones which are to serve us for the coming year. One phone is at the Inn and the other at the Store (not the Yew, as formerly planned). No direct contact can be made by people calling the island. Calls are to be made to an answering service by dialing (207) 784-9161. Note it is 784 not 783, as I reported previously. Calls will be picked up by someone on the island, me at present, but I hope not for long, two or three times a day and posted for return by the callee. It will be probably far better to use the mail than to rely on the phone, if possible, since there is no way to contact people on the island unless they check with the bulletin board which will post the calls. One year to go!

On this coming Thursday there will be a meeting of Monhegan Associates, Inc. trustees, the Plantation and NET officials to decide on where the tower and service building will be placed. They must be within 150 feet of each other, so the tower can't be down by the water towers if the little brick building is to be used. Also, the site must be accessible for delivery of supplies, including oil or gas for the generator. Problems, problems! All the NET engineers who have come out here are loath to put up poles, so we may be able to use ground conduits in some places where poles would be obtrusive, as long as there would be no traffic over them or they could be buried well in such places.

Two weeks ago I went in with NET men after a hurried meeting here, to attend the Mass. Med. Soc. meeting, got mixed up on the dates and missed the day I had paid to attend, also eleven CME points! Elizabeth, Tommy and I went into the Boston Public Library to hear a lecture by Tommy's friend, David McCullough, on Teddy Roosevelt. David is the son-in-law of the late Clare Barnes, an old golfing friend of my father, and Martha's Vineyard summer resident back when we summered at Oak Bluffs. Practically the entire Barnes family was there and it was great to see them after hearing so much about them from Tommy. David is a fine speaker and we all could have sat another hour listening to him. His book on Teddy is "Mornings on Horseback".

Rita White went in early in the month to visit June Day and to see Henley, who is a patient at Togus. While ashore she decided to get rid of her pigtail and is now sporting a permanent and short hair, quite a transformation. Henley has been getting increasingly confused and is probably going to stay much of the time at Togus. It will seem strange not to stop on the road and discuss the meaning of obscure words or talk about the weather.

I must get ready to go to the Inn to pick up the morning messages and greet the boats arriving around eleven. Both have new engines and are making their runs in shorter time, so are still arriving about the same time to wage the battle of the dock. I think in good weather the Balmy Days will arrive first, and she should, because it takes only minutes for her to discharge passengers and what little freight she carries.

It is now Monday morning. The rain and fog of the last two days are still with us making holiday festivities a bit soggy.. The party at the dock for the Balmy Days was only moderately well attended. Many didn't get the message because of not being out and about in the rain. Passengers coming over on the boat were few. But those of us who did go had a great feed and liberal libations. The party for the Trejanians, held at the Boody's in the evening was well attended. Every family brought desserts - rhubarb pies, brownies, cheese cakes, cookies of all descriptions and other fancy desserts. Heidi Boody had made a banner of farewell which all attending signed and everyone, including the children, had a great time.

More people are leaving the island. Steve Rollins has his house and boat up for sale, the Mersfelders will spend winters in shore. On the other side of the coin, John Murdock brought back his new (to him) boat yesterday. She is well equipped with all kinds of electronic devices and mechanical equipment to help him capture the ever increasingly elusive lobster.

From all reports the wedding of Kelanie (Tinker) Cundy to Shawn DeRepentigny in Boothbay Harbor was a grand affair and the reception at the Tugboat Inn a merry one. All who went from here stayed over rather than trying to come back in two kinds of fog!

The other day I picked up a few new masters for my ditto. They were different from the old ones and when I went to roll them through the machine I found I had typed on the wrong side and everything came out backwards! So now I am having to retype both pages!!!

Speaking of weddings, Buddy Murdock wants to make sure you all know that the "biggest wedding in the history of Monhegan" will take place on June 6th when daughter Katy marries David Boegel. The wedding will be at Patty Miller's and the reception at the Lothrops' Ice Pond House. Both Ruth and Gil are looking forward to the festivities. Ruth recently underwent abdominal surgery and Gil had a heart attack the day he was to pick her up at the hospital. They are both fine, now.

While I was ashore the last two weeks Phoebe was in the care of my tenants, the Rejanians. She behaved far better than I had dreamed she would but Rita says she was down at the dock every day hoping to find me. By the time I returned she had to renew acquaintance with my scent before she would come to me! I spent three days in So. Hadley at my 55th reunion of Mount Holyoke '27. Eighty-nine and twenty husbands filled Prospect Hall. The weather was none too good. The rains held off until the alumnae parade and assembly in the amphitheater but it came down in torrents on Sunday. I sang in the alumnae choir - lovely music. After lunch I drove Roger and Eleanor Ussher Baker to Amherst, then over Route 2 in the pouring rain and on to Prout's Neck in Maine for three days of medical meetings and more brownie points. If you can believe it, I took a Casco Bay cruise on the "Longfellow" on the one day the sun shone. The bay is beautiful with its hundreds of islands. Because my dentist was ill I was able to catch the boat back home in time to see the installation of the interim phone system and make the first call. Since then I have been the one to pick up calls from the answering service, one has been for me, all the others for either the Inn or the Store!

It is now June 7th. On the 3rd there was a meeting of the Monhegan Associates, Inc. trustees, Jerry Crimp and Peter Gorman of NET, some Plantation officers and interested home owners during which NET officials presented their proposal for leasing the oil storage building and enough surrounding land for the radio tower and possible photo/voltaic energy system. They feel a 100'foot tower, will be necessary on which to mount the two micro-wave antennae needed and, although the islanders preferred using the area near the water tower, this site was considered unsuitable for many good reasons. After much discussion the trustees voted to acceed to NET's wishes, realizing that the tower will be less visible than one placed by the water towers. NET can now begin final engineering of the full service system. One year to go!

Dr. Rothstein was taken back to Maine Medical last week and is supposed to have had a quintuple by-pass this morning. Gil Lothrop had a second attack early Sunday morning for which I was called. He, Patty and I went in by Coast Guard, ambulance and my car to Pen Bay where he is now doing well, but will probably not return to the island until after his prospective by-pass surgery. It was the morning of Katy and David's wedding, so there was quite a to do. John Murdock had to come in for Patti and me and pick up the minister because Jimmie's trip to Matinicus, which he thought was cancelled because of the weather, went off on schedule. John's boat made the run in record time and we got back only a little late for the wedding which took place at the Ice Pond House with fifty invited guests. Five hundred were invited to the reception, also at the Ice Pond House. Both festivities had to be indoors because of rain, so you can imagine the crush. Stan Haskell from Sea Coast Mission was the minister. He rode out with us in John's Sea Hag.

George and Ruth Colton are here in their former cottage for a few weeks until Carl Hiller arrives. The Haussers are enjoying their greatly remodelled cottage, which is really beautiful. Peter Boehmer did himself proud. Vernon Burton is back and looks better than he had all winter, but no carpentry for him for a while.

Bobby Mersfelder is selling his gear and has his house up for rent. Steve and Jeannie Rollins are selling off everything and moving inshore. The Sprouls have also gone in shore. It looks as if Monhegan is going back to the fifties when the population dropped to 45 and then to 29 after the Store fire in '63. But it will spring back as soon as people can afford to buy houses. There is work for carpenters and all round handymen, even if lobstering is none too good right now.

My cottage is still not rented for June, the first three weeks of August or the first half of September, much due to the poor telephone service as well as unforeseen happenings to my regular tenants. Let us hope next year will be better.

At a special Plantation meeting tonight a five member school board was elected, but the salaries asked for were cut down to that budgeted for the three-member board in March. No action was taken about covering the phone calls.

In the confusion on the dock on both Friday and today, with all the people coming and going, Katy Curtis Donahue's suitcase went back on the boat the day she arrived and was taken all the way to Connecticut, before the flood, and those of Patty Miller and Peggy Gillespie were left on the dock today, instead of going with them to Port Clyde. The dock looked more like mid-summer than early June and today a NE gale swept across. Dave Sleeper, here to finish up several jobs, including a P/V system for Jamie Wyeth, decided against bouncing over to Port Clyde, but nearly one hundred people went in anyway. Let us hope the sun comes out for more than a moment or two, before our batteries are depleted.

My garden has recovered from the dryness of May and is producing tomato sized radishes, which I planted overenthusiastically. I'll never have peas by July 4th, even tho' I planted them on Patriot's Day. Mary and Bob Burton have beautiful flower gardens around the Inn front yard and a new loading platform by the kitchen door, making it easier to reach the phone just inside the door. Victor Lord had to go in today because of a painful shoulder which has kept him from finishing work on the Monhegan House. The Yew had a huge shipment of new mattresses and box springs, apparently enough for every bed.

Toward the end of the month I hope to join John and Rue Riddle, Emma's son and daughter-in-law, in Center Tuftonboro to help them settle things at the Farm. Emma's estate has finally been probated after nearly a year. I have missed my trips over there and seeing the many friends I made through Emma in the town and in Wolfeboro. Also, I want to pick up some maple syrup before it is all gone.

Phoebe is doing well as long as I give her cortisone every few days. She sends a tail wag to all her friends.

Now to send this through the ditto machine. Here's hoping I've typed it correctly.

Goodbye for now,
Alta Ashley

Box 87
Monhegan, Maine 04852
July 15, 1982

Dear Friends:

Believe it or not, the temperature is now up to 87. It is muggy with almost no wind although a breeze is just beginning to come up this afternoon. It is the first really hot day this summer. The garden should be glad except that we have had no rain since the first of the month. I've caught fog and dew to put on the garden and have watered a couple of times. But the peas and berries aren't filling out very well and one by one my pepper plants have succumbed to the heat and the slugs, which flourished during June's cold and wet. In May it was dry and we cried for rain, which came almost every day in June, then the drought. Thank heaven I don't make my living from gardening.

The front garden has finally come to life and produced petunias, lilies (except those the deer demolished last night), anemonies, and portulaca. The lawn is a mass of mustard, poppies, and forget-me-nots where it was dug up for the leach field. Lupine and flowering shrubs have been more beautiful than usual, including the rugosa roses which should furnish us with plenty of big rose hips in the fall.

This seems to be the year of visitors. Jean and Ben Johnstone, cousins from Florida, came by following Jean's 50th reunion at Radcliff. Then came a first cousin twice removed, Tim Breen from California, who had come east for a wedding. While he was here Scotia and Fred Waite stopped off for a couple of days after Fred had attended his 55th at UNH. Last week my college classmate, Hallie Patey

Long, came by with her sister Phillippa on an overnight stay from camping trip to Mt. Dessert Island and now today Dorothy Dolliver Foster Dorr, who was in the class of '24 at Girls' Latin School, stopped in to say hello while on a day trip from Boothbay Harbor where she now lives for six months of the year.

In between all this I sandwiched in a trip to Concord to attend the funeral of Gil Lothrop in Reading. Gil died unexpectedly while waiting to be discharged from the hospital the day I mailed my last letter. Ruth is back on the island and making a good adjustment. I later dashed off to Center Tuftonboro to the Farm while John, Rue and Susie Riddle were on from San Francisco. It was good to see them again. The stay was all too short but I had to get back here for a meeting with telephone and town officials about the new system.

Tom Masterson from NET has been out here frequently and will have to spend a lot more time with his measuring device for determining the possible position of poles and cables, overhead and underground. No one wants more of either, but we'll have to have some in order for the system to be put in. There is no question about ever relying on the underseas cable again and no system can be put in unless there is a minimum of fifty phones. Already the requests for services have topped that figure. We on this end of the island have had to live with poles and overhead cables for the past twenty-one years in order for any kind of telephone service. If you ask most people where the lines are few can tell, they just don't notice them. Incidentally, the pole and cable were not there when I bought my house, are plunk in view from both my livingroom windows, but are (were) necessary for the island to have telephone service over the old system.

If we are forced into another hearing with the PUC this could stop all engineering and possibly delay work so long we will have to go through another year with only radio phones. No one in business on the island looks forward to that. I certainly don't want the responsibility of picking up calls twice a day for another year.

It is now Sunday and we have been in a steam bath for several days. The temperature has been in the upper 80's and the fog horn has blown with little let-up for three days. Yesterday a beautiful yawl from Falmouth, MA pulled into the harbor at boat time and I remarked I wish I knew the owners. Well, I do! They are the Hathaways who sang in the choir in Falmouth, so they took me aboard for a sail around the island and dinner aboard. Great fun. On the lee of the island we were almost dead in the water at times. In fact, the tide made us run backwards around by Whitehead. But as we came by Seal Ledges toward the harbor the wind picked up to 20 knots and we heeled over at about 30 degrees. Great sailing.

Mr. Hathaway is a geologist, I believe with the Oceanographic Institute at Woods Hole. Recently I was asked about petroglyphs on Monhegan, similar to those on Manana (which I have always felt were the result of natural weathering and rock fractures). I took him to see markings I have found in various areas on the island and he agreed with me. So we may have to discard the notion that the Norsemen landed here and wrote on stones.

Things are pretty slow on the island. There have been so few people at the Monhegan House the Periwinkle has been closing at 3. There don't seem to be as many Sensitivity Awareness people here, although Mae Cass has assured me the classes are full. But people have been coming out in droves on the boats and the harbor is full of yachts, as usual. But there have been few wind-jammers so far this year.

Anne Hubert is expected out sometime this week and Ed is supposed to follow her after a few days, having made a good recovery. Dan Rothstein is doing well now and Henley is here, at least on furlough, from Togus. Bob Stewart and Mary Stewart Martindell were here for just overnight. Mary has moved back to Portland to care for their mother, Edith Sterling Stewart, now well into her nineties but "bright as a button".

I still have three weeks vacant in the cottage in August, and three in September. Someone asked for a "secluded and quiet" cottage for the second week in August, but I'm afraid my place isn't exactly secluded, right on the Lighthouse Hill Road and the generator out back. But now, with my new pressure tanks and pump control I can manage with only two or three pumpings a day. What a relief. Last year the pump often went on during day and in the night because of the bulge in one of the tanks which caused the pressure to drop erratically, and for me to wonder what the tenants were doing with the water. Apparently it went back down the well!

My next job is revamping the brochure on the Museum. About five years ago I ordered some 5,000 of them printed and had the wrath of the gods fall on my head and now the supply is about exhausted. Carl Hiller will be helping me sometime next week.

At the end of the month Ed Dunn from Cincinnati days plans to come out for a visit after thinking about it for twenty-one years! I hope the island will measure up and that the weather will be good so we can do some hiking.

The Tams have returned to Kauai where they have built a new home; the Werbys again have their house. Amy and children are in the Bates cottage and she is selling produce from her garden. Danny and Barbara are on Manana again. The Horns are back and planning to stay the winter but the Sprouls have gone, back to East Boothbay. The Mersfelders have sold their Green Point property and have the Horn's Hill house up for rent. I believe the teacher will be there this winter, or at Clara Burton's. This may be the last ditto letter. I'm getting a mimeo and have a small copy machine, too expensive for 100 copies.

Box 87
Monhegan, Maine 04852
August 1982

Dear Friends:

A lot has happened since I last wrote, but I don't want to mail this until we have had the hearing before Land Use Regualtion Commission (LURC) which is to be held on the 30th.

We have stumbled along with the radio phone system which works for a while, then gives up. So, if any of you have tried to reach the island and have expected a call-back, we just haven't been able to return the call nine times out of ten. The mails are still the best means of communications at present.

Soon after my last letter I had a very nice visit from Ed Dunn, who had planned to come out for almost twenty-one years. Last year a trip to China intervened, which it should have, but he finally got here in a drenching rain and obliterating fog. I went in and picked him up at the Owl's Head airport, then took him on a museum tour between sheets of water from the sky. We stopped first at the Transportation Museum in Owl's Head with which John Hosmer is so involved, and sure enough, John was there that day. From there we went to the Farnsworth where Indiana had his show, He is the one who devised the LOVE and MANY OTHER SUCH (whoops, I hit the shift lock) contrivances. It was a bit overwhelming and was diluted by a climb to view the three generation Wyeth exhibit on the third floor. Between raindrops we made our way to the Shore Village Museum on Limerock Street, a combination GAR and Coast Guard Museum.

The trip out on the Laura B was in thick fog, so Ed didn't get an idea of what the island looked like. We arrived just as the heavens opened and rode up in my golf cart in a driving rain, reaching the house soaked to the skin.

The weather did clear enough for us to explore most of the island the following week, although we came home from one hike again soaked to the skin. We missed Burnt Head and Black Head, but did pretty well as far as the rest of the island is concerned. I had planned to visit artist studios while he was here, but nature walks took precedence. We did visit the Monhegan Museum, however.

The day Ed left it was cloudy and dismal until we reached the airport, when the sun came out assuring him a good trip along the coast en route to Boston, a scene he missed on the way into Maine because of dense fog. The next day I went to Augusta to attend a meeting of LURC about the site chosen for the micro-wave tower. The upshot of that meeting was that so many people had objected to the site, which is in an area already designated as Historic and Scenic area and a public hearing must be held, which is scheduled for tomorrow. This may mean a long enough delay to put the completion of our full service system beyond the target date of May 31, 1983 and has split the island down the middle between winter and summer residents tempers have raged, the summer people disturbed about the possibility of the radio tower destroying the view of the Lighthouse, the winter people feeling they can't afford to go through another summer with the inadequate telephone system we now have. It is functional only about 50% of the time and no direct calls can be received. Now there is even concern over the trampling of few fringed gentian plants, even tho' they can be found all over the village side of the island, and of birds in migration, even though no significant number of dead birds have been picked up around the Manana tower which is over twice the height of our proposed tower. And so it goes.

Sept. 5, 1982 There is good news, so I'll try to finish this up today. Kim Kenway from the PUC came over for the day today and said that NET is ready to go ahead with plans for using the alternate site for the micro-wave tower, a spot on the edge of the ball field, about half-way between the Lighthouse and the water towers. They have asked local contractors to bid on building the necessary access road and a cement block building in which to house switching and power equipment. They feel this work can be done this fall and the actual tying in of phones can go on in the spring and be ready on May 31, 1983. On the strength of that I think Phoebe and I will start off on a long wintertrek to visit many of you in the southland and west. It will be a good year to do that because there still will be no telephone communication with the island so I won't try to keep up my column while I am away.

August was busy. I spent a week in a course on Forensic Medicine at Colby. The Cooks were in the cottage and their twins were only too glad to take care of Phoebe. Laurie MacDonald and a friend were in my apartment the weekend I left and I expected Elizabeth Gardner almost any day, but she couldn't come.

A photographer from Country Life was here taking rolls of film of me and my photovoltaic system. Yesterday a Mr. Macrow, consulting engineer from Buffalo stopped in to get information of P/V for the house he is building in Florida, Charlie DeCarlo came by, then Betty Keller and Jane Varnum while I was writing letters to the FCC on the phone situation and getting ready for cocktails for Marjorie Young, A.E. Benefield and Carl Hiller. I went off with them and the Goldsmiths for Inn dinner. Tonight Charlie and Natalie Johnson, Leonard and Meredith Johnson Allen will be here for cocktails. Yesterday noon I had lunch at the Inn with Mary Hoxie Jones and Miriam von Steinem. August and early September are social times. Yes, Friday I joined Charlie and Jewel MacDonald at the Odoms at dinner. My refrigerator is far from busy these days. And the freezer is crammed with blueberries brought over by the Cooks from their farm in Otisfield.

Larry Winter and Amy Melenbacker have been painting the house and cottage. They ran into rotten sills etc. so rebuilt and enlarged the Mooring sunporch and will have to put new sills and posts on the south end of the cottage. Next spring I'll have a wood shed added to the barn, new shingles on the cottage and where there are still the awful green ones on this roof. That should keep things in good shape for a few years and let me recover some expenses. What a year!

Anne says Edgar Hubert is coming along fine. Anna Partridge, however, has had a heart attack and John is receiving treatment (chemotherapy). They both have been in the Milton Hospital. Sally Murdock is off one of these days for a few weeks in Plymouth. John Murdock's boat sprang a leak last week when they were out near Cash's Ledge and they had to be rescued by Coast Guard which was able to tow them to Portland, so he didn't lose the boat. Henley Day is back in Togus. June tried to have him here, but he got too confused. Rita broke her ankle the first of last month but has been able to work with some help from Mary Burton in the laundry. Phoebe is doing well on stepped up cortisone and I lost the bridge which was replaced on Thursday so will have to go in to have it fixed on Tuesday. It was first put in in 1935! REACH OUT AND TOUCH US ON MONHEGAN IN JUNE!!

As ever,
Alta Ashley

When I was in for my bridge I stayed with the Gambles again and went with them to an historical society meeting in Union and met the fabulous Ruth Aiken from Friendship.

Box 87
Monhegan, ME 04852
Sept. 25, 1982

Dear Friends:

While the washing machine is going I'll start this letter for Sept.-Oct. Each time I write I am in hopes I can give definite news about the telephones, but we are still W A I T I N G! The LURC decision came officially two days ago in which they turned down flatly the tower site by the oil storage building. Meanwhile more surveyors have come over to map out forty foot square areas between the ball field and the water towers, more correctly, the road leading to the towers.

Monhegan workmen have banded together in hopes of doing some of the construction work, but they have been hampered by having only the specifications for the original site and tower from which to work. Also, the quality controls for cement to be used in the foundation for the tower and the service building appear to be out of reach for Monhegan constructors. Peter Gorman from NET and Bob Burton, who had agreed to head up the Monhegan consortium are on vacation so things are at a standstill right now.

A couple of weeks ago a German TV company came over to do a documentary on Photo/Voltaic Systems. They went all over my place and had me in front of the camera for quite a bit. They were back again on Monday saying the film was destroyed in the developing. This time I had a chance to think out what to say and did it so fast there were several blank seconds on the tape. They filmed, including my set-up, Sandy Dickson's, the Post Office, Alfred and Dot Stanley's, Don and

Linda Stone's, Ruth and Shermie Stanley's and Jamie Wyeth's systems to show the various types and uses, sizes, ect. I wish we could see the film, but I guess we should be in Frankfort this week to view it. How many West Germans will be coming up the slip next summer because of this publicity?

The island seems suddenly very quiet. The Inn closed on Saturday and the Periwinkle closed Labor Day weekend. Ade still has the Rockcrest Shop open on request and Zimmie is behind the counter at the Spa with Karen Wincapaw helping out until the first of the month (or is it the last of October) when she moves to the Post Office as Post Master, unless someone from "away" gets the post. If she is the successor Carl will go out of the trucking business because he can't drive the truck and lobster, too. The Post Office will stay at the Burtons' month by month until GSA finds a suitable place. Too bad they can't move back to the old Post Office, but Chris Nicholson has life tenancy there even though he hopes to spend most of his time with the Purintons in Florida. They are due here anytime now to take him back with them.

Tom and Irene Martin are due tomorrow for their semiannual bird photographing session. It is good they are coming early because the fall migration is well along already. Hawks, including peregrine falcons are everywhere. Yesterday one with a warbler in its beak came down in front of me on the road, got a better hold on its prey and flew off to feast.

The Balmy Days quit her run on Sunday and on Monday we began our fall schedules with the Laura B coming once a day, no Sundays, and leaving at 11:30 instead of Noon. Victor says he'll stay open as long as there are enough requests for reservations. The Yew will probably close October 11th or 12th. Martha and Richard will keep the Tribler open until water goes off; Sara and Lexi are taking roomers all winter as is Barbara Hitchcock. I'm closing up on November first, the cottage that is. I'll probably stay here until Christmas, after the bird count. I've asked Davis Finch and Cleo Misfud if they can come before instead of after Christmas this year so I can shut up both the house and the cottage and GO, I don't know where as yet.

The Horns are going to stay here on the island this winter. The teachers have the Mersfelder place. Craig and Judy probably won't be back. Surveyors were out to map out their property. The Cundy house is still up for sale. Larry Cooper's relative from Alaska is now living in the Mersfelder place down at Green Point, provisioning it by boat.

He bought both houses last spring.

Irwin Nelson came today to close up and take Cyril home. C. has been here all summer and alone in Three Chimneys for the past three weeks and doing well with oxygen only for short periods in the night. I can tell when because the generator goes on to run the machine. Nancy Kordak Bergelt came today to help Vi close up the Log Cabin. Harry T. Bone has been here all summer but expects a call back to his ship any day now. He dug out under his shed, raised the building about a foot and made a storage area underneath. The view from the top windows is better than ever.

Jimmie Barstow is giving a big party today (Saturday) with bands and what not. Rumor has it that three bands are coming, but take that as you will. Supposedly all the people in shore who had anything to do with refurbishing the Laura B are invited, plus some islanders. I haven't had an invitation but perhaps it is a blanket one. I may mosey down to the Red House later and find out.

Tom Martin arrived today without Irene who, he says, developed phlebitis and couldn't make the trip, 'tho she hopes to come next week. Doug and Harry moved out of the fish house for them and will be in the Underhill house when they return from Doug's Massachusetts reunion tomorrow. The Nesbitts left the Mooring Thursday and Jean Hayes came in today. I just bought some new material for fresh curtains for the master bedroom in the cottage and hope to have at least one pair done while she is here.

Sandy Barker Davis is having a birthday today or tomorrow. Her new husband, Bud Davis, as well as her son Bill and his son Bill, are here for the occasion. Ralph Hinchman and Lulu Lange are here for the first time in ages. Sarah and Cot of the Mooring Chain arrived today. Ruby Court and Kiki Winship came over the first of the week so the island is filling up for its last few weeks before chill winds drive everyone away. The Haussers left today after spending two weeks working on their house which is just beautiful. They have practically rebuilt the entire place and with wonderful imagination. They said that Penny is taking a course in Public Health at Yale.

Chris Nicholson is on the outs with Doug and Harry so they won't be staying in the Old Post Office this winter. As yet Doug hasn't obtained permission from the Planning Board to build his house so they plan to stay what time they are here in the winter up at Harry's - the Underhill house. It is too cold in the fish house and they don't want the Mooring sans water. Danny Bates hopes to go up country where

Barbara Johnson is in charge of a guest house; Amy will be living in the Vaughan cottage down by Swim Beach entrance. I guess that is all the news I can give at this time. The party was quite something. There were two combos, Tom Judge's and Craig Rogers'. Among the notables present were Lou Colby (with whom we watched man's giant step onto the moon down at the Store, remember Elizabeth?), Bob Campbell and many others I didn't know. Most were people who worked on or helped with the Laura B. Sonny Remick and his son, in marina business in Rhode Island, came too. The air is full of bluejays.

Enough for now,
Alta

Monhegan, ME 04852
October 28, 1982

Dear Friends:

It doesn't seem possible that it was more than a month ago that I started the last letter. I want to start this one now, although I'll hold it open until after the Hallowe'en-Retirement Party for Winnie, which will be on Saturday, followed the next evening by visits from hobgoblins - there are sixteen this year, not counting some adults who may want to use their party costumes twice.

The island is pretty quiet now. Nearly all our human robins have left. The water will go off early Monday morning, driving most of the rest onto the decks of the Laura B.

The biggest news at the moment is that construction on the new telephone system **has begun**! People from NET, Harriman Associates (architects) and Colwell Construction came out on Monday to discuss the blue prints. Later three men from the construction company came out on the Laura B, along with some equipment which was taken to the site and work of clearing the area was begun shortly after lunch. The plan to use the wood and slash for fuel for their construction building on the site (later to be removed). We were all rejoicing and I sent write-ups to several newspapers. Then yesterday, when Jimmy came in with the Laura B he said that the fully loaded barge, which he had planned to bring out the day before but couldn't because of high winds, had sunk at the dock in Port Clyde. On it was a load of gravel for the foundation, cement, a bulldozer, and other equipment. They were able to get the bulldozer ashore and wash it down with fresh water, but it remains to be seen how much it was damaged, or how much the accident will delay the project. About every available barge is now in Portland being used for the new BIW drydock.

The construction company elected to use my cottage in which to house the workmen, so they moved in Monday afternoon. They are three nice young fellows, are cooking and keeping house, using sleeping bags and their own towels, so there is little I have to do. They expected to be here four nights a week until mid-December, but are going in today (Thursday) until needed materials and equipment can be landed on the island, hopefully sometime next week, but maybe not for a couple of weeks, which will knock my time-table, and theirs, into a cocked hat. With my luck in renting the cottage this season I guess I should never have undertaken this. Misfortune seems to have dogged my tracks ever since last March.

But there have been some bright spots, although at the time I was a bit overwhelmed. First John and Rue Riddle from San Francisco came out a day later than I had expected and stayed a day longer than I had hoped they could. The Saturday they left we had a horrendous storm and I begged them to stay overnight, at least, not knowing that the Kellys accompanied by Elizabeth Gardner were aboard and that Jean Hayes, who had been renting the Mooring, decided to stay until Monday. We managed. The Kellys had the lower room, Elizabeth took my room and I my winter bed in the new living room. It all worked out well. In fact, one or two more singles could have been accommodated in this part. Too bad John and Rue didn't stay over, because they had a terrible trip. The Laura B took two seas right over the pilot house before she passed Nigh Duck; and half-way to Burnt Island the life raft blew off and had to be retrieved. The trip took two hours and nearly everyone succumbed to mal de mer, except John and Rue, intrepid voyagers!

Elizabeth left on Monday, the holiday, the Kellys and I went on Tuesday to Tuftonboro after being the speaker at the Knox County Medical Society dinner meeting at the Sail Loft, giving stories of the good and bad experiences in medical training and practice. I could have talked twice as long, but the first game of the world series was on and everyone began to get restless. Fortunately I had managed to get my hair done and teeth "scaled" before the meeting and Phoebe to the vet's for flea spray. Our Laura B trip was great, in warm sunshine over a following sea, but Jimmie did take us to the westward of Allen's Island because the surge through the channel between the Georges was so great.

From that angle the surf by the bent spindle was spectacular.

Instead of staying only overnight in New Hampshire I stayed three days. We organized a cocktail party the first afternoon and everyone made it except the Phelps who had gone shopping for the day. It was good to see Thelma Phillips, Lois and Ted Drowne, Ginny and Howard Huxtable, Margaret and George Pettey, especially after such short notice. The autumn colors were just beautiful all along the way, but particularly along the coast where they were at their peak, passed to some extent in New Hampshire.

On Saturday I was back on the island and on Tuesday, Nancy Peters and her roommate at Wheaton (back in the forties) arrived on the Laura B for an overnight stay. The weather was perfect so we set off for White Head. A couple out there wanted to go to Cathedral Woods, so I led the way down to Little White Head, along the Long Swamp Trail, through the woods and home. Going down the side of White Head was almost impossible for me, but the others did well. We found the witch hazel trees on the Red Ribbon Trail on the way out to be in full bloom, more than I have ever seen them. Hobble bush leaves had turned and several maples were also in full color. Mount Washington graced us with its presence that evening. I hated to see the girls leave the next day.

Early in the month I gave a reception for Doug and Betsy Curtis. He is running for re-election as State Representative. I thought maybe five people would show up but over twenty did. My munchies had to be spread thinly! Two Sundays ago Jackie and Julie Hudson, Zimmie Brackett, Rita White and I joined Jane and Will Curtis and their house guest Ruth Bailey for our now annual fall picnic. As usual, the wind was too cold and brisk for us to picnic outside so we all gathered in the Curtis (Pitkin) cottage and feasted on a delicious spread. Then we hiked out to Jackie's studio in Calf Cove where she showed us the several paintings and lithographs she had there. The view from the studio is beautiful, from Manana to the Georges.

A few weeks ago I ordered a small tractor for the garden. It arrived, fortunately, when Phil Baker and his son-in-law Todd Gifford were in the Mooring so they set it up. It is now in the barn waiting for Bob Boody and proper tools to check the engine oil level. Now I'm wondering if I can get it up to the garden and through the fence gate. If so, there should be no excuse for clumps of grass scattered about and my yield should improve over this year's, which was terrible.

Lance Burton decided Albuquerque was not for him so he set off on his motorcycle and made it cross country in five days. He is now at Kent's Hill where Chris Rollins is; Jeff Rollins is at Hinkley, Zoe Zanidakis is at MCI! Doug Rothstein is back at Brewster, Margaret at Stoneleigh-Burnham, and Timmy Boehmer at Proctor. Kathy Wincapaw is sailing through Morse High in Bath.

Rickie Boehmer's book "A Foraging Vacation" is off the press and she and Peter are off for a six week trip to Japan, all expenses paid, to gather information about fishing and the use of fish and sea vegetables. How does she do it?

As many of you know, Winnie is stepping down from her postmaster job on October 29th. Karen Wincapaw has applied for the position and will be acting postmaster until the permanent replacement has been appointed. Needless to say, we all are rooting for Karen. The post office will stay at Winnie's for the time being on a monthly basis, but the postal service is scouting the island for a suitable, or even just an available, spot for a new building. Let us hope it is near the center of town and not at Lobster Cove or Green Point! It would be nice if the previous post office could be reinstated, but Chris Nicholson has that for life tenancy and Jeanie Lord is reportedly not interested in selling it or renting.

Jimmie Barstow says the Prock Marine will take over the job of barging materials and equipment to the island. The other barge sank by the Fish Co-op pier in Port Clyde. How lucky it wasn't en route to Monhegan where it could block a channel or endanger the Laura B as it sank in tow. There were 60,000 yards of gravel and a front-loader, no other equipment. The loader was washed down after being lifted ashore, extent of damage not yet known. It was Prock whom the Monhegan Company had hoped to use and were turned down because of cost!

It looks as though I'll be going back to East Falmouth after the first of the year, although I still want to take a trip to the Galapagos Islands and Equador in February. I really must get on with my Monhegan history and find a publisher. Nancy has been editing some of my write-ups, which is a great help.

The nose count for winter on the island is, as far as I can see, 76, counting Harry T. Bone and Billy Payne plus Mort, Doug, Harry and me who will be off for a few weeks, plus 10 away at school and college. There will be nine in school here and seven preschoolers, all of whom will be stopping by for treats on Sunday. I'm going to pop corn and offer apples as well - no razor blades etc! The winter count doesn't include the four of five Coast Guard or the construction workers.

Last week there was great activity on Lighthouse Hill. Three men were busy clearing the site for the telephone system, Coast Guard men finished up insulating the generator building for the light, and Larry Winters worked on the inside of the museum. All is quiet now until the telephone workers return next week or the week after, whenever the front loader and gravel can be delivered to the site. Stay away storm!

Sally Murdock returned from her Cape Cod vacation and took several paintings in to be exhibited at the Courier-Gazette in Rockland. My friend, Louise Webber of Augusta, is to have a show at Harlow Gallery (KVAA) in Hallowell all November and there will be a reception for her on Sunday, the seventh, at the gallery. So Phoebe and I are planning to go in on Wednesday, go down to Concord and return for that reception Louise (and I) is one of the twenty or so charter members of the Kennebec Valley Art Association formed back in 1956.

Winter season is approaching. Fishermen are busy with their gear Don Cundy has brought in two loads of bait, and all are supplied with salt. Soon the island will be redolent with the pungent odor of bait as bait-bagging takes place. There is less hammer tapping this year than usual because most have gone in for wire traps. So the wooden trap will take its place with wooded buoys and glass toggles in this modern age of plastic. The wire traps are plastic coated.

Bags of coal have appeared on the dock all week, fire wood came in earlier. I still have a good supply left over from last year, surely enough to get by this winter if I take off eight or ten weeks.

Danny took the children in this weekend but will be back. He is not going north for the winter. Doug and Harry plan to lobster. Bryan Hitchcock has moved to Richmond and is continuing his aerial photography while Barbara is turning her house into a place for winter visitors. Margie is here, as are Bill and Nancy Thompson Brown (Shumaker cottage) and a few other guests so the party at the Inn, really for Winnie, will have at least 60 adults, even though Doug and Harry are down on Prudence Island for the weekend and Harry T. and Billy Paine aren't here. Everyone on the island is planning to attend, costumes or no. There will be sixteen hobgoblins, including guest children, Sunday, after the school play.

Sunday. I've just finished resetting my chime clock. It didn't occur to me I could stop it for an hour until I was through listening to 11 hours of quarter hour chimes. At least I was able to study the chimes and found, at this late date, that there are five which are chimed in order. Always I had thought there were only four and each chime started with the same pattern!

The party last night was a great success. It was billed as a dinner dance for Hallowe'en, but was really a celebration of Winnie's retirement. She was taken completely by surprise and was literally moved to tears. There were all sorts of munchies, drinks, entrees and a huge carrot (or punkin?) cake. Music from a hi-fi was excellent and the dancing enthusiastic. Only a handful of people on the island failed to attend. Several guests, among them Dave Sleeper and Rusty Mallery, Margie, and Lisa Dickson, swelled the number to over sixty. Even two year old Joshua Freeman was there for a while, and Alice wandered around underfoot. People said she was Phoebe in an Alice suit and wanted to award her the prize for the best disguise! That probably should go to Margie who came in a KKK outfit. I thought the person underneath had blackened his/her face! I resurrected my July 4, 1976 costume and found my waist was four inches larger now!!! That I can't understand because my weight is the same or less, but the rolls are there. The traditional retirement gift was a dainty digital wrist watch with which Winnie was very pleased. We'll miss her behind the P.O. window. Best wishes to Karen in her bid for the position!

Rather than keep this open until after the school play and hobgoblins visits I am going to close so envelopes can be stuffed and I can get these letters in tomorrow's mail. We start out three mails a week schedule tomorrow.

Last night Dave Sleeper, probably in jest but maybe not, asked if I'd like to go with his company to Colombia in January on a Solar Energy conference. So, who knows where I'll be this winter or what I'll be doing? Probably I'll be watching the swans on Eel River and singing in the Falmouth (Mass.) choir. At any rate, I shan't be leaving here until probably after Trap Day, except for short stays off island for Thanksgiving and Christmas. But Colombia and Galapagos are calling! At least I'm sure I won't be trekking through the country. Maybe next year. So, all good wishes that you have enjoyed Hallowe'en and that you will have a fine Thanksgiving.

With that Phoebe and I will sign off.

As ever,
Alta

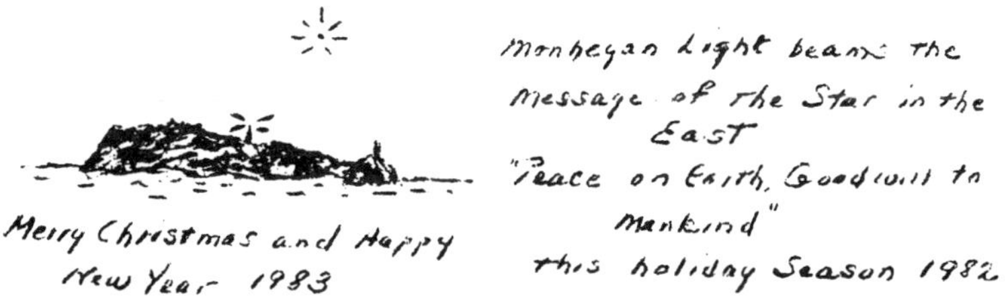

Monhegan Light beams the
message of the Star in the
East
"Peace on Earth, Goodwill to mankind" this Holiday Season 1982

Merry Christmas and Happy
New Year 1983

Dear Friends:

Winter has descended upon us at long last with temperatures close to zero, after a week or more of springlike weather. Christmas is but two weeks away and no further delay in getting out this letter can be had, no matter what.

1982 has been a busy year. It seemed as though each day brought a visitor, another need for letters concerning the telephone system, or an interview about photo-voltaics, and desperate effort to pick up "Brownie Points" (C. Med. Ed. credits) by pouring over journals or taking in a symposium, or lecture on the mainland.

About the middle of January Phoebe and I took off for what we thought was going to be a relaxing time and a chance to catch up on writing and editing materials for my Monhegan history. We were to house-sit for my niece Nancy Mead Peters in her cottage in East Falmouth on Cape Cod for a couple of months. The cottage is on Eel River so we watched the tide come in and go out bringing swans, geese and ducks, until it froze solid early in February. A heavy coating of snow kept it frozen for much of the time we were there. Cape Cod didn't prove to be the semi-tropics I had jokingly said it would be. But it definitely was better than Monhegan with its sustained below zero weather and worse chill factor, which did a job on my plumbing. At least twice I travelled from Falmouth to the Cape Cod Hospitals over

Digging for telephone tower bed rock — none found!

Looking over plans for telephone service building and tower

icy and snowy roads to attend lectures for my CME's.

When I left the island I thought the telephone situation would need no more help from me, but almost immediately things began to happen, needing telephone calls, innumerable letters and a trip back to the island in early February for a special Plantation Meeting to vote on a Stipulation of Agreement which had been drawn up by Plantation and New England Telephone lawyers because of the delay in the hearings examiners decision from Public Utilities Commission. In the agreement NET would give us full service micro-wave system if we would furnish a suitable site for the micro-wave tower and building to house necessary equipment. Plans at that time were to use the Lighthouse for the micro-wave dish and the little oil storage building close by for the service building. The Plantation accepted the agreement and a great load was lifted from my shoulders. Work on the system was expected to begin in summer or early fall.

Phoebe enjoyed Falmouth. She fell in love with the Scullys' Archie, who came over to play with her every morning, sharing bones and dog biscuits. I joined the choir at the Congregational Church, led by Mr. Heremann who, incidentally, was the organist for the Baluchi funeral, too. It was great to sing good music under such fine direction. I was tempted to stay for the spring concert and for Easter but telephone business began to pile up and I returned to the island in late March.

Our return in mid-March was a great mistake. We were greeted with frozen pipes and, what was worse, frozen drains, of which we weren't aware until everything from upstairs came out the toilet downstairs! And the dishwasher spilt its contents all over the kitchen floor. It took Bob Boody about a month to straighten things out - new water tank, leach field, and generator. Paul Horn has just installed a gas heater for the cellar in hopes of keeping the pipes happy this winter.

Class reunions played a big role in this year's activities. I went off to South Hadley for my 55th, gave up going to my 50th at Cincinnati and won't take my 45th at Vanderbilt. As usual it rained, but we did have a good time. Hallie Patey Long came out to the island with her sister Phillippa on hearing more about the island. Fred and Scotia Waite came by after he went to his 50th at New Hampshire and my cousin Jean Woodbury Johnstone and her husband Ben made their first visit to the island en route to Canada after her 50th Radcliffe reunion Ed Dunn spent nearly a week here after promising to do so for many years, and my cousins Rue and John Riddle paid their first visit to the island when they came on to settle things at the Farm in October. My niece

Nancy Peters liked the island so well on her first visit last fall she brought out her college roommate, Mary Pierce, for an overnight stay. Elizabeth Gardner, Joe and Elie Kelly came out the day the Riddles left, so for a while I felt I had wall-to-wall company. But it was great seeing everyone and introducing new-comers to the island. Fortunately most had good weather, although John and Rue took off in an horrendous storm which sent seas over the pilot house of the Laura B before she had passed Nigh Duck.

It seemed that every week or so there was someone wanting to get a story either on the telephones or on photo-voltaics. A TV crew from West Germany came out twice to film the P/V story and Mary Denison, who worked at the Inn this summer, saw a movie on photo-voltaics when she was travelling in Europe and nearly jumped out of her skin when several shots of Monhegan were shown and she saw me telling how great it is to take power from the sun - all dubbed in in German. The people she was with couldn't believe she knew the people and places shown. Little Monhegan is coming in for a lot of publicity between our electricity and communications problems. Sleeper has now installed fifteen systems on the island with more to come. A third of the year round cottages are so equipped already. For the past few weeks, however we have had to use our generators it has been so clousy and drismal, as Carl DeSuse says. Have you seen the story in Dec. Country Journal?

In June the Monhegan Associates Trustees accepted the Stipulation of agreement concerning the phone system, agreeing to lease the oil storage building and land around it for the tower - NET had decided against using the Lighthouse for the antennae because they felt two would be needed. As summer people came on the island much controversy arose when easements had to be signed for the tower site as well as for erection of poles and stringing of wires. The lighthouse area had been listed in the roster of Historic and Scenic Sites, which brought in Land Use Regulation Commission and the Commission for Preservation of Scenic and Historic Sites necessitating several hearings here and on the mainland and a delay of construction until a second site, off the southwestern corner of the ball field, was approved. Finally, on 10/25 a crew came out to clear the site. A barge loaded with gravel and a front loader to be used in excavation was delayed by storm a day, then sank at her dock in Port Clyde on 11/2 and another barge didn't get here until the tenth. Fortunately cold weather held off until Dec. 10th so that footings and the foundation for the utility buil.ding and the footings for the tower were poured and two-thirds of the brick

Telephone tower completed

Telephone poster by school children — May 1983

facing of the building were set by the time near zero temperatures arrived. According to the construction crew work is progressing on schedule and the target date for completion is still May 31,1983. Then you can call the island and not have to rely on the radio phones now in use with almost no possibility of receiving in-coming calls at present.

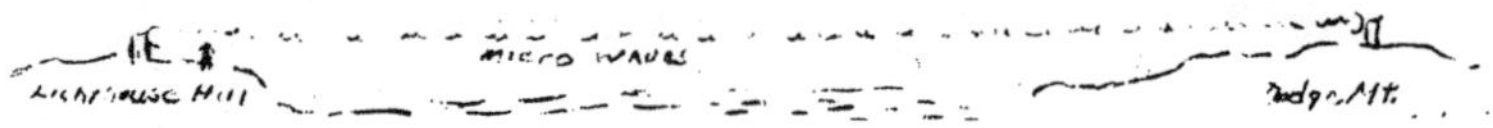

You'll be able to "reach out and touch someone" on Monhegan after 6/1/83.

No longer are dogs welcomed at the Samoset, but Fred and Irene Gamble in Tenant's Harbor have come to our rescue. By now I think they probably think they have welcomed the "man who came to dinner", especially when we stayed a week, instead of overnight last Eastertime because of the big snowstorm in early April. It has been wonderful to be able to stay there, meet their friends and join in their family festivities. Now that the Farm in Tuftonboro is closed up, we have been spending more time with Elizabeth in Concord. We spent a week at Thanksgiving and plan to be there for Christmas, bird-watchers, construction workers and weather permitting. The Monhegan Construction Company lost out to Colwell Construction of Augusta for the telephone contract and the crew is staying in the cottage for the duration. After they leave NET personnel will come in, so my hopes of going to Falmouth, the Galapaggos or even Colombia in January and February look pretty dim at present.

Now for some island news. The Sproul cottage has been sold with Bill Sweetland having an eight year rental option. David and Katy Murdock Boegel are living in the "shop" bought from the Cundys, that house is still for sale. David and Katy were married in June, the day we took Grandfather Gil Lothrop to Pen Bay Med. Center, where he died two days later while waiting for discharge orders. Clara Taylor died during the summer at her home in Port Clyde after a long illness. Paul and Barbara Horn are back on the island as year round residents, but the Mersfelders are in shore for the winter; their cottage is being occupied by the Freemans, our teachers. Both Debbie Bates and Anne Court have moved off. Courtney Day's Deronda is the newest boat in the fleet, but with Hitchcock, Sproul and Mersfelder no longer fishing here there will be only twelve boats, as far as I can count.

All last winter we rode the Prudence A or Penobscot Gulf (looking for all the world like a spy ship or smuggler) while the Laura B underwent extensive repairs...When she finally went on the run in late May she looked beautiful - two new heads, refurbished lower cabin (all set for cocktails on moonlight sails). She was welcomed with open arms by all islanders and summer folk.

Winnie Burton retired as Postmaster in late October. A combination retirement part and Hallowe'en dance was held at the Inn with just about everyone, plus Dave Sleeper, Rusty Mallory and other visitors present. Rusty is now working with Dave. They recently put up a P/V system for Lexi Krause - I believe the fourteenth for the island.

Lance Burton left school in Alburquerque and is now at Kent's Hill and happy to be back east. Bob and Mary planned to stay here for the winter, but when Bob wasn't needed for the phone construction they decided to take off for New Mexico and left last week. We'll miss them.

We'll have our community Christmas dinner and tree celebration on Sunday afternoon. The school play will be held on Saturday. Having everything on one day was too much, especially for the little ones. So, on that thought I'll close with another wish for happiness and health for you and yours for the coming season and new year.

Alta Ashley and a tail wag from
Phoebe Starling

Monhegan, Maine
December 12, 1982

Box 87
Monhegan, Maine 04852
Jan. '83

Dear Friends:

HAPPY NEW YEAR! (A week late).

As many of you who read the local papers already know, Trap Day was New Year's Day and happened with almost no fanfare. No traps were moved to the dock until Friday, but all that day trucks labored up and down Dock Hill until there was scarely an inch of surface left on the dock. In fact, no aisles were left for anyone to reach the bollards on which, to hang Laura B's lines when she came in that morning, which caught Bob Campbell by surprise. He has been captaining the Laura B since before Christmas while Jimmy Barstow was in Rochester visiting with Judy's family and nursing a bad back. Jimmie was back on board yesterday well recovered.

Conditions on New Year's morning were just about perfect. There was a coming tide, full at noon, no wind and a calm sea. Some of the younger fishermen wanted to start at 12:01 a.m. in the light of a full moon shining in a cloudless sky, but saner heads changed their minds at a "lobster meeting" the evening before. But, rather than waiting for all boats to be loaded and ready to leave their moorings before opening the season it was decided that the signal would be given at seven a.m., no matter who was ready, since there was no reason for anyone not to be ready by that time.

Many boats were loaded late Friday afternoon and fortunately all survived an unpredicted snow squall which passed through the harbor in early morning, probably because most traps are now of the wire type which can withstand wind pressure better than can the wooden ones. Things went along well and many boatloads of traps were taken off the dock by noon. But, just as the tide was of ideal height a brisk wind sprang up making a heavy surge at the slip and slowing the loading of the boats, as well as making trap setting difficult. By mid-afternoon several of the boats returned to their moorings and called it quits for the day.

It took only a few more days for the dock to be cleared, even of second loads of traps brought down on Sunday. Now, the following Saturday, there are only a few "trailers" left on the dock and scarcely any traps can be found on the island. Gone are the great piles one sees all summer which form many a background for tourist photographs.

Charlie MacDonald was here to help Doug and Harry. He got a big kick out of his spending three days at hard labor trucking traps to the dock and loading the boats while on "vacation" from his job as vice-president of a bank. We finally had our ritual "first-of-the-catch" dinner on Tuesday night, and did those lobsters taste good! I took some in to the Gambles on Wednesday and we had another feed that night.

After giving up hope of getting off for Christmas because of being loath to leave the cottage too long with the telephone workmen in it, I was able to go in with Lexi Krause on Thursday and reached Concord just as the snow, first encountered in Haverhill, began to accumulate. The Sutherlands came up on Christmas Day, but didn't arrive until evening so David, Tommy and I drove over to see Aunt Helen at the Baptist Home of Massachusetts near Lake Street. She is very happy there saying "it couldn't be nicer", which is a great relief to us all I'm glad we didn't try to get her to Concord for the day, she doesn't want to go out anymore. She'll be ninety-seven on the 19th of this month. Maybe we all will want to stay put if and when we reach that age.

Christmas was one of the pleasantest days I have spent in a long time, due in large part to the clever "stocking stuffer" presents David managed to get for everyone. Christopher and Ian Sutherland acted as Santa's helpers, insisting no gift could be passed out before the preceding one had been opened, shown and duly acknowledged. It was fun, particularly when some bore appropriate and funny messages.

Our island festivities took place the week before Christmas. The school "play" was presented on Saturday and dealt with both the nativity and also Chanukah, as well as the story about the writing of the carol "Silent Night". At the end the children read their letters to Santa, which they had not expected to be made public. Is was interesting that younger children asked for things for themselves - 65 foot boats with 85 hpw motors, no less - while the older children asked for peace and an end to nuclear warfare madness, very touching.

The community dinner and Santa's visit came on Sunday. Having everything come on the same day has been too tiring for the smaller (and older?) members of the community, and the dinner had to be held after the play because of the scenery, which could not be in place at the time of the dinner. Courtney was Santa, readily spotted by Sammy.

There are twelve boats in the fleet this year, plus skiffs and dories from which Mary Beth Dolan, Danny Bates, Victor Lord, and one more person I can't at the moment recall, will tend their few traps.

Wednesday I went in and picked up my new glasses, the "veriflex" lenses which have no demarkation. I hardly know I have any glasses on. It is supposed to take up to three weeks to become accustomed to them, but I found no trouble after the first few minutes. Great! I wish I had had them years ago.

The telephone situation still has us guessing as to what next can happen. Why did Murphy ever invent his law? I haven't heard just what will be done about the tower in shore. Here they have dug down at least seven feet and have given up trying to find ledge - the only place on the island with that much soil, we all swear. Mild weather has been with us most of this fall and winter, ideal for pouring cement, and none can be set until the engineers come up with revised plan. The constuction crew hopes to come pack with a solution to the problem on Monday. Peter Gorman hopes to get over again on Tuesday. He says already 54 requests for telephone services have been received to date with down-payment. We should have our numbers assigned soon. Electricians are due in mid-week or early next week so there will be no chance for me to get away to Falmouth, the Galapagos, or even Columbia. Too much goes wrong with my equipment when I'm not here and the workmen are in the cottage. So far there has been no talk of having to postpone the target date, but there is Mr. Murphy leering at us.

Patsy Farrell is back after spending Christmas with Pru and Don. Don is here right now doing chores about her house which is being kept warm by one of my big oil stoves. I'd like to get rid of the other and buy a smaller one for downstairs, setting it up on the other side of the chimney so that the room could be utilized to better advantage, as well as having better circulation of heat to the upstairs. The gas stove in the old cellar and an electric heater, which goes on only when the generator comes on and doesn't keep it going, heats the new cellar where all my P/V system batteries are. One of these days I'll have things working right, and then I'll decide not to stay here in the cold weather.

There is no frost in the ground. Today I picked parsley not cropped by the deer. But a NW wind is blowing and mercury is expected to fall into "the single numbers". We have had two or three really cold days and nights, but then have come 40-50 degrees days, and the phenominally warm Christmas weather in the 60's. A funny winter so far. Lobsters seem to be fairly plentiful, but small. Doug brought a lot of crab claws last night which I picked out, getting a bit more than a pound of meat for an evening's work.

I overshot the previous page, so several copies are missing the last line "of meat for an evening's work". The blurred word is "lobsters". Sorry about that.

There is a little other news to relate although rumors as big as eagles keep flying around. While I was in shore last week I did some more dickering for a copy machine. The big Sharp I bought last January in Falmouth has been bought by Rockland Business Machines at last. But the small copier was found to have a defective part which is not readily replaceable, so now I'm hoping for a desk model Sharp which doesn't need special paper and can print on two sides as well, a great savings, **if the thing will work without going into tantrums**. I have a big basket of material ready to be copied just sitting there for months, putting me way behind in filing news items in the albums I am making for the library and my own files.

Thank you all for your wonderful notes included in your Christmas greeting and contributions to the cause! May '83 be a good year for all of you. Phoebe and I are doing well, she is even back to jumping up on people, a thing she couldn't do when her hips were weak last winter. And I thought she had finally learned to docorous in her greetings!

Don't forget to put Monhegan on your agenda for '83.

As ever,
Alta

P.O. Box 87
Monhegan, Maine 04852
(207) 594-2158
May 7, 1983

Dear Friends:

Here it is about time for another of these letters. As the end of this month draws near we are growing increasingly excited about the approach of the time we can make a satisfactory call out and even receive calls directly, something we haven't been able to do since Sept. '81. I wonder how long it will take us to cut off our telephone bells! Most of us have had our phones installed, some of us, including me, have visited the phone center at the Maine Mall and picked up our new phones. Mine is a trim-line with a long cord between the base and the receiver so I can wander a bit around the room while talking. We can't have push-button phones because the Rockland exchange isn't set up such.

Wednesday. I went to a special meeting of the Lincoln County Republican committee to help nominate a successor to Helen Weston who recently died. Someone proposed my name! But it didn't take long to convince people I couldn't fill the bill very well because of travel difficulties & lack of phone service. Thank goodness.

Yesterday I had lunch at the Methodist Home where Mother had an apartment and where I stayed one winter. How glad I am I had enough sense to move out. I used to think that such places were great for older people, but to me they are deadly, unless you just can't fend for yourself, then they are fine, if they have some kind of activities to keep the mind oiled, which few of them have. In the afternoon I went to Emma Field's funeral. She seemed to recover from the fall she had last month, but took a turn for the worse on Sunday and died Tuesday.

It seemed as though all April I planned to go ashore for one thing or another - medical meetings, dentist and doctor appointments, driver's test, shopping and what have you - and each time some New England Telephone official would plan to come out just that day. Fortunately they often chartered the Laura B for the day or for special afternoon trips so I went back and forth frequently without having to pay a fare. Not bad, because the fare has gone up another $2 round trip.

A week ago Thursday I went in to a briefing session in Bangor for our People-to-People Mission trip which will leave from New York September 17th and return to San Francisco October 8th, a round the world trip! I'm so excited, even more than over the phones. When I picked up my car at Ivan Simmons' I found him ill with the "flu". That night I drove from Bangor to Waterville, had a breakfast conference with Peter Gorman, our NET man, and went on to Portland for another medical meeting.

That night I picked up Dave Sleeper and we joined Phil and Ginny Baker at dinner at the Galley at Falmouth YACHT Club, great fun. Phil and Ginny have the cottage in October, and were the ones to bring Dave and his photo-voltaics to the island. When I returned my car I found Ivan looking awful, took his pulse and realized he was having heart problem and sent him off to Pen Bay Medical Center where a team met him at the door. They put him on an emergency pace maker, IV's and the works. When I saw him Thursday he was much better, with the pace maker only on stand-by, but he had a close call. I almost didn't stop in that day to check. Had I not he might not be here now. I can't get along without Ivan!!

While I was at the hospital Jim Elliott, who is a Kennebec Valley Art Associ. member and once headed the Portland School of Art, came down the corridor so we had a good chat. His wife is the daughter of Ralph Cline, Andy Wyeth's "Patriot", who lived in Spruce Head. He supplied my first load of wood. It is such fun to live in Maine. One makes so many contacts it makes you feel you know the whole world.

What a spring we have had. It has rained for the last six weekends, plus most of the days in between. My batteries have taken a beating, especially since I have had telephone people in the cottage since last October. The bunch that has been here since early March goes home every weekend, leaving me with all their bedding plus a turkish towel daily for each person. That has meant draping the wash around my rooms to dry (I have no dryer). Today the sun is shining **for the second day in a row, a record.** So the sheets are out now and towels will follow shortly.

I ordered a TV color DC/AC the other day from Sears. I had to put a different end on the cord to fit my outlets. The thing ran beautifully for two days, drew hardly any amperes, then the DC side quit. I lugged the thing in, but didn't take in the DC cord, thinking Sears repair shop would have one. They found a loose antenna wire, said that was the problem, so back I brought it yesterday and it still doesn't work! Dave came out in his boat yesterday but left early today so I haven't been able to get him to check this machine out. I found he handles the same model (and for less!!).

Right now I have two telephone poles side by side in my front yard. I can't get my golf cart out of the barn until they take the old one down. Let's hope they haven't forgotten it. Most of the phones have been installed, but new orders keep coming in, more than sixty now.

A painted bunting has had everyone excited. I haven't seen it nor have any orioles come to my oranges, but Rita had Baltimores and orchards, plus a dicksissel, summer tanager and black throated sparrow. Why are we having so many western birds, I wonder.

It is now May 10th. The wind has died down so **maybe** the barge will come out with gravel for the roads, lumber for dock repairs and to take back an enlarging amount of metal trash - refrigerators, old culverts and anything else you can imagine, plus surplus telephone wires, cable and conduits - which is taking up about a third of the dock area. It seems that everyone has a truck and all jockey for position on the dock at boat time. The two telephone poles in my yard keep me from taking my cart to the dock!

People coming out to the island this spring have accepted the tower with little comment. Vince Dostie came out yesterday to see about the erection of the chain-link fence. Many of us have suggested planting barberry bushes along it. They will soon grow to cover it and prevent anyone's climbing over, also they will look beautiful in the fall.

Teco came out a couple of weeks ago but had to be taken back that afternoon because of shortness of breath. After a few days at Pen Bay he returned to Portland. Zimmie hasn't come out yet. He had eye surgery and is now supposedly doing well. John and Lynne Hultberg moved onto the island a couple of weeks ago as permanent residents. No one has heard from Newt since he left late last fall. Anna and Laury Cooper are here. Laury will plant the Lewis' garden with the help of my Troy built Pony tractor which I haven't yet had the courage to use. Wilma and Ernest aren't returning until July, too late to put in the garden themselves. Winnie Murdock is back in Reading awaiting her baby. Her pregnancy has been complicated by the appearance of a cyst which is growing about as fast as the baby. She is due in early June.

Joe Pommerleau and Amy Melenbacker will be doing the garbage pick-up this year, Steve Rollins will have the garbage boat as usual. Carl feels he can't truck now that Karen is in the Post Office full time. By the way, Kathy is valedictorian and has earned a scholarship to Colby College, for which everyone is very glad.

Work on the telephones is winding up. All cable has been hung or laid. Most new cable has been buried. I wish some of the old could have been too, so I could be rid of the pole and cable which go by my windows. But NET feels I've lived twenty-two years with them so I can go on doing so. All phones will be installed by tomorrow. Only the very time consuming work of setting up the switching relays remains to be completed. That was delayed first by non-delivery of the equipment from Port Clyde to Monhegan because Jimmie took the boat off the run to paint her and get her ready for the summer run. Then, the cramped quarters in the utility building make it difficult for more than two, or even two people to work simultaneously. The grand celebration is now scheduled for June 9th, although we may be able to call a few days before that (but maybe not until a few days after!) There will be a big reception at the Inn on the 9th, special boats are to bring out special people, but details are lacking because Peter Gorman hasn't been here as he had hoped, he may come out today or tomorrow with some dignitaries from NET and Augusta. The cottage is engaged by NET thru June 9th.

The room downstairs is now fixed up - all culch moved to the new cellar! A new ceiling will go in and new carpet laid as soon as the boat makes her two runs a day. I'm also getting a new carpet for the loft in the cottage and later plan to replace the big rug in the cottage livingroom - after the pocketbook gets fattened a bit.

Our phones are dial, not pushbutton because Rockland exchange is not set up for such at present, maybe in three to five years.

All the rain we have had has made dandelions grow tall and tender. I dug enough for a mess for Doug, Harry and me by just digging them out of the brick walk to the back door on Saturday. I've had two cuttings of asparagus and more is ready for today. Also have feasted on Bamboo sprouts. What fun it is to eat off the land. The forsythia is more beautiful than usual and the magnolia tree will have about twenty blossoms. This was the first year I haven't wrapped it, too. The deer have nipped the tops of many of my lilies already because I didn't get around to sprinkling them with deer blood. They did it when Phoebe and I were away. All tulips were eaten, too. I think we should emulate New Zealand and corral the deer and raise them for meat. It would save our gardens and preserve our tempers.

Victor Lord is holding fire drills each Sunday. The last was a simulated fire at Mary Beth Dolan's in which they used water from the tank and the well nearby. Things went well. This was all probably inspired by the fire at Joe Pommerleau's (Grilly's cabin on the Meadow) which fortunately did little damage and no injuries but gave everyone, especially Danny Murdock who was living there, a good scare. Vic is very conscientious about his job, a good man as Fire Chief.

Heidi Boody has been accepted at Gould Academy. She is our only eighth grader this year. We are hunting for a school teacher and lodgings again. Mort's house is up for sale but is not considered adequate for a teacher's family. I am **not** renting the Mooring!

I guess that is about all the news. I am still having trouble with my mimeo and haven't found a suitable copy machine, although Sears may have one which would fill the bill. So you'll have to bear with this mimeo for now.

Phoebe sends a tail wag. She feels better rid now of the tape worms she showed up with a couple of weeks ago. What next? Oh, the Bernstein fund is still growing. It will probably be used to buy an encyclopedia and other volumes to bring the school library up to standard. Thank you all.

As ever,
Alta Ashley

P.O. Box 87
Monhegan, ME 04852
June 10, 1983

Dear Friends:

This may be the last mimeographed letter you will receive. Last time I was inshore I invested in a copier, but it uses treated paper (one

Telephone tower in the making

Jack Kearns — NET engineer

side only) and I think this letter will be a bit long, to relate all that has happened since my last letter, and especially since May 27th.

All May installers, cable stringers and micro-wave engineers worked long hours to meet the deadline of May 31st for the phones. They did so well, in fact, the system was operational by the 27th, just after midnight. The islanders went ecstatic and placed calls even to London that night. I lifted my receiver at 12:00:00 and found the dial illuminated so placed a call to John and Rue Riddle in San Francisco, knowing anyone in the east wouldn't appreciate a call at that time of night.

Many islanders were down at Billie and Jackie's at Deaman's Cove, celebrating with Jack Kearns and Victor Fuller, NET radio engineers. They tried until four o'clock to raise me. I had not turned up the bell in my phone so didn't hear the ring until then.

Because of various problems of logistics and the possibility that the system's initiation might be delayed, formal dedication was put off until Thursday, June 9th. A committee had worked hard to set things up, trying hard to meet without my horning in on the discussions.

The Balmy Days began her excursions that day, Dick Wall brought out some NET people. Kim Kenway came to represent the PUC, Peter Gorman the NET. Reporters and camera people were here from The Courier-Gazette in Rockland, the Portland Press Herald, Channel 2 in Bangor, as well as representatives from all the legislators' offices - both State and Federal. Through Jamie Wyeth, who arrived that day, arrangements had been made for me to place the first official call to Vice-President Bush and then to Governor Brennan. Billie Boynton helped with the second call. Then I gave the toast to dedicate the telephone system and to render thanks to all concerned in its creation.

It was all overwhelming for me. First I was given a corsage of red rosebuds and baby's breath, then a plaque which reads: "To Alta with sincere gratitude and appreciation for your devoted time and effort. Your Monhegan friends". Then I received a gold telephone charm to hang around my neck and lastly a T-shirt with an outline of the island on which is superimposed a drawing of the tower and the words "We (Heart) our tower. Monhegan Island, Maine. Finest kind!" Designed by Peter Gorman, who presented similar shirts to several others, including Billie Boynton, Mary and Bob Burton. Monhegan T-shirts were given by the islanders to Peter, Alan Stone, Kim Kenway. The celebration was held at the Island Inn, preceded by a luncheon for all those involved with the project and their guests. Fred and Irene Gamble came over to share the day with me, and what a day it was. That night there was another party down at Billie and Jackie's with many of us showing off our T-shirts.

Today's Portland paper carried the story and a great picture of me on the dock with my plaque and T-shirt. Channel 2 had the story on the six o'clock news, but it didn't come on Channel 6 until eleven o'clock. We can't get Channel 2, unfortunately, but Jack Kearns called from the Bangor area to tell me how good it was. On the lawn of the Inn was Rusty's big whale saying "Alta. You did a whale of a job!" Tomorrow I'll feel like an ex-president or a mother with an empty nest!

Some funny things happened. First of all, Vice-President Bush was asked to "hold on a minute while I call Dr. Ashley to the phone" as Peter Gorman handed me the microphone. Then Billie told Gov. Brennan "Goodbye, we need to get on with the celebration". Both men spoke highly of Jamie Wyeth and asked if he could be brought to the phone. Jamie's face was scarlet. A special phone hook-up was used so the entire audience could hear both sides of the conversations. Rickey Boehmer taped the whole thing and presented me with a copy today.

Later.

The Courier-Gazette carried two stories in the Saturday edition and another great picture. My mail is full of copies sent by thoughtful friends. I trust my hat size hasn't changed too much.

While everyone was celebrating at the party down at Deadman's Cove Sally received a phone call from John telling her that her new grandson Ben Neilsen Murdock had just arrived, weighing 8#1. Kristina is ecstatic, John and Winnie are doing well!

Formal graduation exercises for Heidi Boody were held at the schoolhouse a week ago Saturday. She will enter Gould Academy in the fall. All the children received an award of some type. Ed Freeman made a very touching speech and all was very impressive. Tomorrow Kathy Wincapaw will graduate with highest honors and be Valedictorian at Morse High School. She will enter Colby College on scholarship in September. Carl and Karen are off this afternoon for Bath.

I forgot to say that the party Thursday night was in honor of Kim Kenway and Alan Stone, who helped so much in getting the phone system. So many people helped, I wish all could have received plaques, etc. Jack Kearns and Victor Fuller worked into the night often and were able to put the phones in operation four days ahead of the deadline. There are still a few bugs to be worked out, but they have nothing to do with the phones per se. The system is an experimental one for NET

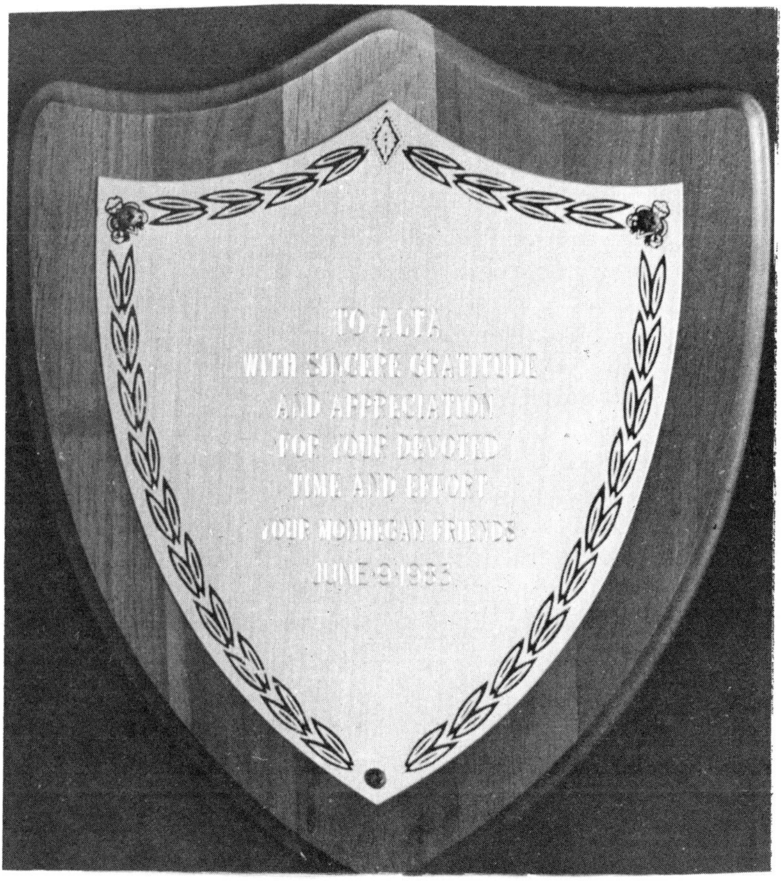
TO ALTA
WITH SINCERE GRATITUDE
AND APPRECIATION
FOR YOUR DEVOTED
TIME AND EFFORT
YOUR MOHEGAN FRIENDS

- the first one installed where there is no public power. I think solar panels will be in place before long. At present there are 67 phones in place, a far cry from the 40 we hoped for, and almost daily new requests are coming in.

Newt Searls arrived this week with his brother Richard who spent a few days fixing up the house and yard. Newt seems glad to be back. Most of the winter was spent with Margo in Rochester, then with Richard and a short stay in the hospital because of a blood clot in his bad leg.

Rickey and Peter Boehmer arranged a very nice memorial service for Mort Bernstein on May 29th. Hardly an empty seat could be seen in the church. On May 28th we learned that Helen Maxcy Bates had died just short of her 100th birthday and my Aunt Helen Woodbury had passed away in her sleep the same day. Phoebe and I took off on Monday for Aunt Helen's funeral and spent a few days in Concord, even though Elizabeth was in Charleston, SC visiting Bruce, but David was there and Tommy Ashley came out for a dinner. The services were private both at the funeral home and at the cemetery in Arlington. Robert and Helen Woodbury and I went over to the Allston church for another service in the evening - quite a concentration of funeral services.

That Thursday evening New England Telephone hosted owners of ATT stock at the Samoset and presented information about the disvestitures of Ma Bell and what it will mean to us as stockholders and as customers of New England Telephone. By the way, Peter Gorman introduced me to Vice-President Bush as "the Ma Bell of Monhegan"!

The island is in its summer mood. A lot of people have arrived for the season or have been here to open cottages for their tenants. Margery Pratt left today, the Curtises last week, also Ruth Lothrop and the Huberts to return later. Bob Burton has done a monumental job on the Pierce cottage, including installing new bathrooms and repairing the porch and front steps. Lumber is lying on the lawn of the Library ready to be used for the new addition which will take shape as soon as lobstering comes to an end on the 25th. The Monhegan House has a new front porch and several rooms have been refurbished. Dint is busy moving his workshop out and making the place ready for the Plantation Gallery. Cynthia Krusell, who will be the museum curator again, has not arrived yet but is expected soon.

George and Loretta Green are in their place for three weeks. George will retire soon from the New York Telephone company. He has been quite interested in our system and may be a valuable resource person once he comes here to stay for the summer or longer.

The newest enterprise on the island is the North End Pizza which Katy and David Boegel have set up where Don Cundy had his shop. They fill telephone orders, deliver after 7 PM and serve from the shop directly. Steve's Fishhouse Frankforts cart has not made an appearance as yet. Maybe they are waiting for Jeff to get out of school at Hinckley.

Lance is driving the Store truck, Holden Nelson in on the Inn truck, Amy Melenbacker and Joe Pomerleau are doing general trucking and picking up garbage, Sherm Stanley is doing gas delivery and Paul Horn gas service. Paul Krause is chef at the Inn again.

Rick Mathieu has his captain's papers so substitutes for Jimmy quite often on the Laura B while Jimmy takes care of his new real estate business. By the way the Cundy house is still for sale, Rusty Court has put his place on the market, the Leow cottage is on the market again. I'm not sure about the Blankenship place down by Jamie's. Bob Mersfelder is manager of a hardware store in Milford, NH and has his place up for rent summer and winter.

Rickey Boehmer had breast surgery last week and is in now for a bone scan. We all are praying for her. Bob Burton is in, too, for a check up. He has given up his coffee in hopes that has been the cause of his problems. We hope so, too.

The island has been beautiful for the past week or so. There never have been so many apple and cherry blossoms, and lilacs are gorgeous. The pictures on TV taken at the time of the phone celebration were particularly lovely. It was a gorgeous day, with bright blue sky, so the island showed up at its best. Lupine and snow-in-summer are in full bloom and the air is sweet with their scent. Soon the rugosa roses will be out. Wild strawberry blossoms are everywhere and my cultivated ones should have a good crop. But peonies and lemon lilies are yet to arrive.

I've probably forgotten some very important items but I can think of no more. I must get back to my Russian and Chinese language tapes in preparation for the trip which will begin September 17, from New York. We return to San Francisco on October 8th. I'm hoping to take a few days to visit relatives en route home. This summer I'll be going to Colby again for the course in Forensic Medicine August 21-25. Meanwhile I am preparing an island telephone book to be sold at the Spa. Gotta keep busy! We are hoping Zimmie can sell some of the new T-shirts, too.

Best wishes to all. I hope many of you will come by during the summer. Another tail wag from Phoebe and a trunk wag from Andrew Elephant!

As ever,
Alta Ashley

Alta & Phoebe & the whale May 29, 1983

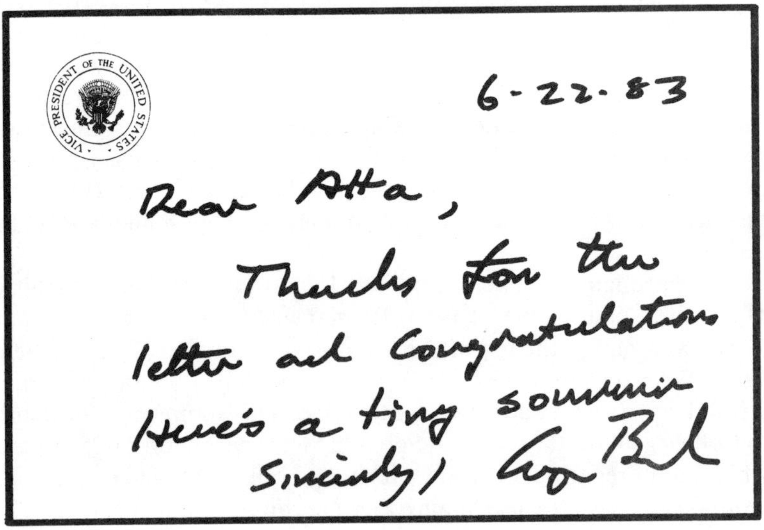

6-22-83

Dear Alta,

Thanks for the letter and Congratulations
Here's a tiny souvenir
Sincerely, George Bush

80th Birthday

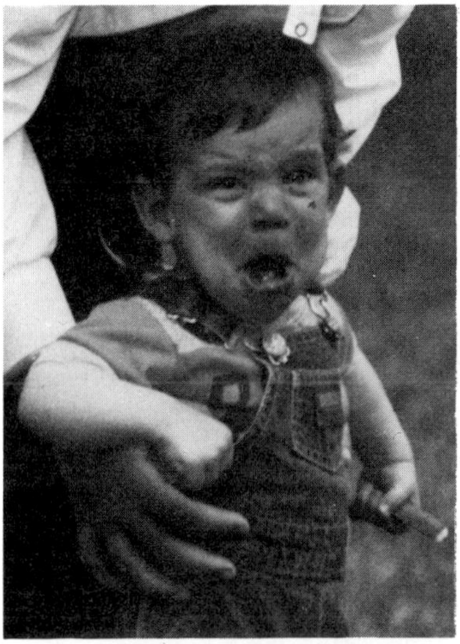

What happened?

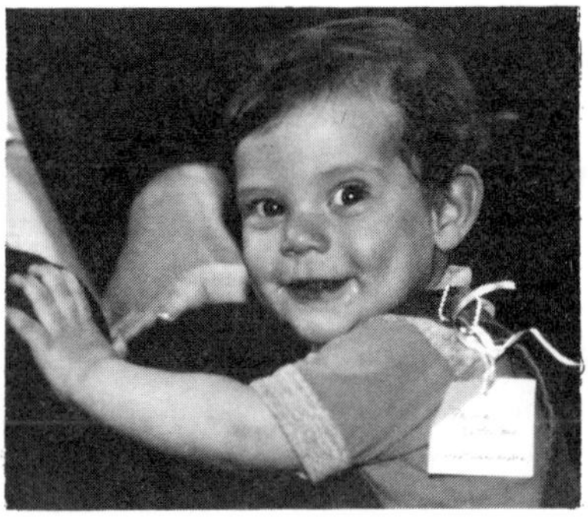

Happy child

Birthday Cake
fields of interesting decorations

Box 87
Monhegan, Maine 04852

Dear Photo-voltaic Devotees:

So many requests for information about my pioneer **domestic** photo-voltaic system have come in, especially since the article in Newsweek Magazine, I am having to resort to my mimeograph machine, poor as it is - and hand cranked, to respond and still have time for gardening, writing and just plain living. So please forgive this impersonal approach.

The energy system for my home on Monhegan Island, where there is no public power, has gone through many changes in the past twenty-two years, from kerosene lamps and **listening** to the **Today Show** on my FM 12 volt battery powered Zenith radio to electric lights and a color TV powered by solar energy.

In 1969 I had a deep well drilled which necessitated buying a generator for a deep-well pump, 3/4 horse. With this generator (run on LP gas in 100 # cylinders, we have no facilities for bulk supply) I was able to have electric lights. Gas was then $8 a cylinder. I used the generator only after dark and not between 11 PM and 7 AM, and for running the pump, vacuum cleaner and clothes washer. Even so, the generator burned 50-55 cylinders of gas a year, plus several quarts of motor oil.

In 1977 the house was wired for a 12V DC system. Two 200 ampere hour batteries were charged when the generator was running to furnish power for lights, 25 or 50 watt bulbs; and an inverter changed the DC current to 110 AC for my electric typewriter, egg beater and a color TV. A B/W ran on DC 12V power. Generator usage was cut almost in half and now I could read in bed or get up in the night with light, without turning on the generator. Gas consumption for the generator fell to about 30-32 cylinders annually. Gas by then was around $20 per cylinder, so the wiring and batteries were paid for in gas economy.

In 1979 David Sleeper came out for a visit and introduced me to Photo-voltaics, asking if I wished to gamble on a domestic system, never done before. It sounded good and, since I was adding a room to my house, this room was wired for it. The system also supplied DC current to the rest of the house and to an adjoining cottage, through the old battery supplied DC wiring system.

Almost never do I have to run the generator just to charge up the batteries, even with last summer's radiation reduction because of volcannic ash or the almost constant cloudy and rainy weather we have had in 1983. Gas consumption has fallen to 15 cylinders, or less, yearly. Generator time is about 20 hours a month (in the beginning it was 4 to 8 hours a day). Gas is now nearing $40 per cylinder. The generator is used "to keep me clean", for pumping water, running the vacuum, clothes and dish washers. With savings on gas plus the tax-write off the system has just about paid for itself in three years.

Cost of my system was high because of the cost of special batteries and the solar panels. Since then some costs have come down. The system has been most satisfactory, it is clean, quiet, requiring virtually no maintenance.

My house is heated by passive solar energy from a window wall, plus a air-tight wood stove and kerosene space heater. Refrigeration, cooking and hot water uses gas. We in the NE don't find solar heat or hot water cost effective. I have no freezer as yet.

Monhegan, Maine 04852
August 31, 1983

Dear Friends:

Now that I have **nearly** caught up with day-before-yesterday I'll begin this letter, long overdue. But since most of you have been or still are on the island you haven't missed much news. I should be working on four "Reflections from the Bird Bath" to send to the papers to tide over the four weeks I'll be travelling around the world, beginning two weeks from tomorrow!

The Chinese said they couldn't accommodate us in October so our whole trip has been turned around and the dates slightly shifted. We are to fly from San Francisco to Shanghai and end up in Helsinki. We do have a chance to spend a bit more time in China and will have a trip to Tian-jin (?Tien-sin) added to the tour. I had hoped we would see the strange limestone pillars of Kwei-li, but I guess that will have to be on the next trip - whenever.

The summer has simply flown. It was hot, dry, but really quite bearable except for a few hours now and then. In fact I have slept with a blanket nearly every night, and even in my Lanz flannel nighty as well.

In my last letter I spoke of how well Joe Walsh looked. Scarcely had I mailed half the letters than I learned Joe died on July 31st from a massive hemorrhage, which was a shock to everyone. The day before he had told me how happy he was that the skiff he had worked on much of the winter had arrived and that after he got through playing with it this summer he wanted to have it auctioned for the benefit of the cemetery fund. The auction was held on August 20th and June Day bid $1015 for it, having been egged on by Dint, who withdrew at $1010. It is now in her yard. She is also having a continuous flea market on her lawn, clearing out much of her accumulated treasure gathered in over the years.

The cemetery fund is growing and is now nearly $1700. Tower T-shirts have been so popular I've had a hard time filling orders, especially since there are never enough medium and large sizes. Right now I have extra-large and small only, but could get rid of ten or more had I the other sizes, which I hope Peter Gorman will send out before everyone leaves for the season. Several people have sent in contributions to the fund and island telephone directories have been very popular. I've sold over 150 and have run out of materials, so plan not to make more until next year when probably ten more phones will be added.

Speaking of the phones, we have had trouble three more times, involving the same phones each time, although this past weekend only twelve, including two of the public phones, were out of order. Although I reported the trouble three times to the phone company neither Victor Fuller, nor Jack Kearns who came out today on a routine check knew the phones were out! I guess the strike messed things up a bit. Right now everything is in order, but I have several fingers crossed - you can see that by the number of mistakes I'm making in my typing.

Joe Walsh's son Joe came out this weekend and he, Charles Brooks and I went to the cemetery and buried Joe's ashes. A beautiful memorial service has been held a couple of weeks ago at the church so there was really no interment service. Young Joe saw the condition of the cemetery and was glad that his father's friends, through their contribution, would see to it that the cemetery will be restored. There is a man here who was a cemetery superintendent and wants to advise us on how to go about the restoration. That is a great relief to me.

Martha Yandle is also interested and probably many more. Letters will be sent to relatives of those already buried, especially of those whose graves and headstones are in poor condition. I hope by next year those visiting will see some improvement, long overdue.

At the annual meeting of Monhegan Associates Ted Cochrane announced that Ted and Ann Edison would like to donate a large tract of choice land to the association but that they would like to see the

Museum divorced from the parent association. The trustees agreed so now we shall have another organization whose name hasn't been decided upon but officers have already been appointed. Ed Deci will be president and I Vice-president. I don't know who the others are but most of us feel the museum will benefit by the move. According to present sentiment, the Associates will come forth with a large donation should the museum need it. I would like to see the cemetery added to the museum organization but don't know just how we can manage that since the cemetery belongs to the town. (plantation).

The Library addition has been closed in and work already started on the inside. They had to take down the Angeline Nunan room because they couldn't get an easement from abutters so that the addition could be extended farther to the south. The new reading room will be known as the Elizabeth Ebert Reading Room.

I just ran off the first page of this letter and it looks as though **maybe** it will be legible in its entirety. I just fed the machine a drink of ink.

The middle of the month I spent five days at Colby in a course on Forensic medicine. Night before last I wondered if I was being put to the test when I heard shots ring out and found Phoebe a trembling mass of dogflesh on my bed. It turned out that Ted Kennedy was here and celebrating with Jamie and a few others with left over July Fourth firecrackers.

Elizabeth is in Europe with the Minute Men and their families reenacting the ceremonies around the signing of the peace treaty in Paris 200 years ago. She'll be back on Labor Day so I'll get a chance for a first hand account when I go to Concord prior to taking off for San Francisco en route to Shanghai. What a great experience that must have been for her. She has been very involved with Concord's celebrations all along. Priscilla, her sister, and I will be staying with the John Riddles a couple of days in San Francisco before our trip and have a full round of activities mapped out for us for the two days we'll be there.

It seems as though most of my old pals arrive on the island these days before Labor Day so my calendar is filled with dates for drinks, dinner and what have you. The garden has suffered from drought and complete neglect. I have a potato barrel I'm hoping will be full of potatoes on my return. No red tomatoes, carrots or beats as yet and the fava beans aren't anything like the delicious flat climbing beans I had last year. I hope my return on October 12th won't be too late to harvest. No time is a good time to go away when one has a garden!

Trefethern family members continue to keep me busy with genealogy information, and now some Humphreys and Albees have turned up for more. Olive Trefethern Moore of Kittery donated a lovely wedding shawl to the Museum. It was used for her wedding which took place in the Mansion. Olive made her first revisit to the island, followed shortly by Leona Trefethern of California who came armed with a huge genealogical diagram which conformed generally but not exactly with one I have been able to make. Hers filled up some holes in mine and vice versa.

It is now September 1st. Yesterday's heavy rain, which came of course just at boat time, has cleared and Willard Scott says the day will be perfect. Helen Moller has returned to the island in good health with no more need for a pace-maker. I haven't heard from Tyler Bartow who had a slight shock a week after their arrival at the Inn. Rickie Boehmer has completed her post-op treatments with a good prognosis and is gearing up for Kris' wedding to Dee Meisler in Kingston. Right now she is off island to take Louisa back to Bowdoin College.

Dougie Rothstein is returning to Brewster Academy as a "proctor" and Timmy Boehmer will be off to Proctor Academy. Margaret Taylor, Barbara Rothstein's daughter, is entering Colby-Sawyer and Kathy Wincapaw will be at Colby. Heidi Boody is entering Gould Academy.

Our new teacher is a recent graduate and I feel for her. There are eight children again this year: Becka and Matthias (Tigger) Hitchcock, Heather Boody, Sam Rothstein, Greg Rollins, Orca Bates, Dwight and Adair Stanley (if I have counted right).

Things are winding up. The museum goes on shortened hours next week, the dog wagon closed its parasol about the middle of August when Sonny Remick's daughter left. Sonny, who had planned to stay here, had to return to Rhode Island because of a fire at his marina he had just sold but whose papers hadn't been passed. David and Katy Murdock Boegels' North End Pizza is still going strong and are they good!

More publicity. About twenty assorted media personnel came out in mid-month, led by a director of the Department of Energy Resources from Augusta, to do a documentary on solar (P/V) power so several of us appeared in print and on TV. We are still waiting for a copy of the documentary tape, as usual. That is always the bait held out but never delivered. We'd like to show them during the winter, or summer, at the schoolhouse. Cynthia Krusell did a tape on me for Mount Holyoke and the museum so I sent her off to interview Mary Hoxie

Jones class of '26 now staying at the Inn. Then some woman from Worcester came out and took steen pictures of my "workshop" which was in complete disorder, as usual. Mary Hoxie's tape should be a good one.

I have now thought of four subjects to write up to send in serially to cover the time I won't be writing my column. Copy would have to go by diplomatic pouch ever to reach the USA in time.

The wedding of Ruth Stanley's sister Moira to John McCrea in Owl's Head was lovely. Thick fog lifted and lowered during the ceremony which was held on the lawn of her parent's home built right on the shore. It was Rockland's Seafood Festival time so some couldn't attend, including Doug and Harry, because no rooms were available in shore. I was lucky to be able to stay with Fred and Irene Gamble in Tenant's Harbor that night. The newlyweds are now in Alaska. They held an open house at their home in Anchorage on the 28th. I wonder who from here was able to attend.

It is about time for me to turn my thoughts toward the four subjects I am hoping to report on. While I'm gone Marion Watson will be here. She has offered to clean the house for me. She just doesn't know what she has let herself in for, but I'll be happy to see it done. I need a caretaker! Oh, by the way, the big elms have died and are to come down any day now. I hope to get two cords. Wood I ordered in May hasn't arrived yet.

Next installment will be post-around-the-world adventure. Let us hope this latest apparent Soviet aggressive act doesn't foul things up.

Best regards to all from Phoebe
and
Alta Ashley

Box 87
Monhegan, Maine 04852
October 24, 1983

Dear Friends:

This letter is long overdue but I haven't been able to settle down to do the many things which have piled up since I left the island in mid-September. One HUGE job is the planting of over $100 worth of bulbs from Breck's. If I ever get them in the ground the back hillside should be full of blooms, come spring. Right now it is raining hard to I can start working on this letter.

The trip around the world came off as expected, even though it was touch and go as to whether or not we could reach or leave the Soviet Union. Because of booking difficulties the trip was turned around so that we left from San Francisco and flew to Shanghai, rather than going from west to east. It turned out to be best as far as jet-lag is concerned, but not too good for our luggage, which became overburdened by our Chinese purchases, including throw rugs which many of us dragged from Tiensen and all through the USSR.

It was a most interesting trip but one none of us wishes to repeat. The non-stop flight to Shanghai tourist class was murderous. They handed out toothbrushes and footsies, as well as blinders for our all night (in daylight all the way) flight. Toothbrushes and fans were passed out in all but one flight on Chinese planes, and on that one we all received a beaded child's purse. Candy was also distributed as well as full meals and snacks about every two hours, just to keep our spirits up, I guess.

Shanghai was a mass of humanity and bicycles and men polishing their buses and taxis. We were housed in one part of the Jin Jiang Guest House where Kissinger and Nixon stayed, but in far less pretentious quarters. Our dining room was reached after a long walk by public toilets by an inside route, or by the dumpster and garbage trucks, if we walked outside. We were near Nanjing Road and #1 Department Store, so saw plenty of activity. Those who ventured out in the early morning saw hundreds gathered for morning exercises. Later in the day one could watch older men in their rhythmic, aesthetic exercises carried on oblivious of an audience (spectators, to be exact).

Living quarters for the Chinese were a study in contrasts. Older houses were nothing more than one-story rabbit warrens, often with a pile of new bricks in the yard. Newer houses were twelve story Soviet type apartment buildings stretching one after another along the roads in newer parts of the city or in reclaimed areas. Laundry hung from lines on the porches or along the street in older sections. Also in the older sections there was much living right on the street. Women cooked, washed and cared for children on the sidewalks while men and boys played games or just squatted. No one in China owns a private car and bicycles are rationed, one to a family. What cars are to be seen belong to the government and are used only for officials and party members

- officials have curtains at the windows. Buses drive only with parking lights at night "so as not to blind the cyclists in the road". The streets in Shanghai are lined with Sycamore trees whose trunks are painted white to a level of about five feet, clearly marking the sides of the roads. The trees give needed shade in summer and shed their leaves in winter, letting in the sun.

We visited pagodas, gardens, industrial exhibits and, of course, the Zoo where the pandas put on a marvelous ham act. We went to the ballet (and slept through most of it) and exchanged dollars for yens at the Friendship Stores and were glad to move on to Xi'An in three days.

My mistake, we went to Beijing for three nights and two days before moving on to Xi'An. We were housed in the Diaoyutal State Guesthouse, a series of buildings with sumptuous quarters. One building had a Friendship Store, and another was host to a group of dignitaries, including Sec. Weinberger, of whom none of us caught a glimpse. The bathroom has two sinks, a great shower, and a phone by the john! We stuck silver luggage tags on all our bags, pocketed the sandalwood soap and took pictures of each other phoning, from you know where.

The meals were delicious and colorful but we were beginning to miss meat-potato-gravy meals, thirsted for pure water and coffee served at the beginning, not the end, of breakfast. The guesthouse was set in a beautiful park, the former "angling terrace" of Emporer Zhang Zhong. We were honored to be considered "distinguished foreign visitors", as the brochure states.

While in Beijing, which the guides continued to speak of a Peking, we had a marathon race through the Imperial Palace and Forbidden City with a guide who seemed to want to get the whole thing overwith in a hurry, but not before she had given us her sore throat and cough with which all came down before the trip ended, prompting a humorous cartoon from Loraine Wales' pen of a bus from which is heard hacks and coughs while two Chinese on the street corner are saying "Engine trouble?" Our trip to the Great Wall was scheduled for Sunday morning. After a long bus ride we arrived at the wall just as a train discourged thousands of Chinese while thousands more tried to make their way back to the station. The result was a frightening melee of Chinese, tourists, buses and carts all trying to push their ways through narrow passageways under the wall. And the wall was so crowded with people pushing and shoving making one's way over the rough surface was pretty scary so Priscilla and I went only as far as the first turret.

The view from the wall, though, was awe inspiring. How many lives were sacrificed to build that mammoth structure?

There was also a day spent at the summer palace with luncheon at the restaurant of the Singing Orioles (we heard none) and a ride on the lake where we saw fishermen catching freshwater shrimp by pulling strings of cans along the lake as others poled their boats over the three foot deep water.

I forgot to say that one of the most beautiful sights in Shanghai was the sitting jade Buddha, larger than life and of exquisite color. In fact, the whole pagoda, or temple, was magnificent.

The Peking Hotel was fabulously ornate. When we returned to the city on October 1st, a national holiday, we went up to the roof to look over the city and Tian'anmen Square where every government building was outlined in lights, celebrating the founding of the People's Republic of China on that day in 1949 by Mao Zedong. It was an impressive sight, but we weren't impressed by the Peking duck served at dinner. Another People to People delegation, this one from Idaho, shared our restaurant. They cheered Idaho and we sang Maine's Stein Song. I took greetings from one potato state to another and received an Idaho potato pin, some other pins and a card from Gov. Evans. He took home a copy of Monhegan Ice Pond. That delegation, slated also for USSR, changed their plans and went to Hong Kong and Tokio instead, "for political reasons".

Rather than make this letter too long I'll reserve tales of the rest of the trip for another letter to come off soon, I promise. The island is quiet, nearly all seasonal people have gone. There may be fifteen boats in the lobster fleet this year if both Bryan Hitchcock and Don Cundy return to lobstering.

As ever,
Alta Ashley

Alta Ashley
Nov. '83

Monhegan
On to Xi'An!

Nov.7. The night before we left Beijing for the ancient capital city of Xi'An we ran another marathon from our restaurant to the theater, through the rain and over humpy sidewalks, mud and puddles as we traversed the sidewalks above the new subway extension. The going was so difficult and the sidewalks so crowded we kept losing members of the group until George Bostwick held a light tan pocketbook high over the heads of all as a standard for us to follow I can't now recall what we saw, I think it was acrobats, because jet lag still put most of us to sleep: that added to the strenuous days we put in following our marathon runner local China International Tourist (CITS) Service guide.

We flew in to Xi'An and were put up in the Remnin Hotel where some of us enjoyed a suite while others were crowded into room barely larger than a good sized bathroom. We also found that the beds were made up with a bottom sheet but only a large turkish towel for a top sheet! In the lobby we found miniature Snickers bars, which we quickly were snapped up by chocolate-hungry tourists, even at 35 cents a piece.

We slopped though a commune where we were shown cook stoves heated by the burning of hay. Electricity was manufactured with the use of methane gas produced through the fermentation of night soil and hay. Heat warmed large vats set into the stove and serving as the source of domestic hot water. There were also deep-well cookers. The Chinese know well how to get the most energy out of the least resources.

At the door of a nursery school were parked brightly painted tricycles decorated with multicolored ribbons. Inside children in colorful clothes (and split pants) willingly recited and sang for us, all the while wondering about our pale faces. At the commune hospital we saw a child in traction with the use of a folded bed sheet tied to the bed frame, a paraplegic whose shoe tips were tied to his crutches, overcoming his toe-drop, and aiding him in swinging his legs forward in walking. We visited an arts and crafts store, saw people painting wall hangings and making terra cotta figures - miniatures of the buried warriors and horses which we saw next day at the Ban Po Museum.

This museum was fantastic. Thousands of figures are known to be buried here. Each warrior differs in facial expression from the next. They were apparently set up to guard the burial site of Emporer Qin Shi Huangdi and were discovered only in 1974. We also saw relics of a neolithic village.

Outside the gates of the museum were dozens of Chinese hawking their wares, cotton jackets decorated with dragons, etc and ornamental animals. They kept up a chant of "Hello, hello" until we bought and how we bought!

One afternoon we set off to find a Friendship Store, got lost and wound up in a street market where all kinds of foods were for sale, flies and all. Before we left Xi'An we visited the Big Goose Pagoda and met up with Mrs. Weinberger and her party. She was a classmate of one of our party, Kay Sager, at Oakland High School and had stopped in at High Mooring to see me a year or so ago, so we had a nice chat. Then we were off to Beijing by plane and the Tianjin by bus, a long bus ride in the dark.

I forgot to say that before we left Xi'An we stopped to see Hot Springs a beautiful Daoist monistary where Chang Kai Shek was captured and held prisoner before his escape to Taiwan.

After the night ride to Tainjin from Beijing by bus we were put up in another sumptuous guest house set in another park by a lake, very much like the Daioyutai Guest House in Beijing, except that our entrance to it was through the back door!

While in Tianjin (Tiesin) we visited the Number 1 Carpet factory where we watched carpets being made - 128 ties to the inch and the same number of rows to the inch - and then sculptured with scissors. Many of us couldn't resist buying a rug, mostly throw rugs, but some bought full sized ones for shipment home. Those of us buying the smaller ones had to drag them all the way back home, with quite a bit of difficulty and inconvenience. We also watched the artists designing the rugs. It was fascinating and the rugs hung on the wall were unbelievably beautiful.

We also went to a crafts shop where kites, among other things, were being made. We saw cloissone pieces being made, jade being cut and, of course came away with some more treasures. Most of us spent almost all our money in China, knowing there would be little to buy in the USSR.

We went to the #1 Hospital of Tianjin where we saw a "very serious" patient on a respirator in an ICU with about ten patients, several receiving intravenouses of herbal concoctions. The "serious" patient was said to have tuberculous meningitis.

That night three of the doctors from the hospital joined us in a farewell banquet at the Guest House, an elaborate affair with colorful platters of food and hors d'ouvres.

From Tianjin we returned for one night to Beijing. It was October 1st so everywhere in both cities there were crowds of celebrating people and red flags, it being the anniversary of the founding of the People's Republic of China. Our trip was by train, very leisurely with tea served from small tables set between each two double seats. It was a nice way to travel.

Bejing was in a festive mood, even though a drizzle had arrived. All government buildings and many public buildings were outlined with lights making Tian'enmin Square a veritable fairyland when viewed from the top of the Peking Hotel.

One afternoon we set off to find a Friendship Store, got lost and wound up in a street market where all kinds of foods were for sale, flies and all. Before we left Xi'An we visited the Big Goose Pagoda and met up with Mrs. Weinberger and her party. She was a classmate of one of our party, Kay Sager, at Oakland High School and had stopped in at High Mooring to see me a year or so ago, so we had a nice chat. Then we were off to Beijing by plane and on to Tianjin by bus, a long bus ride in the dark.

I forgot to say that before we left Xi'An we stopped to see Hot Springs, a beautiful Daoist monastary where Chang Kai Shek was captured and held prisoner before his escape to Taiwan.

On October 2nd we were off to Moscow via another Chinese aircraft after a gala evening in Peking. Moscow was cold but clear. We went to Red Square both during the day and at night where we marvelled at the beauty of St. Basil's at night. We had a wonderful meal at a Georgian restaurant where we joined with a group of Italians in singing and dancing until called to our tables to eat our ice cream, a special Russion treat since Kruschow's visit to the USA. We toured the Kremlin, saw the opera Anna Karanina in the new Soviet Building, bought a few things in "dollars stores", and had trouble cashing travellers' checks. Moscow was a bit less drab than it was fifteen years ago on my first visit. The subway was as clean and beautiful as before.

We then flew to Kiev where we housed in a comfortable hotel. The city is beautiful, filled with Ukranian treasures and parks. The war memorial and Catherine's palace, St. Anthony's Cathedral and the cathedrals near the catacombs were all beautiful. Wherever there was any unused space a park was built. Leningrad was rainy, its architecture impressive. The Hermitage was filled with gorgeous art treasures and the summer palace at Puskin staggeringly beautiful. Our stay in Helsinki was all to short. We did see the Church in the Rock - fantastic. On returning to New York we passed over coastal Maine and could identify our own towns. It was a great exhausting trip and we were all glad to be home.

Congratulations to China and the USSR for their art and architecture and historic treasures. But THREE CHEERS to the USA for its portable water, two sheets on a bed, cold drinks and individual freedom!

Monhegan, ME 04852
Nov. 16, 1983

Dear Friends:

We are having another NE storm with no boat today. This is the second time this week and the fourth time in two weeks that the boat hasn't come out. The winter that is said to be a hard one is starting out to fulfill that prophecy to the fullest. Both Bob Burton and Doug Odom are starting their buildings and freight is greater than usual. Jimmie had hoped to pull the Laura B out to have some work done on her this month but instead, he has been making extra runs, weather permitting, then losing out on regular runs.

My bulbs are still waiting to be planted. All during the fine weather we had after I returned to the island I had to spend time organizing my photographs to be ready for the reunion we had at Pemaquid the first of the month and again Sunday in Augusta.

Phoebe and I took off on Hallowe'en (not the first) and visited Lorain Wales in Chamberlain (near Pemaquid). Margaret Baker and Fran Lash came to dinner and Barbara Briggs came over later to share slides and photographs of our trip. Children showed up periodically at the door for handouts and to show off their costumes. It was a great night. The next day we mosied around Pemaquid, called on Ruth and Ken Brookes, John and Barbara Hanson showed up, too. All four were friends of mine in Augusta when Ken was our minister. Then I took off for my medical meeting at the Sail Loft, spent the night with the Gambles and headed back to the island. On the boat was a couple from Pittsford, NY, who heard there was to be a special boat the following day, so they decided to stay over. I found they had recently been to China so invited them to drinks and dinner and we had a great evening together.

On Thursday the Griffin went in to get Dan Rothstein and take in the Rays (the NY couple), broke down and had to be towed back to the island. Jimmy didn't come out with the Laura B and no one else would make the trip because of weather predictions.

Friday the nor'easter struck and dumped more than five inches of rain on us before the storm was over. Jimmy didn't come out until Sunday, stranding the Rays, Boehmers, Teco, Winnie and Vernon, and Hary T. Bone who had been called back to his ship. Dan Rothstein was still on the other side waiting to come out.

Tuesday was voting day and I sat at the poles that afternoon so didn't know the Kelani and Donna had came out with three cords of wood, two of them mine. With the help of Doug and Harry and Jeanne Rollins we got all the wood off the dock. Laury Cooper took one cord, I stacked a cord but the second is still in a pile on the lawn. It was lucky we did that because on Thursday the Kelani and Donna came back with three tons of coal and supplies for the two new buildings. The Laura B came in about one-thirty loaded with cement, shingles, lumber and other materials for the buildings. WGBH was due Friday morning on the Freeman Clark with Dave Sleeper to do a segment of the New "This Old House" using my place as part of its setting. They planned to leave at seven, giving me time before I had to leave on the Laura B to go in for our reunion in Augusta on Sunday. Well, Jimmy said there would be no boat Friday because of a predicted storm. What to do?

I decided to go in on the Laura B, to pick up my car, then drive around to Boothbay Harbor to meet the WGBH crew, come back with them on the Freeman Clark and return with them and go on, as planned to spend the weekend with Bob and Margret Baker in Newcastle. It took so long to unload the Laura B we did not leave until dark and low tide. It meant jumping onto the rail in the dark, something I almost lacked courage to do, but finally did.

Magee Barstow and the Coopers went with us. The only individual who didn't have qualms about boarding the boat was Phoebe who got all four of her feet well planted on the deck. It was pitch black when we left, which made backing off the dock a bit of a maneuver to slide by Smutty Nose and not land on the rocks in front of the Rockcrest. The ride in was a sideways rollercoaster, much to Phoebe's distress.

Our weekend with the Bakers was delightful. They also have a Phoebe, a golden retriever puppy, and a large yellow cat. The two Phoebes wore blaze orange ribbons on their collars to protect them against hunters who were all around. We first spent the night at the Gambles after sharing a supper with Sandy Dickson who said the WGBH program had been scratched - no need to go to Boothbay. The next day I called on the Kennisons in their new home, a farm house in Warren which they are busy remodelling, then stopped to see Gib and Jo Strout, also in Warren. They have a house which Gib built so Jo can get around in her wheelchair (she has MS and can't walk). Gib used to be our State School Agent. That evening the Bakers, another couple and I set off for Monmouth where we had dinner at Curley's and then went to see the Yoeman of the Guard. Bill Kennison played in the orchestra, Alfred Senter sat behind me. The Aults (tennants of mine this September) and the Bibbers who went to Spain with Elizabeth and me several years ago were also in the audience.

The next day Margaret and I went shopping for her huge dinner party that night, ate at the Samoset restaurant in Pemaquid at the same table with Irene Davis. It was quite an old home week. I thought I'd remember who was at the dinner party but right now I can't recall a name. We showed slides of our trip and talked.

Sunday morning Bob, Margaret and I were off to the Cong. Church in Newcastle where we all sang in the choir. Such fun! Who should be there but Vera Varney from old Augusta days. Then Margaret and I were picked up by Peggy Hornberger who had Loraine Wales and Fran Lash with her and were off to Augusta to meet with others on the trip and share pictures. Others there were Frank Stred and his wife, Arthur and Marion Moulton, Hilary Ervin (Ed was on call and couldn't come), Phil and Mary Thompson, Dick and Nita Shaw, Warren Strout, George and Mary Ann Bostwick and Anne Hunter. Everyone asked for Priscilla. There were a lot of slides and photoghraphs to look at, memories refreshed. Anne's slides and pictures were particularly fine and Phil's movies brought back a lot of things. But everyone had great pictures to show. Thirteen of us went to a nearby Chinese restaurant and most of us had poopoo platters. I'm sure not known in China. Some used chop sticks, but most of us used our fingers. It was great fun getting together. We all hoped to have more reunions. One person I forgot about seeing was Alice Chadwick who is now a patient in the nursing home section of Miles Memorial Hospital in Damariscotta. Alice lived here for many years and was at the Methodist home in Rockland when Mother was there. Everywhere I went people told me that they read my column so now how can I quit writing it, as I had thought of doing come January, the beginning of my thirteenth year at the job?

The Obermans came out over the weekend to close up their cottage before the water was turned off which I believe happened over the weekend. I never know because I have my own well which, by the way, overflowed after our five inch rain a week ago and is probably doing it now with this rain.

I have a new pot burner downstairs which keeps me almost too warm upstairs. I'm going to see how comfortable I can be before making plans for the winter. We'll be off for the dentist Monday then on to Concord and New York state for Thanksgiving with Priscilla and family. Tommy will help with the driving. Elizabeth and David will stay in Concord. Happy Thanksgiving.

As Always,
Alta and Phoebe

Box 87
Monhegan, Maine 04852
Dec. 16, 1983

Season's Greetings!

Sorry to be so late with these greetings. Many of you have read in previous letters and in the local newspapers the accounts of the world trip I made in September and October - Round the world in twenty-one days! Early in the year I was invited to join a contingent of Maine health workers in a People-to-People Medical Mission to the People's Republic of China and to the USSR, including a trip to Kiev in the Ukraine. it was too good to pass up, especially since my niece Priscilla Jackson wanted to go, too, and be my roommate.

Our itinerary finally settled down to leaving from San Francisco and returning to New York from Helsinki. That was fortunate in more ways than one. The Korean plane had been shot down only a few weeks from our departure date of Sept. 19th, thus giving time for the air travel embargo to be lifted by the time we were scheduled to leave the USSR, and the jet lag going from east to west is easier to handle than is the reverse.

We had two days in San Francisco, which we spent with the John Riddles who showed off their beloved California. We stayed at their home which is perched on the side of a canyon in a very interesting part of the city. Much to my disappointment we didn't see Muir Woods or Mt. Tamalpaice (?sp).

The overseas air trip to Shanghai was gruelling, despite the footsies, toothbrushes and paste, candies, and night eyeshades handed out.

My legs were swollen to the knees by the time we arrived in Shanghai.

Our first meeting with oriental culture was memorable. Memories of Shanghai are of people everywhere, white shirts, bicycles, platters of strange and often tasty food, orange drink and beer but no potable waterer. I was chronically thirsty because I didn't drink beer or tonics. Then, of course, there were the beautiful pagodas, wonderful purchases of silks, cloisonne and other Chinese trinkets in the ever present Friendship stores (all over China). I had brought a package of velveeta to give to a Chinese friend of Harry and Lucia Miller, then couldn't find her address. Getting it delivered into the arms of a Chinese doctor took over an hour on the phone by our tour guide, Tom Manning. Then there were the piles of bricks, houses being put up or torn down, and every scrap of unoccupied land under cultivation. The tender care with which taxi drivers and bus drivers treated their vehicles was amazing. There are no private cars in China and even bicycles are rationed, although one would never guess it by the number seen parked or on the road - maybe one per family per year can be purchased.

The economy in Beijing and Tianjin seemed a bit better than in Shanghai. There were more official cars in Beijing, party members' cars had curtains at the windows. The Great Wall was a disappointment because of the unruly crowds there the day we went. The surface is rough, even has steep steps in places so one has to watch one's feet, hard to do in a crowd. Pekin duck was also not the feast we expected. But the summer palace and its Long Walk was spectacular.

We could have had more time in Xi'An to see the buried warriors and horses, as well as the neolithic museum with its wealth of artifacts, but we had enough of the Renmin Hotel and its live stock and turkish towel top sheet.

Tianjin was a thriving industrial city, smoggy as all China seemed to be - we saw few people with face masks, however. The carpet factory was most interesting and few of us resisted bringing back one of those beautiful rugs. We spent most of our money in China, knowing there would be little in the USSR to buy. I couldn't resist a silk jacket which is reversible - red and royal blue, also some cotton vests decorated with serpents, frogs, pandas etc. bought from hawkers at Ban Po museum.

We visited Moscow, Kiev and Leningrad in the USSR. Things seemed more prosperous than they were fifteen years ago. The Moscow subway was just as clean and impressive as before. Kiev was beautiful and the architecture and cathedrals all over were gorgeous, breathtaking. St. Bsil's at night was spectacular. We visited the Hermitage and I took a photo of a Rockwell Kent painting there. We also went to the summer palace in Pushkin. Both places were filled with treasures, beautiful walls, ceilings and floors (we had to wear special shoes in Pushkin).

Our visits to hospitals in both countries showed that their medical art is far behind ours. In China much herb medicine is still being used along with acupuncture and moxibustion and massage (through the clothing). Little more than expedient medicine is done either place. We were able to fly out of USSR to Helsinki where a half-day was all too short.

I bought a chop in China. It is the character for Alta. With this I must say best wishes to all.

Alta Ashley

MERRY CHRISTMAS 1983 HAPPY 1984 from MONHEGAN

Dear Friends:

December 25th is barrelling down upon us with presents to wrap, holiday cooking to do and Christmas greetings to write and **mail!** Ever since I returned from my world whirlwind trip I've had to make many trips in shore and have almost invariably been stranded there or had to leave early because of bad weather. The Laura B has been off the run for a month now, complicating passenger and freight transportation as well. So forgive me if your greetings don't arrive on time.

This has been quite a year for me and Monhegan. As you all know, the phones went in on May 28th, two days ahead of schedule despite all sorts of set-backs. Two days of gale winds didn't delay the erection of the micro-wave tower, which was completed in seven days. But the dishes had to be remounted because the correction for true north was made in the wrong direction and we were beamed to Friendship instead of Rockland's Benner Hill.

With the telephone system taken care of (so I thought) I decided to put my energies toward repairing damages to the cemetery from time, weather, vandalism (?) and just plain neglect, so a fund for that purpose was started, initiated by the sale of "Tower T-shirts" following my gift of one from NET at the great party we had at the Inn in celebration of the completion of the phone system held on June 9th. I made special island phone books and sold nearly 150 for the benefit of the cemetery fund and received several sizable donations, we well. The day before Joe Walsh died unexpectedly he told me he wanted to auction off the skiff he had worked at building all winter "as soon as I am through with it" with the proceeds going to the cemetery fund. $1015 dollars was raised in the auction, and money raised at the Spa for a memorial to Joe was added, so now the fund stands at just under $2000. I must get busy on plans of how to use it.

All summer I tried to learn Chinese and Russian from tapes in preparation for a People-to-People mission to the People's Republic of China and the USSR. It was time poorly spent because I couldn't remember any word for more than fifteen minutes. I got way behind on my self-assessment quizzes for my medical "brownie" points and have had to do twelve quizzes this past month, with one to go by Dec. 31st, in order to retain my license.

Karen Wincapaw was sworn in as Postmaster on Feb. 2nd and Bob Burton is building a place for the P.O. down by the Store so Winnie and Vernon can have their home to themselves. Progress is slow because of boat and weather problems. The Odoms' cottage is coming along better on the other side of the Store and should be closed in by Trap Day. The Library ell is finished.

We now have close to 70 phones. At first one batch of phones kept going out of order but finally in September the reason for the failure was found and things have gone along well since.

Monhegan boats have had their problems. Rusty Court's boat went on the rocks in the harbor, Bob Burton's broke from her mooring and landed on the rocks in front of the house, the Laura B hit the reefs between Allen and Burnt Islands, and Doug's Chris was fired on when she was in Friendship Harbor having work done on her. Others have suffered the usual mishaps of boat engines. There may be fifteen boats in the fleet, in 84.

Among the many old friends on Monhegan who died this year were Mort Bernstein, Emma Field, Iris Gibbs, Dan Shaper, Vincent Tibbetts and Joe Walsh. Tyler Bartow suffered a stroke but is improving and Zimmy Brackett has regained some sight in his operated eye.

There were several weddings, none on the island, however. Moira Paddock married a man she met in Alaska, Chip Richardson also was married and now lives in Alaska, Ricky Cundy married Elise Burnham and Kris Boehmer was wedded to Dee Meisler.

The three elms which stood in Ida Proper's yard are no more, and the Balm of Gilead tree at the Influence bit the dust in a storm this fall. Among Trefethen relatives to visit the island were Olive T. Moore of Kittery, Leona Trefethen of California and Ashton T. Welch of NH.

The island received much publicity from the telephones, solar P/V systems and the Laura B. Two documentaries were made on P/V, one for Disney World EPCOT and one for the Maine Energy Conservation agency. Old friends dropping by or staying a while were Pepper and Floyd Milbank, Nancy Sulides and Gert Black, Dick Nakashian in his Kismet, Anne and Stu Barber as well as many new friends who signed the guest book. Elizabeth spent several days here during the summer.

Our spring was wet, cold and windy, summer superb but record rains and winds since mid-October. I am planning to house-sit in Nobleboro on Pemaquid Lake for ten weeks beginning early in January. Sorry, no time for personal notes. Best of the blessings of the season to all.

From
Alta
and a tail wag from Phoebe

Box 87 Monhegan, ME
Jan 2, 1984

Dear Friends:

While I wait for Charlie MacDonald to come for my **nine** pieces of luggage to take to the dock I'll try to get at least one page of my first 1984 letter typed.

Today I leave for Nobleboro - the address is RR 1, Box 52, Nobleboro, ME 04555. I plan to stay there at least until Town Meeting, which will be the first or second Monday of March.

Phoebe and I left the island the Wednesday before Christmas in the Chris, knowing that the Laura B would be very late that day because of all the accumulated freight she would bring out, her second day on the run. Luckily we did, because Victor Fuller and Jack Kearns from NET came out to put a heater in the utility building and the Laura B waited for them. Also, it was lucky we didn't plan to drive to Concord the following day because snow and freezing rain made travelling very hazardous.

We got about six inches of snow in Concord before it turned to rain and then froze. What a mess! The temperature went below freezing on Saturday night. My nephew Tommy Ashley came out for a Christmas Eve dinner and exchange of presents - Phoebe had been taken to Pat Palmers kennel before everything froze up.

On Christmas morning, with the mercury below zero, we tried to start both Elizabeth's and my cars. She finally got hers going, but my gas line was frozen. Her gas feed was frozen open making the engine race dangerously until she had driven around the neighborhood several times. We finally were able to take off for Naugatuck, CT only a half hour late.

Cindi (Sutherland), my grand niece, had a colorful table set for us and a tasty souffle. We then exchanged more gifts and were off to the elder Sutherlands' for more tree festivities and Christmas dinner. There were fourteen at the table! What a day, what hospitality. Young Christopher and Ian (my great grand nephews!) held up better than did we adults.

Next day we were off again for Concord. It was still very cold and my car still wouldn't start. Finally, with the aid of an overnight light bulb under the hood, a hair drier and a boost from Elizabeth's car battery the thing got going and I was off with Phoebe for Maine. We stopped in Nobleboro to meet the owner of the place we are to stay in (Mrs. Bickford) met her mother and son. The cottage is great, built like a schooner, with all space well used. There is a large living-dining-kitchen area with fireplace, wood stove and electric cooking; an oil burner down cellar, wood piled to the ceiling, and an elevator to bring it up on. Phoebe had a great time exploring the woods and the lake. There are neighbors, but none too close. We should be quite comfortable and happy there for the next ten weeks.

Yesterday was Trap Day. By mutual consent they didn't set on New Year Day because it was Sunday and the last time they set on Sunday was in 1978 the year of the great storm with loss of the telephone cable, boats damaged and gear lost in record amounts. Too bad, because the wind and tide and temperature were ideal. Courtney's boat had not been fitted to its new engine and Alfred's had sprung a leak. They decided not to wait for Courtney - it might be a week or more - but if Alfred was back by noon on Monday they would set. News came that he had left Port Clyde and was expected here around two, so they decided to go as soon as he was back and loaded with traps. Doug and Harry were at the dock when he arrived, helped him load and slipped the first trap over the side at 2:33 PM in front of the Rockcrest. Charlie MacDonald is here to help. He also has helped with the building, which may have its roof on by now. He also has just taken my luggage to the dock.

Kathy Rhoda and her POSSLQ, Buddy, are staying at Newt Searls' this winter, Joe Pomerleau is living in the Bunk House back of Mary Beth's; Danny Bates and Daphne are in the Bates' cottage; Amy Melenbacker has bought the Cundy house and she and Bill Baker are busy fixing it up. There will be twelve boats in the fleet this year plus the dories of Danny Bates and Victor Lord. Victor is fishing thirteen traps. He says he isn't superstitious! He is also helping build the Odom cottage. Bob Burton has laid the floor on the new building by the north side of the Store where the P.O. will be by spring. Must run this through before boat time. Best of 1984 to all. May add a bit after getting to Nobleboro.

As ever,
Alta

Box 52, RR 1
Nobleboro, ME 04555
Feb. 2, 1984

Dear Friends:

Happy Ground Hog Day! Any self-respecting ground hog in the north-east has seen his shadow and retreated for a six week sleep. This morning it was ten below here. The lake is an almost unbroken expanse of white. Only two snowmobile tracks have been made since 13 inches of powdery snow fell on Tuesday. Although it was powdery it is still sticking to the pine trees which surround this house. Every so often they let go their load in a cloud of white and a thud. Then the ice booms, the baseboard heat knocks and tinkles and way in the distance can be heard the traffic climbing the hill on Route One, so this house in the woods is not deathly silent. I'm enjoying being here immensely.

Last week Margaret Baker (one of the persons on the R-t-W trip) and I went over to New Harbor and picked up shrimp - 90 cents a pound. All the next day I spent peeling the 15 pounds I bought and got just about six pounds of meat. But are they good eaten fresh like that. You can imagine the mounds of heads and shells there were. Fortunately the boy came to pick up garbage on Sunday - the dump is open only on Sundays and then for only a few hours, at least in the winter. Doug and Harry dropped off some lobsters later that day which Margaret and Bob Baker and I feasted on that night. They live right on a salt river so kindly took the shells home for disposal. The lobsters were well worth waiting for. Last night I regaled Doug and Harry with a boiled dinner and a blueberry pie, with enough left over from both to feed me for several days.

At last I have gotten back to working on my Monhegan articles and have three in shape for printing. Now to get quotes and some printer who wants to take on the job. I've also had 100 copies of the Monhegan telephone book run off, with corrections and additions. There are too many B's!, making the first page too long. I thought I had spaced it right. This time I had them run off at Rockland Business machines where this will be run off also, at such a low cost it is silly to try to do it myself. this new typewriter makes better copy, too, so **maybe** every word of this letter will be legible.

Last Saturday I had lunch with Doris and Bob Slayton in their charming home in Alna. Bob will be ninety his next birthday and looks about seventy. Doris is as perky as ever and looked lovely in a lavender outfit, with her lovely white hair. This noon I am going to Barbara and Arnold Briggs' home in Damariscotta Mills to lunch. They, too, were on the PRC-USSR; seven of the group came from this area.

The Newcastle church choir singing has been lots of fun. Bobbie and Ruth Ives work in that church, he as assistant pastor and she as a Sunday school teacher, so I see them every Sunday. They are now living in Pemaquid and conducting their Carpenters' Boat Shop for boys in social rehabilitation. They make beautiful small boats - skiffs, punts, dories and the like, some canoes, I believe, as well. The Bakers are in the choir, too.

Saturday night Margaret and I drove to Augusta to attend a performance by the Monmouth Players at Jewett Hall at UMA. Naturally there were some Augusta Players people there and other people I know, including Alfred Senter among the ten. He seems to turn up everywhere I go. He said that Zimmie's last report on his eyes was good.

Ruth Shaper is back at Miles Memorial in Damariscotta after receiving chemotherapy at Maine Medical Center. Daphne lives just across the lake from me so she goes in to see her mother every day, a great help to Ruth. Ollie has been up, too. I believe he is now living in Marblehead, or maybe in Boothbay Harbor by now.

The announcement of the wedding of Judy Kurzer and Jon Bogdanove in Rockland in December arrived this week. They should be back on the island any day now. Dyan Fitzpatrick has been living in their houseboat on the Hudson this winter while she attends the Art Students' League (or maybe Cooper Union). Marion and Tom Watson brought Mary Stanley by for a short visit last Sunday. Marion has had lazer treatment for her eyes and is much improved. Mary looked fine.

Severe weather and lobstering has halted work on Doug's cottage and the new P.O. Chris Nicholson is back in the old P.O. Lobstering is only fair but the price is very good - it always evens up.

It is now Friday morning and I want to finish this so I can take it with me to Rockland to pick up some more Brownie CME points this noon. Arnold Briggs came over and took me to their home yesterday. Well he did because I would never find the place by word of mouth and doubt if I can even now. Their house is built in the woods right on Damariscotta Lake, against a hill so they have window walls on two floors. It is a lovely spot and a beautiful home. Loriane Wales, another fellow traveller (in the good sense), was there too. As usual I found that I was acquainted with the other guest, Lise Aubry, who used to be a kindergarten teacher in one of the schools where I did tuberculin testing. You guessed it - we had shrimp casserole for lunch - delicious! (Arnold & Barbara Briggs were also on the China-USSR trip.)

Last night at choir rehearsal we did a lot of the music to be sung during Holy Week and Easter. I'm sorely tempted to extend my mainland stay until after Easter, which is the third, no fourth, Sunday in April. A lot will depend upon what plans have been made for this cottage, however.

Sunday I hope to get over to Boothbay Harbor to a Republican cocktail party, provided the messy weather which is closing in on us doesn't make travelling too hazardous. We are expecting rain this afternoon, on top of all this snow!!

Let us hope Presidents' Birthday weekend will be good. Nancy and Elizabeth, my nieces, are hoping to come here for the weekend and I'm hoping we can go over to New Meadows Inn to dinner that Saturday. I haven't been there since the twenties and before it burned and was rebuilt.

The Sing and Supper at the Kennisons' a couple of weeks ago was lots of fun. The group has been singing together for some time and are quite accomplished. I'm afraid my contribution was not musically valuable, to say the least. We sang a capella all evening, first from an Appalachian hymnal and then from a collection of Bach chorals, quite an exercise for me.

I can't seem to scrape up much island news. Each Saturday I call Rita, then have scarcely a page to send to the papers. But people still say they look forward to reading my column, people who don't even know anyone on the island!

It is time to get the buggy on the road. So, enough for now.

As ever,
Alta

P.S. Phoebe is growing a nice crop of fur to cover the spot where the tumor on her back (benign) was removed. She has made tunnels in the snow all over the back yard in her travels to her toilet and to visit the goat next door. A red squirrel has taken up residence under the stoop at the side door, giving off confusing and enticing scents. Thank goodness it isn't a skunk.

My Chops

"Alta" in

Chinese

RR 1 Box 52
Nobleboro, ME 04555
April 11, 1984

Dear Friends:

This will undoubtedly be the last of these letters sent from the above address, at least until fall. Eleanor Bickford, who owns this place, will be up on Easter weekend and I'll know then how long I'll be staying and if I'll be back here in the fall. I do know that I'll be in the area come next January.

The galley proofs are back at the printers, my income tax forms are ready to mail this afternoon, I've just finished three monthly quizzes and there's nothing pending except to clean up the house, get the kitties out from under the beds, and write this letter. I'm setting here looking out at the lake, hoping to see the ice "go out", which I have never observed. There is free water around the shore and over most of the lake except right in front of this area. A tongue of free water is reaching down from the north and the remaining ice is turning very black. But the sun is hiding, as it has for several days, and the wind is cold, so maybe the ice will stay until tomorrow by this house. No more ice fishing or dogs playing on the lake. It will be nice to see ripples on the water and the reflection of the far shore, which I haven't seen since a few days after my arrival in Jan.

Yesterday Marilyn McLoon (she is also our choir director) took me around to see some houses "which would be just right for you". One is right in down town Damariscotta but on a side street and has a big yard, part of which is well fenced in. But it is one of those square houses with all bedrooms and bath on the second floor, and has olive green painted shingles. We didn't go inside. The second house is in Bremen, a couple of doors from Lon Gibbs' house and overlooks the lovely Medomak River, or maybe it is Muscongus Bay. It has brand new appliances - dishwasher, huge refrigerator, counter-top stove and elevated oven, and wood-burning stove in the kitchen. Everything is highly varnished. Upstairs is a green, a rose, and a lavender bedroom(s) with linoleum rugs to match - cabbage roses, fern leaves and I've forgotten what went with the lavender room. To me the place has the advantage of being a marvelous conversation piece. Maybe I'll settle for a mobile home or cluster housing, a sensible thing at my age, I guess, although I still would like to build that solar-efficient house.

Rita says that work is finally progressing on both Doug's cottage and the Store addition which will also house the Post Office and both should be ready by the summer season. Last report was that the boat is still off the run. Jimmie decided that as long as she had to be hauled out to have her engine re-installed he would paint and refurbish her for the summer season. I hope she is back by the time we appear on the dock with all the accumulation of the winter, plus 10,000 new "Introductions" in four cartons. I took one carton down to the Campbells in Boothbay Harbor to use on the Balmy Days, last week. They are fine. Their house looks right out on the harbor toward Squirrel Island. Bob is making the run (work run) out there this spring, taking workmen out to open cottages, make repairs, etc.

Time for choir rehearsal. More later.

Thursday 4/12

There is still ice spanning the lake about two hundred feet wide in front of this house, otherwise the whole surface sparkles whenever the sun comes out, which is precious little. It would seem that this spring the sun is going through phases like the moon, right now it must be a "new sun". My hand healed well. Had a cancer removed near my left thumb a while back. So much skin had to be removed that there is quite a bit of tension but it should ease up soon. It has slowed me down a bit, plus our recurrent snow storms, because it is hard to drive when the roads are slippery. Then my front end got out of line and I ruined a tire before I realized what was wrong. It was after a fruitless trip to Falmouth to pick up the "Introductions" the car began to shimmy.

The Women in Medical Care project is bumbling along. Phyllis MacNeil, a retired surgeon from Groton, MA. now living in Newcastle, and I went to Pemaquid Point last week to see Marcia Fite, another retiree and quite a gal. She helped develop and try out medicines for the control of intestinal parasites, carrying out projects in Central and South America and the West Indies right after WWII.

Had a nice luncheon with Marion Watson and her daughter Georgia at Alfred Senter's "Greenery" restaurant the other day and saw Alfred who said Zimmie's eyes are continuing to improve. Teco is now staying with Zimmie and can't be left alone, which makes it hard for Z. He, Teco, is pretty well wedded to his oxygen, but still smokes!

Don Cundy is now back lobstering on the island. Bryan Hitchcock is picking up his traps, delayed by weather and skiiing vacations with the children. You probably know by now that Ruth and Shermie Stanley have separated. Ruth will stay until the end of school. Next week I am going to the Paddock's in Spruce Head for a get-together. Bridget expects Ruthie will be there for the Easter weekend. This separation-divorce fever has certainly run amok on Monhegan. Ade Mersfelder says, though, that he won't worry until Doug and Harry decide to separate!

As May first draws near my calendar is getting very full. I keep planning a dinner here for these friends and those, then get invited to their houses instead. All through Lent we have had church suppers once a week so I'm about out of casserole materials. Last night I baked a blueberry pie with the last of the berries I froze in August. I still have some raspberries and strawberries to put on ice cream before I leave.

Before leaving here I have to firm up arrangements with the monument man who is to work on the cemetery restorations, pick up or arrange for delivery of the tree to be planted in memory of Iris Gibbs, check on the progress of the book as well as rehearse and sing the Maundy Thursday and Easter music - I'll have a small solo part in the Maundy Thursday program.

I think the name of the book will be "Under the Grey Gull's Wing" taken from a poem I wrote:

A seagull wheeled and glided overhead
Mewing and crying as he passed.
And with his cry came memories
Of childhood days spent long ago
On this small isle,so misty and remote,
Happy to play carefree and safe
Under the grey gull's wing.

When I went to Brunswick I stopped at bookstores in Bath and Brunswick and then went to Camden, Rockland and Thomaston. Everyone was helpful and all agreed to handle the book. I have decided to donate a percentage, if any, of the profits for the benefit of the Monhegan Museum Association which was formed following the splitting off of the Monhegan Museum from Monhegan Associates, Inc. I trust this won't be an idle gesture. The book should retail for around $6 or a little more, depending on the number of pictures and cuts to be used. Olive Metcalf has supplied me with several of her drawings of which I'll use at least one for this book and others for the next book,if I ever get it written. They say the first is always the hardest and the longest aborning.

Ad Winship is retiring from Dartmouth this spring and the administration has given him a showing of his painting with an opening reception tomorrow. I had hoped to go, but my hand won't stand up to such a long drive. Good luck, Ad! He and Kiki were on the island getting their place opened for the season last week.

The ice is out. Just before I sat down to type this there was still a fair sized floe out in the middle of the lake, but it disappeared in seconds. The patches of open water began to widen about noon. The a brisk NE wind sprang up and the ice began to move rapidly in huge ragged rafts, pock-marked with pools of black water. They floated down the lake, spinning as they hung up on the shore, leaving small thick floes along the bank, then taking off in ever decreasing dimensions. I ran out to take some pictures of the process and hope they'll show something of this interesting phenomenon. It is good to see the water again. Some ducks flew in this morning but were so far away I couldn't tell whether they were golden eyes or buffle heads. Hope they come back and closer to this side of the lake.

Glen Layman sent me in today's mail a nice plaque to put on the organ he gave me that was in Elsie's cottage. It is very nice and goes on with miracle glue - no nails. That instrument should stay on the island, if and when I move off, but I'm not sure just where. The Museum already has a parlor organ. Any suggestions?

Don Grindle who runs off these stencils said I ran over on the last ones, so I had better say enough for now. Phoebe's fur has grown back well. Both of us have grown pudgy and need walks on the Red Ribbon Trail.

Thanks for your notes and subscription donations. Preparing 130 of these twelve times a year is getting costly.

As ever,
Alta and Phoebe S.

R.R. 1 Box 52
Nobleboro, Me. 04555
March 1984

Dear Friends:

What a full month this has been! The longer I stay here the more involved I become. There was, of course, the hours spent on Income Tax preparation. I took all my papers to Mr. Strout in Rockland with the notation that I couldn't find the exemption paper about the People-to-People trip but would get a copy if need be. You guessed it, the stuff was sent off to the computer, then I got a note saying my tax could be recalculated if I had that paper! I'm still on its trail.

With the tax off my mind I settled down to the material on Monhegan and took it to Lincoln County Press here in Damariscotta and am now waiting to see the galley proofs. After I pick up some photos on my next trip to Monhegan (which I hope will be Monday for Plantation Meeting) the book should go to press. It will probably be about 100 to 125 pages, paper-back, which I hope won't be too expensive. Anyone interested? Next will come trips to bookstores in the coastal area. Zimmy's Spa will be one.

Lincoln County Republicans caucused at Lincoln Academy on Lincoln's birthday, but because no one else from Monhegan was there a caucus couldn't be held so one was arranged for while the boat was at the dock on the 21st. of February. The day before I couldn't stop the toilet from flushing here in Nobleboro, the well drained dry and mud got into the intake pipe fouling up the pump and I **had** to call the plumber, who came just when I was to take off for the boat. Three people held the caucus, but I had all the papers to sign, so must get that done on my next visit.

The following day I did get to the island and had time to spare because the Laura B took off an old jeep of Shermie's and they had to wait for the tide, but I forgot the papers!

There wasn't a speck of snow on the island that day and things were beginning to poke up through the ground. We had had a premature spring for about ten days before the cold, and it has been cold - down to zero and not getting much higher than the teens this past week. The island looked a bit drab and the house most uninviting, with water pipes all disconnected, no plants (they are at the School house for the winter), so I was glad to get back on the boat. Progress on Doug's house and the new Post Office were at a standstill. Maybe there will be more to report after my trip on Monday.

My nieces, Nancy Peters and Elizabeth Gardner, came up for the long weekend. We went to the Lincoln-Sagadahoc Counties combined Republican dinner and they had a great time. Elizabeth even found a classmate, the wife of the president of the Lincoln County delegation. She is following in her aunt's footsteps, apparently. The dinner was a delicious one served at the New Meadows Inn in West Bath.

With my writings at the printer's, my IRS materials at the computer I turned thoughts to other things. There is a monument dealer in Newcastle who is familiar with Monhegan cemetery and is willing to come out later to work on the broken stones, add needed dates and reset those needing positioning. His costs seemed reasonable.

Yesterday I called at Pinkham's Plantation, a local nursery and found they have potted hawthorn trees which can be easily transported and planted so I ordered one to be placed by the Library as a memorial to Iris Gibbs, who loved the library, was active in Arbor Day plantings throughout the country. She died last May after a very short bout with cancer. Lon still lives in their home in Bremen and hopes to return to the island with son Geoffry for a simple ceremony when we plant the tree. This has been cleared with the library trustees, except for the exact location on the lawn.

An interesting project I've been involved with here is one concerning women who have made a significant contribution to the field of medical care either as residents or as people born or educated in the mid-coast area but working elsewhere from 1850-1950. We are having a brown-bag meeting on Wednesday which I hope to attend, if I can find transportation back from the island on Tuesday or early Wednesday. A Jean Palmer Walbridge sent me a nice resume of the service Louise Connaway gave to the island the sixteen years they lived there. If anyone has any other information about women in medicine I'll be very glad to add it to the materials we are assembling. She need not be a physician.

Margaret Baker has been a wonderful friend and has introduced me to a lot of different groups. One day (in the pouring rain) we went to an S & B (Stitch & Bitch) luncheon at Betty Noyce's down on Keene Neck in Bremen. The house is gorgeous and Betty's needlepoint magnificent but the drive down through the woods is not one I would relish in winter. The same group met at LaGarage in Wiscasset Wednesday. Fun!

I drove around to the house the Hudson girls have purchased. There was a workman there, but I didn't try to go inside. The house is high on a hill, next to the old jail and commands a great view of the Sheepscot River and Wiscasset.

On Valentine's Day there was a progressive dinner in Damariscotta-Newcastle area attended by 150 patrons of the arts, and called "Give Your Hearts to the Arts". The houses we went to were all gorgeous, and the last one was large enough to accommodate all 150 for dessert which consisted of cream filled pastry swans served from silver platters. The family had been in the foreign service so the house was filled with exotic things chiefly from Indonesia and the Far East. I went along with Bob and Margaret Baker. They are also members of the choir where I sing - and am to do a solo next Sunday - "Oh Rest in the Lord" from Mendelssohn's "Elijah", the same one I sang in Cedar Rapids several years ago.

This part of the state is so lovely and the people so interesting I have taken a big step and have listed my Monhegan property with a real estate agent, with the proviso of a life tenancy in my part of the "estate" in hopes of getting enough from the sale to be able to build my energy-efficient dream house to use winters and when I no longer feel I can manage trips to the island at all. It's a big step and I'm a bit scared. Meanwhile I am busy drawing plans again and searching magazines for ideas. Better now or never!

I'll save the rest of this page for a report on Plantation Meeting. It is now Friday afternoon and I'm back from Monhegan. The meeting went smoothly. Katy Murdock Boegel replaced Ruthie Stanley as Second Assessor and Winnie Murdock was elected Third Assessor, as well as serving as Moderator, and did a great job. Joe Pomerleau will be the truant officer and Newt was named Truant Officer Emeritus at Rickie Boehmer's motion. I left all my notes on the island but I believe the tax commitment will be about the same.

It began storming Tuesday afternoon, dumped over a foot of snow and then turned to rain. The thermometer went nuts, as did the weather vane. Wednesday at 2 the temp. was 38, at 2:30 it was 26, it had been below twenty in the morning, was 25 the next morning and by boat time (we had no boat Wednesday) it was up again nearly to forty. the wind howled all day and night, Wednesday, and my bed down in the big room at the NW corner of the house was noisy; shingled rippled all night in the over 45 knot gale. By boattime the winds had gone back into the south so the crossing wasn't too bad. I ate dinner Monday and Tuesday with Doug and Harry but didn't attempt to go out on Wed. Managed to find something to tide me over the extra day so lost no wgt.

When I got back here I found the road had been plowed but the snow had been piled up against the garage and house doors. Getting in is like scaling the Alps. Let's have no more snow! Rickie Boehmer fell on her way to the boat and dislocated or broke an ankle. She didn't have to stay at Pen Bay so I don't know what did happen. She fell in the Cochrane's yard, taking a short-cut and wasn't heard for about half an hour!

Doug's house still has no shingles on roof or sides. Bob is having trouble with LURC so work on the new P.O. is at a standstill. By the way, LURC is taxing Monhegan at least $12,000 yearly whether or not we use its services. The high tax is because all our property is considered "shore property".

Just called the island. Rickie dislocated and broke her ankle and has a nasty situation. She and Peter have an apartment in Rockland so she is there, on crutches. Bryan Hitchcock is picking up his traps. I wonder what he'll do now. Don Cundy is still planning to fish from Monhegan and Rusty still has his traps in even though he spends much time in Boothbay. There was a general exodus from the island on Thursday. Vacation was to start on Friday but a "snow day" was declared for Thursday - walking was impossible. Ruthie took her two to her parents' in Spruce Head. A party is scheduled there for tonight, but I'm staying here because of possible freezing rain and I am to sing a solo in church tomorrow and don't dare risk the driving.

I hope I can get this done before Monday when I'm having an operation on my left hand to take care of a "sun spot". I trust all that will be needed is a bandaid.

I've made new telephone books, new Introductions are in press and the book is progressing. I'll probably stay here until about May 1st. Thank you all who have written, will try to send individual regards. Phoebe is fine, a bit too fat, as am I.

Happy St. Patrick's Day!
Alta

May 11, 1984

Dear Friends:

The long awaited off-spring of my fingers and (?) brain has arrived! A few copies will be available by one o'clock today so Sally Haines, one of the choir members, and I are off for coastal bookstores from here to Brunswick, and maybe Falmouth, to deliver copies and hopefully collect a bit of money to fill in the hole in my exchequer occasioned by this venture. You'd think it was a huge tome, I've done so much talking about it. Instead, it is only 144 pages, including all the indices, front pieces, etc. A dollar from the sale of each book is to go to the Monhegan Museum to help keep that new enterprize afloat.

Mr. Stevens from Page Memorial Stone Company here in Newcastle will be coming out during the summer to work on the restoration of the cemetery for which more than $2,000 has been raised so far through the sale of T-shirts, phone books, and the auction of Joe Walsh's skiff. I hope to have more T-shirts for sale this summer but haven't heard from Peter Gorman as yet. He has probably fallen in love with another community by now.

I've just come in from digging a mess of dandelion greens which I am going to take to the Baker's for supper. Margaret and I are then going to the Gordon Bok concert in Damariscotta. I shall miss all the activities here and my many very good friends but it will be nice to be back on the island and seeing everyone there. I plan to leave here on Monday, visit bookstores to the east of here and go out to the island on Tuesday. I promise!

May 12

Sally Haines, one of the choir members, went with me yesterday afternoon while I delivered books in Brunswick, Bath and Boothbay Harbor. Everyone was most interested. I had to go back to McBeans in Brunswick and found both clerks busily reading the book with enthusiasm. (It should take about an hour to go through the whole thing!) The Huckleberry Bookstore in Boothbay Harbor was very glad to get a book on Monhegan. The proprietor, Jared Holden, said the last time he had seen me was when he sprained his ankle on Monhegan and I fitted him to a knitted ankle brace!

Last week Alice Fleming and I drove to Augusta so I could obtain a sales tax number and find out about donating a copy of the book to the State Library. They have a whole room devoted to Maine authors only. We took in the State Museum and the State House; Alice was quite impressed, especially by the museum. We stopped by Louise Webber's and I bought one of her paintings - yellow day lilies, which she did for the Wiscasset show last year. She is only 90! Then we drove to Alna and had tea with Bob and Doris Slayton. The rain stopped before we got under way so we had a good trip. The countryside was lovely in its new spring dress. Augusta is way ahead of us. Magnolias are out and some spring flowers already gone by.

It is pouring today after three lovely warm days. I autographed 50 books and took them to the Maine Coast Bookstore in Damariscotta - I think now they wanted only 15!!! Well, they'll be ready for the autographing party on July 13th. I'm to have another that day at the Personal Stamp and Bookstore in Thomaston. I can't get over all this attention. Let's hope the book sells. I am donating a dollar per volume to the Museum so I have to sell a lot of books, best through the bookstores. The price is $6.95, 35 cents Maine sales tax and $1.00 for postage and handling. Grey Gull Publications, Box 87 Monhegan.

I had a great birthday on the 5th. I decided not to do any work and even forgot to phone Rita for island news. Margaret Baker, Sally Haines and Alice Fleming took me to dinner at Backstreet Landing and then to see "Terms of Endearment". We got in late so it was a bit confusing, but what good acting.

Eleanor Bickford came over Easter Saturday and said she would like to have me stay here each winter, which I have decided to do. This is a wonderful place in which to spend the winter. Maybe I can get out the next book while I'm here next winter, before May 1st, I hope.

Much of April was spent on the book and Republicans. Jock McKernan spoke to the Lincoln-Sagadahoc Med. Soc. early in the month. He is a bright and attractive young man. The Republican Convention was held in Augusta on two of the four sunny days we had in April. Ed Myers (the mussel culture man) and I have decided we've had enough of politics. In four years I for one will be too old, anyway.

Lon Gibbs just called and read me his review of my book. He writes beautifully, but I fear I can't live up to his review.

The sheriff's office has discovered me and I was called twice in one week for medical examiner cases. If I get enough of those I can make expenses, but I'd just as soon not have too many, especially if they are homicides requiring court cases.

On Easter Margaret, Sally and I left Damariscotta by 5:30 a.m. to go to the "sunrise" service at Pemaquid Point. It was lovely, but cold, clear morning and Monhegan sparkled on the horizon. There were two to three hundred people there. After the service we had a ham and scrambled egg breakfast at one of the New Harbor churches, all very nice.

One reason the book has been so delayed is that after the galley was printed I had to write a preface, a piece "About the Author" and make an index. What a job! Then the printing was held up for one thing or another. What am I going to do with nine cartons of books? plus two cartons of "Introductions", plus twelve pieces of luggage to carry back to the island. Fortunately the Laura B is back on the run.

The Tuesday before Easter Barbara Barstow Mann and her family came out expecting to spend a week at the Red House. But that night they had a chimney fire which spread throughout the second floor. Fortunately hoses could be lowered to the harbor so the fire was brought under control in short order, but there was a lot of water and smoke damage. Rita says they have had to remove the entire second floor and two or three walls of the front room. Jimmie hopes to have repairs completed by the time the Smith's arrive, but this weather isn't helping. The wiring for Doug's cottage has been completed and Peter Dirks is now painting the outside trim. The P.O. addition to the Store is still a long way from being completed. Bob has had to make a lot of trips inshore.

Jo Day arrived last Monday and should have the Yew open by the middle of the month. Dyan is back. Steve Rollins came in and got Zimmie on Monday so the Spa is open full time. Harry T. bone is back and Peter Dirks (Mary Beth's husband) is planning to stay all summer. Ruth Lothrop is on the island to plant her gardens and the George Cabots came out to open their house.

Teco is back at Pen Bay but is doing better on new medication. He plans to return to the island, where, I don't know, because his place has been sold, as has Ralph Hinchman's. Ralph's daffodils are particularly beautiful this spring.

Sandy Dickson took some time off from her busy schedule of P/V promotion and spent a couple of days on the island. Rickie Boehmer is managing with a walking cast and is conducting programs on fish preparation for the Extension Service. Martha Yandle has now qualified for Notary Public so someone can function when Karen is off island.

Judy and John Bogdanove are back for the season. Jeanne Rollins is in New Jersey to welcome her father who is returning from the Faulklands on the vessel which went around the world without modern navagational equipment, the Naustar. They encountered a severe storm so were some-what delayed.

Phoebe has just nudged my elbow, telling me it is her suppertime. So I had better bring this to a close. By the way, it is the plumbing, not the wiring, that has been completed in Doug's cottage. We hope that Iris and Lon Gibbs' son Geoffrey will be back from Europe and can get out early in June for the memorial tree planting in Iris' memory on the Library lawn. I hope the Haussers will be there, too.

Jimmie is planning to run Sunday boats through October and the Isle au Haut is planning to run daily trips from New Harbor later known as the Hardy Boat. I haven't learned when, how long on the island, departure times and fare. That will help my getting back here on occasion.

Steve Rollins finally got the water on last week. He was hampered by a broken valve up at the tanks, and then by a lot of breaks which showed up after the water went on.

If you know of anyone who might be interested in my house do let me know. The agent hasn't been out to see it yet so it hasn't been advertised as yet, but it is on the market. I still hope to be able to use my part as long as I want it, but would settle for a rental else-where if need be.

Now to feed Phoebe. Last greeting from Nobleboro until winter.

As ever,
Alta Ashley

I'm dewy-eyed about leaving here. The lake was beautiful in the morning mist.

June 1984
Monhegan, Maine 04852

Dear Friends:

It has been more than a month since I sat down to type out the last letter, and what a month is has been! That time I sat looking out over Pemaquid Lake from the cottage in Nobleboro and this time I am facing Lighthouse Hill, with the Gulf of Maine behind me. The lupine and "snow-in-summer" are in full bloom, lilacs are fading, but the rugosa roses are coming out to perfume the air. The past two days have been perfect September days, cool and dry, after several days of hot and humid weather (but not above 85 degrees) which came on after six days of torrential rains. I haven't planted my garden yet and doubt I'll get much in this year.

As most of you know by now, the day before we left Nobleboro I began to run a fever and landed at the Gambles the following afternoon with a good high one which seemed to respond to antibiotics overnight, so I set off for Monhegan, never dreaming the Coast Guard would have to be called to take me to the hospital the following day. Ten days later I went back to the Gambles, tried out my pins for a couple of days and came out here before the rains came - it began raining as I reached the Spa. A stiff neck, cured by a neck brace for a couple of days, set me back a bit and I lazed about for nearly two weeks more. But for the past several days I've been just about back to normal. It's kind of handy to say I'm tired if I don't want to do anything, though!

Marion Watson came out from Five Islands on Tuesday and stayed until yesterday (Friday), helping with setting up the cottage for renters. There was little to do because Ginny and Phil Baker left the place in good order. Their daughter. Linda Gifford, took Phoebe with her to Readfield while I was sick where she had stayed while Priscilla and I went on our world trip last fall, and she didn't get back to the island until this Wednesday, due to Sandy Dickson's kindness in picking her up in Augusta and putting her on the boat after an overnight stay in Port Clyde. She misses her two retriever pals, but is happy to be back on the island. She looks great. The stay with the Giffords' did her a lot of good.

The summer season is upon us full force with the opening of the Inn yesterday. Barbara Hitchcock has been busy all winter taking in tenants. Amy Melenbacker has bought the Cundy house and has made three or four efficiency apartments in it and they are full most of the time. Marion says they are very attractive.

Don Cundy has been fishing out here all spring. He is living at the Cundy Cottage, which is good for Patsy. Progress on Doug's cottage, which looks like a miniature Monhegan Store, is slow at least until all traps and gear have been hauled out. Lobstering comes to an end a week from Tuesday here on the island.

Vi Sturner arrived Friday with Mary and John-Paul, George and Loretta Green left today after opening their place for the Marders. The Curtises were here with Katy and her three children, getting their place ready for the Casses.

Nothing has been done on the Red House, and it looks as though there is little to save. Ironically, the new South door was finally put in after two years and it wasn't damaged a bit, but it leads into a void!

Jackie and Billie are moving their "Monhegan Framers" shop into the shed behind Newt's house where Ida Proper used to paint. Joe Grilley's little place has been enlarged for Shermie Stanley. Barbara Rothstein has converted the garage on their house to a shop where she is selling Maine and Monhegan crafts. She has made a very attractive place and has some very fine things in it - including Under the Grey Gull's Wing.

The book is selling very well both here and on the mainland. I'm developing quite a brisk mail-order business, too. By the way, should you want to know, the book sells for 6.95 plus Maine sales tax of 35 cents (not charged to out-of-state sales) and $1.00 postage and handling. If sales keep up at the rate they are now I'll begin to bank some money soon, right now I haven't met expenses.

Ray Orne has had to undergo extensive oral surgery and may not be able to come to the island this summer. Too bad, because he has worked so hard to improve that cottage the past few years. Jo Day is still running the Trailing Yew.

There is to be a reception for the opening of the Wyeth show at the Portland Museum on Monday. A bunch of Monhegan people are going. I have a speaking engagement in Bristol on Wednesday, so decided to take in the festivities. Margaret Baker will go with me. This afternoon I bought a blouse and skirt to go with my Chinese jacket at Zimmy's and am all set.

Becky Brown from Marilyn McLoon's real estate office came out on Tuesday and was quite taken with the Mooring complex. They will advertise in Down East, which should bring some inquiries. I'd like to have the sale completed by fall, if possible. I hope the person buying will be compatible so I will feel like staying on in my "High Mooring" during summer and fall. I do love this new room.

Sunday, June 17.

Kim Kenway, who was one of the attorneys for the PUC during our struggle for telephone service, came out with his family for overnight. With them were his in-laws, the Parkers, who are now retired from foreign service and looking for a place to retire to (in good weather) in Maine. They seemed interested in the Mooring, but have to sell a place in Italy before they can think of acquiring more real estate. Amy Melenbacker's mother, Dora Kendall, is also interested. I took nearly a full roll of B/W film of the place in hopes a good picture could be sent to Down East. Maybe I should apply for "house of the month" in Yankee.

Last weekend two TV crews were out here. One filmed a segment for PBS program "A House for All Seasons", developed by a Colorado station. This segment will air sometime in January '85. The second crew was from Channel VII in Boston and will air on June 24th at 9 p.m. on CBS. They interviewed me for both programs. The first I had a ball with, the second was a bit more difficult because I am not one of the Wyeth worshippers. They asked such questions as "What effect has the Wyeth family had on Monhegan", "What do you feel is the contribution the Wyeths have made to American art", and "What do you like best about Wyeth style of art"? I kept appealing silently to Mary Beth Dolan to get me out of the fix and she just sat there enjoying every minute! Apparently the interviewer was quite satisfied, but the whole segment will probably be scrapped.

Dana Pratt and his wife are here now. Last night I walked over to their cottage and asked for help making out my copyright application. Dana is in the Publishers section of the Library of Congress. He said that not long ago he served on a panel with David McCullough, and spoke very highly of him. (Tommy Ashley, take note.) Margery was here earlier, but I missed her because of my sojourn at Pen Bay Med.

I forgot to say that Colorado TV crew was interested in alternate energy, particularly in photo-voltaics, so I was in my element. Both Sandy Dickinson and David Sleeper were very pleased with how all the interviews went. There are now fourteen systems on the island, including those home-made ones of Danny Bates, Lawry Cooper, and Bob Boody. They have certainly cut down on generator noise around the island.

Tuesday I'll be tearing around in the Damariscotta area delivering more books to stores. The Lincoln County press is having a hard time keeping up binding to meet demand. And both Yankee, Down East, and even Courier-Gazette said the book wouldn't sell!

Another thing I'll have to do is check up on Page Memorials to see when or if they intend to come out to fix the cemetery, which is really in awful shape after winter. As for the memorial tree for Iris Gibbs, some people are now wondering if the children will get hurt from the thorns on a hawthorn tree. No one has gotten worried about the wild hawthorns that are all around, or the japonica bushes that are covered with thorns, so I may go right ahead with previous plans and see about the tree when I'm Damariscotta Tuesday.

Ruthie Stanley's house is up for sale. Shermie is living in the old but revamped Joe Grilley place. Sarah Burnham has set up her kiln and is producing nice pottery which Barbara Rothstein is selling in her shop located in the garage at her house. She has assembled some fine examples of Maine and Monhegan craftsware and is doing well. both winter and summer craftsmen are represented, not restricted to winter people as Winter Works was.

Amy Melenbacker and Bill Baker have put in four efficiencies in the former Cundy House and are filled up much of the time. Barbara Hitchcock has been busy with her apartments all winter. Bryan has gone to Alaska. He pulled up all his traps early in the year. The Werbes are back in the Tam cottage - Reuben and Gerry aren't expected to visit this island for a year or so; they are happy with there Hawaiian house.

The Lewis house is not on the market at present. The Levys are in Nigh Duck cottage and luxuriating in the plumbing after so many years in the Lobster Pot, which had no facilities of any kind.

Orders for four more telephones have gone in and more are expected before the summer is out. We'll reach the 96 lines capacity before we know it. Hi there, NET and NYNEX!!

Best Regards,
Alta Ashley

P.O. Box 87
Monhegan, ME 04852
July '84

Dear Friends:

Where to begin! This has been a busy five weeks or so since I last sat down to write you a letter. Don Grindle from Rockland Business Machines came out on Monday with a new copier, took back the other one which required coated paper, and the mimeograph. Coated paper sticks in damp weather and since that is about all we have, the copier just didn't work more than a tenth of the time. Then, I realized that to keep a mimeo in good order it must be used at least once a week and mine was used about once a month. So--I decided to take another shot at a copier. This one is about one-fourth the size of the Sharp I had in Falmouth. It requires generator power, but will do 99 copies in one setting. We'll see how it works with this letter. Sending in masters to have Don run them off and then getting the copies back here didn't work out too well once I was back on the island. So here goes!

The big news is that the books "Under the Grey Gull's Wing" are sold out. I ordered 1000 more back in June 27th and haven't had delivery yet. I also made application for copyright and Lib. of Congress registry even before that and have heard nothing, so that data won't be in this printing, either. The Press is correcting the many mistakes, some mine, some theirs, so the second printing should be better than the first. This little venture has taught me how to go about things, so my next book should be better organized as to distribution, etc.

The summer is marching on. So far Elizabeth and Tommy, David Gardner and his girl friend Rose have been guests. Having the old office space turned into a guest room has helped a lot. Eleanor Bickford came over for the day, as did Alice Fleming, Betty Allbright Ransom and her family (in their own boat), and the Furmans in theirs. Having the Hardy coming over from New Harbor four times a week has brought a lot of people over from that area.

The weather has been pretty foggy. When the Hardy started her run it was so thick of fog that week she gave up. But on good days all three boats are filled to capacity and the shops are doing well. The Wyeth reception at Portland Museum was a gala event and quite a crush. I doubt many really saw the paintings; all were too interested in greeting old friends and munching on goodies. The talk at the Bristol Historical Society meeting went ever so well. I enjoyed it, anyway.

The autographing parties in Damariscotta and Thomaston on the 13th were less successful. It was in the 90's and many people just didn't want to stir from their homes. But I think between the two bookstores we sold close to eighty books. It was fun meeting my reading public. Between the two sessions I sandwiched in a luncheon at Farnsworth Museum and saw a lot of old friends. Living in Maine gives you the feeling you know the whole world, there are so many connections wherever you go.

Young Joe Walsh stopped by in June to see about setting a stone in memory of his father, so I sent him to Page Marble in Newcastle. Two men were out last week and in two days had straightened and cleaned more than half the stones. Three had to be taken back for repairs and I'm hoping to get local help to clear away the weeds, alders and spruces that are growing out from under some stones and toppling them. The town has voted to get fill and to renew the fence, this time enclosing the entire cemetery. That might help keep rowdies out.

There is more activity as to phone services. More and more people are applying for phones, some who were **adamant** against having a phone in their house. But now that messages aren't delivered and one can make only credit card calls, even locally, the need for one's own service is great. The company may soon install Photo/voltaic power, especially since cost of materials is coming down - none too soon for me. I'd like to expand my system and make the two houses independent energywise.

I've had no real bites on the house, but quite a few "interested people" making contact. Now that I have come back to the island and sat in my front room gazing at the water, the sunsets, and the general activity of the island I am very reluctant to think of giving it all up for at least five years. I may settle for not renting the cottage and using it only to accommodate friends for the next few years and be relieved of the hassle of landlordship.

The library addition was dedicated on June 27th. It is in two levels. The lower will be known as the Elisabeth Ebert Reading Room and the upper as the Angelina Tribler Nunan Children's Room (the original had to be removed to build the addition). Jackie Boegel rearranged all the books, made and hung the draperies (tan with a faint darker brown design), and she and Billie Boynton finished the chairs and tables. It is a very airy and pleasant library now, and no longer crowded to the gills with people trying to browse or hunt for books.

Many of you apparently saw the article in the Boston Globe on Monhegan since the telephones written by Christine Tree with photos by Merry Farnum. Both girls interviewed me at length and Merry took about 50 pictures of me in various poses. Zimmie says she took 75 of him! I wonder where they all are. It was a good article and I'm glad not to have to see myself again. Thank all of you who sent clippings. I have now filed the story in the albums for the library and museum.

The Museum opened on the first with a reception for the Alice Kent Stoddard show, the special exhibit for the summer. It has been the most successful season (financially) so far even though the boats haven't had as many passengers as last year. This is a good omen for this, the first year of independent existence for the museum. The money from "The Grey Gull's--" will help, too.

Newt Searls is back. He was in a rehab program in Rochester and will return to it in the fall. Teco is now, I believe, in a nursing home near Portland. Togus wouldn't take him. Carl Wincapaw has lost weight and looks like a million dollars. He was so spiffed up the other day to go to his 24th H.S. reunion we hardly recognized him. Work on the new P.O. is at a standstill until the Inn closes and progress on Doug's house takes precedence to a dozen other jobs Bill Sweetland has. Charlie MacDonald has begun to clear his lot on Horn's Hill and John Murdock is putting in his foundation close by up back of the Cundy Cottage field. Phoebe finally made it back to the island in mid-June and has been a model dog ever since.

The back of the cottage has been shingled and not to be out-done, I have shingles on the right hip, nineth day. I started right off with prednisone and local cortisone ointment, plus lysine (any straw in the wind) and have gotten along pretty well - no fever, not much pain, some itch, and a good excuse not to do anything I don't want to do! Let's hope I'm over the worst. Doug won't believe it is the shingles because his still hurt after seven years.

Ruth Harlow has left to visit her new granddaughter Ashley. I hope she can stay longer next year. End of paper, so enough for now.

As ever,
Alta

July 29, 1984

P.O. Box 87
Monhegan, ME 04852
August 23, 1984

Dear Friends:

Here it is more than four weeks since I started my last letter, and nearly the end of the summer. The season has marched along through fog and drizzle, some heat, and several gorgeous days, especially this past week. We have had so little rain since the deluges stopped at the end of July that things are drying up, believe it or not. But we are in for another deluge this morning, an answer to my wish for rain. The last time I made such a wish it rained six days straight!

Thank goodness for my well. I didn't put the cottage on town water, as I had planned, so we were spared the days of no water during the past two weeks. Two more well points have been sunk and Barry Timson is due to come out to advise the Water Company on increasing delivery. It may all be due to inadequate pump, because Laurie Cooper sank over his ankles in water the last time he went into the Meadow, so the water is there, if not up in the towers.

The evening of the Monhegan Associates meeting we all got a scare when Stevie radioed from his boat out on the garbage run that there was smoke in the vicinity of Green Point. It was about five-thirty and the volley ball game was in progress. People ran to the beach to board Lance's, Courtney's and Lexi's boats, others piled into the emergency vehicle and the fire engine, and still others ran along the shore towards Green Point. In short order they were back after learning that a man lost on the trail had set a signal fire after not being able to find his way out of the woods for over five hours. He had done so after carefully constructing a fireplace on a ledge so all the anger built up in everyone's mind was quickly dispelled when the truth was known. But it did scare everyone because of the extended drought. It certainly dries out quickly here after only a week or so of no rain.

Many of you will be glad to hear that George Green is doing well. He is on chemotherapy so decided against coming back this fall but hopes to be back in the spring when his course is completed. His tumor was in the bowel, not the stomach and his prognosis is good, according to his daughter Janet.

All sorts of things have been going on this summer. There were several ecology lectures throughout the season. Ruth Faller and Sally Rand conducted a walking tour of the old houses from Harry Bone's to Harry Odom's with Ruth telling of the history of each and Sally telling how each can be dated by its architectural findings. I missed that because of being at Colby for my annual Forensic Medicine course, which was particularly interesting this year. There was a Dr. Odom from Honolulu who was quite interested that Odoms were here on the island, so I gave him a copy of "Under the Grey Gull's Wing", the second printing of which is beginning to sell quite well..I have over $1000 to give to the Museum whenever its tax-exempt status is established.

An appraiser came out yesterday to look at the property, and one was here while I was up in Waterville, so by September 4th I ought to know whether or not the present buyers will be completing the deal. If they do I'll have life-tenancy, so won't have to move out of this part. After spending the summer gazing at the ocean out of my front room window I would find it impossible to give up this place completely.

Both the Monhegan Associates and the Library have held their annual meetings. Ray Werbe is the new M.A. president, Faryl Henderson the Library's. The Museum association is ready to go as soon as we hear from the IRS.

It is now Sunday morning. Bruce and Elizabeth Gardner arrived on Friday's late boat and they have "done" the island these past two days. Bruce will be leaving today to do some visiting around before heading back to Charleston, S.C., Elizabeth has decided to spend the beautiful weather forecasted for the next few days, at least in part, on the island. It is lovely to have such weather but **we need water!** Even my well is beginning to falter.

Pepper and Floyd Milbank of Hayford Heights days and their friends the McCarneys of Owl's Head came over yesterday and were here to look at my pictures of China and the USSR. I'm afraid there was more munching of goodies and sipping of ice-cooled drinks than conversation about the trip. But it was quite a gabfest. Floyd is still at the VA in Albany and Pepper is doing real estate there in Albany, too.

While I was at Colby Phoebe stayed with Ina and Ray Sturgis who are renting the Cameron Cottage on Horn's Hill this summer. The other day I met Ray outside the Periwinkle and Phoebe nearly wriggled out of her skin with delight, then started to follow Ray home! And I thought she adored only me!! I guess I haven't been taking her out to walk enough. I'm beginning to feel insecure on the trails, especially after having the shingles and then a week of dizziness. But all that has gone, even the shingles (except a huge pile from the cottage, now piled up on the back hill).

No more work has been done on the cemetery except for the bushes and small trees David Kent cut down around the fence. Now you can see the stones from the roadway again. The town is trying to find a reasonable way of getting a load of loam delivered to the site. Over Labor Day weekend there will be a service for Cyril Nelson who died last month at the age of 91 or 2. Funds are being collected in his name for either the library or the museum, he did so much for both institutions, especially for the library, in their early days. Quite a bit was collected in memory of Joe DeMartini who died on his way here from Nantucket early in July.

Some of you may have seen Ramona Barths exuberant piece in the paper about a "flotilla of little boats" filled with gals coming to Monhegan to pay homage to me!!#!! It was supposed to occur yesterday but I told Ramona I was upset, felt I had more than enough publicity about the phones, photo-voltaic systems, the book and what have you, had never made a name for myself as a physician (which they were supposed to be honoring) so she called the whole thing off, even the picnic for those wanting to visit the island and me. Too bad, because it was a beautiful day for such, but thank heaven no reporters came.

My previous tenants were Alan and Ann Bernstein whom I knew in Augusta. Ann came over for Clay Kent's workshop in water color painting and found the session very helpful.

The people interested in the property invited me over to New Harbor for a visit to get acquainted. They have a house in New Harbor which once belonged to Zimmie Brackett's grandfather and is next door to where Jo Davis Townsend (Manville's sister and who, with her husband Fred, used to run the Store winters for Doug and Harry) lived until she died last year. They were so fond of Jo and have made so many trips to Monhegan they want to have a place here, as well as on the mainland.

Charlie Martin has another book out "Summer Business". Zimmie is busy selling all these books by Monheganites. Larry Goldsmith has one on painting. Lee Court has an expensive one out, but I haven't seen it on the shelf. It is more of a reference or cocktail table type of book. A print of the painting of the Red House (before the fire), embellished with a drawing on the bottom of its mat is being auctioned, no, chances are being sold for the benefit of the School Trip fund, Labor Day weekend. More than 1000 chances have been sold already.

More telephones have gone in and the number is creeping slowly to 96, the limit of the system as set up at present. NET is still considering photo-voltaics. I am hoping to add to mine and other people are planning for systems but Dave Sleeper, who brought over his truck a few weeks ago, hasn't shown up. He was due last Monday. Sandy Dickson is planning to have another group of energy conservation people over on the 7th, so maybe David will appear by then. She promises to start the interviews and visits at this end of the island. Heretofore they have started at Lobster Cove and never reached my place after making me hang around all day long.

Speaking of no progress, the Odom cottage has had no work done on it since last spring. Bob Burton hasn't been able to do anything on the new post office but will as soon as the Inn closes in mid-September. Doug said he and Harry might move into their place by 2000 AD! The John Murdock house is coming along and Charlie MacDonald was here to do some measuring for his place which will go up on one of Harry's lots on Horn's Hill. Paul Horn has raised the roof on his cottage and redone the sunporch, so they should have a snug place come winter.

Mike Loew's place is up for sale and Ruth Stanley is still looking for buyers for her place. The Kalurs from Cleveland bought Ralph Hinchman's and have had a great summer there. They even have a phone. Ad and Kiki Winship and the Sullivans are waiting to see what can be done about a telephone pole which needs to be erected for their services. "We Love Our Tower" shirts are being worn again. I decided not to try to sell any more this year, but telephone books have sold so well I have to keep making more. I think about two hundred have been bought since early June. So many numbers had to be added I labelled the last set 1984 (S). Bob Smith finally got his phone installed this month.

Jimmy Barstow has been spending some time here working on the Red House, which will be quite nice by the time he gets it done.

Chris and Molly Black Buscaglia have been living in the apartment Sturge and Eleanor Dexter lived in following the big Store fire that destroyed the Starling-Wincapaw house where they used to spend the summer. On Friday they gave a party to introduce Molly's son and two daughters and their families. It was quite an affair. Perhaps the most unusual offering was baked garlic - delicious! Elizabeth and Bruce came in time to take in the party.

The North End Pizza has had a banner year. They have been making the best pizzas I have ever eaten. There is a crowd around the window each noon and evening and people can be seen carrying large square boxes home to all parts of the island. This August has been a good one for all the stores, inns and rooming places.

I must correct a previous remark. Floyd Milbank retired from the VA last January and is doing consultant work and working on special projects. Can he be old enough to retire? Tempus fugit!

Teco is still in and out of the hospital. He has moved out of his cottage and the people now renting it have option to buy. Monhegan without Teco isn't quite right but he hasn't been able to cope for the past few years.

Phoebe is doing well and running around like a puppy. People can't believe she is at least nine, although her muzzle is quite white. We both should eat less or walk more, or both, however. The Museum has had its best year yet and much to my delight people have voted the portrait of Gertrude and me the best of the Alice Stoddard show, the special for the season, There will be a Jim Fitzgerald show at the Farnsworth with a reception on September 30th which I hope to sandwich in with a speaking engagement in Camden for AAUW and a visit to Tuftonboro when the Riddles come on the last two weeks in September. And so----

Best regards, as ever,
Alta Ashley

Box 87, Monhegan, Me. 04852
September 23, 1984

Dear Friends:

Here it is already fall and time for another letter which I hope to be able to run off before the boat tomorrow when Phoebe and I will be off to the mainland for a busy week in shore. The car needs a 25,000 mile servicing and I a new permanent which I plan to get while the car is in the garage tomorrow. Then we'll go on to Newcastle and the Bakers' for overnight, with hopes of having time to stop in at the real estate office and the Lincoln County Press to pick up more books and to leave the manuscript for a new book to be looked over. There is still no decision about selling the house. The buyer has now said I will have to pay rent, which was not in my agreement, since I came down so far on the price.

The book is still selling well, though nowhere nearly as fast as in June and July. However, we have made quite a dent in the second printing and have gotten rid of all I had on hand. On Tuesday I am one of the speakers on the AAUW "Maine Authors" program and hope to sell a few books then. Me a Maine Author!!

Sandy Dickson brought out a group interested in alternate energy systems, among them the builder I contacted last winter about possibly building me a house. Peter Talmadge was also in the group. He now lives in a house he designed in Kennebunk which is almost completely dependent on solar energy for power. He outlined a system for me, should I build that house, all for little more than it would cost to buy regulation equipment, and that difference would soon be eliminated by the lack of fuel costs. We shall see.

Things were a bit lively around here early in the month. Katy and David Boegel put on a pizza and beer party for the island to meet David's family. Everyone came and twenty large pizzas were needed to satisfy everyone's appetite. Barbara Horn's father entertained with his card and sleight-of-hand tricks.

An 80th birthday party for Zimmie Brackett was put on down at Jamie Wyeth's on the 8th, five days early. Mary Beth counted 110 in attendance! So it was quite a crush. Zimmie thought he was having a dinner with six people and was moved to tears when he saw all the people and began opening his presents. A huge orange Harvest moon rose out of the water to celebrate the occasion. It was so bright we didn't need our flashlights to walk home by.

The Balmy and Laura B are still running daily bringing out hords of trippers. Yesterday two groups of Audubon members came out, some for just the day and others for overnight. Stell Shevis and some other women artists came out for a dinner which Sally Murdock put on for them. I haven't learned who else came out. Phyllis MacNeil whom I met last winter in Newcastle came out with one Audubon group, also Jeff Tourney and his wife. Jeff was Director of the old Knox Hospital. The weather has been superb all month and people still are interested in day trips. The Hardy boat gave up her trips and probably the Balmy made her last one today. The Laura B has been making two trips to bring out lumber for the Murdock house which is going up back of the Brown's place, and bringing fire wood and gas tanks, in preparation for winter.

Ben Odom came out yesterday with cabinets for the kitchen in Doug's new cottage. Work is proceeding, though ever so slowly, but Doug and Harry are in hopes of moving in before winter, even if they have to do some of the finish work inside. Outside shingling is still just as it was in the spring.

More telephones have been put in or applied for, bringing the number to 75 by my last count, and telephone books have been about as popular as my book. I have a hard time keeping Zimmie supplied. With every flurry of new phones I have to type up new masters. Thank heaven for this new copier.

A couple of days ago I dragged out my old letters dating back to 1963 with hopes of getting them ready to show to the people at the Lincoln County Press. Somehow or other they have gotten out of order, what with having been sent here, there, and yon to see about being published so this room is now covered with piles of letters in my vain attempt to find the missing pages. I finally have found all but two or three and can probably get along without them, if need be. Even so, there are nearly 300 pages of double space. Will anyone plough through all that? And that is only up to January 1977.

After the lecture and luncheon at the Sail Loft on Tuesday Phoebe and I will be off to Center Tuftonboro and the Farm to visit a few days with John and Rue Riddle, on from San Francisco. They left there just in time to miss the heat wave and enjoy fall weather in New England. On Sunday there is to be a reception for the opening of a Jim Fitzgerald show at Farnsworth Museum in Rockland, so I won't get back until a week from tomorrow, if then, should I decide to stay in for the medical meeting on the 2nd.

Jean Hays is in the cottage now, the Nesbitts finally made it for ten days in September. Phil and Ginny Baker will come when Jean leaves and will be here during World Series week, as usual. My last tenant had to cancel, and I'm really relieved because the big "on demand" generator needs an overhaul.

David Sleeper is due soon to change over my photo-voltaic system and enlarge it so the cottage is independent of my part of the house. He was to come out a week or so ago but was held up because my new panels didn't arrive until a few days ago. He says all the work will be done in two days, but I don't dare hope for such speed. I just hope everything will be completed in 1984!

The Yew and Monhegan House will close Columbus Day weekend, then only places with efficiencies will be open. North End pizza is still going as is the Spa, but Ade's Rockcrest shop closed today. Most human robins have closed their cottages, but the Hudsons plan to stay until November. Bob and Mary Burton took off yesterday so the big generator no longer supplies power to the Store and other places. Time marches on.

Teco is now in a care facility in Old Orchard Beach, still not doing well. Bill Varnum looked great when they were here in August.

I guess I had better finish this on my return from New Hampshire.

Alta

(Sept. '84)

It is now October 1st. Phoebe and I have just returned from our week in shore. Last Monday I got a permanent while the car was being serviced and we were both ready at the same time, so off we went to the Bakers' in Newcastle for the evening, after dropping in at the real estate office in Damariscotta, being told of a lot of land just a mile from town. We drove over, looked at it and decided to make an offer. Yesterday Margaret Baker and I went over again and looked at it and I plunked down the cash for a downpayment!

The lot is the third in from Bristol Road and overlooks Damariscotta River (when the leaves are off the tree). It is perfect for a solar house so now I must get ahold of the builder in hopes of starting by late fall or early spring. According to the agent the closing of the sale of this house (with my life tenancy/agreement okayed) should take place before the end of the month. I'm a bit overwhelmed but everyone thinks this is a good move and an excellent place for me to build.

The program in Camden last Tuesday was great fun. Howard Scheele of the Scheele keel fame, preceded me on the program and he got the audience in good spirits with his humor. There were many familiar faces scattered through the audience and I autographed scads of books. Twenty of mine were sold for the benefit of the AAUW scholarship fund, all profits of the Owl and Turtle Bookstore going to the fund.

We had a delicious luncheon at the Sail Loft in Rockport and then set off for Center Tuftonboro, NH and the Riddles with four lobsters in a nifty insulated carrying case, received with utter joy by John and Rue. I don't remember much about Wednesday except that in mid-morning a squall blew through, bringing down many of the leaves - the same squall which tipped and sank the Isaac Evans, a schooner which had visited Monhegan only a few days earlier. The Drownes, Huxtables and Pettees were all able to come to an impromptu cocktail party we got up on Thursday. It was so nice to see them all again and to talk of the good times we all had had with Emma.

One day Norma Phillips and I went to Wolfeboro while I peddled my books at the two bookstores, and, with Norma's help in introducing me, was able to leave eight books. We drove back along the lake where the colors were beautiful.

John, Mitch, Jaffy and Zane Allen (Emma's granddaughter and family) came up late Friday and John took the boys for a snipe hunt Saturday night armed with a flashlight, two sticks and a large paper bag which was empty on their return!

Sunday I left early and reached Newcastle in time for choir rehearsal and church, then to the island on Monday. It was great to be back. I probably won't go in to stay for any length of time until after Christmas. The house in Nobleboro won't be ready for me until the first of November, and there is a lot to be done here before I leave for the winter. I'll be in the same place on Pemaquid Lake, if all goes well.

Alice Fleming went with me to the Opening Reception for the Jim Fitzgerald show at Farnsworth yesterday. Everyone I know must have been there. There was a good contingent from Monhegan, including Jamie, Ed Deci and the Huberts, of course. The show is great as is the companion show of the works of Beverly Hallem of York and Abbott Pattison of Camden. Hennig, the author of the Fitzgerald Biography was there.

Next week back to the mainland to the dentist and hopefully to pass papers. Hail to Autumn!

As ever,
Alta & Phoebe (woof, woof)

P.O. Box 87
Monhegan, Maine 04852
October 16, 1984

Dear Friends:

While David Sleeper and David Stevens are on the roof installing more P/V panels I'll take advantage of generator power to start this letter which I'll probably not finish until next week. This system will separate the energy going to the two parts of this complex. Thus one system won't be robbing the other of power when too many lights are on in one or the other houses.

On Friday we will be going in to sign papers for the sale of this property and the lease for the life tenancy for me to use my part which includes the 1969 addition and the 1980 big room in the back, plus the enclosed porch which separates the two buildings (or connects them!).

Ever since I last wrote we have had superb weather, at first a bit too cold but followed by two weeks of Indian Summer, not technically true here because we haven't had a killing frost as yet. Evenings have been clear and sunsets have been glorious pink-orange so characteristic of those at this time of the year, and Mt. Washington has been visible on quite a few evenings. (yes, Bill and Fran, it appears south of Pemaquid Light, as you said).

I should be out planting those bulbs which arrived a year ago, but now I can't find them. The best thing I can think to do with them is to stick them in around that mound of "compost" the telephone construction men dumped on the front lawn two years ago. At present it is hosting a healthy growth of mint and wild roses.

The island is pretty well buttoned up for the winter. Zimmie left yesterday, turning over the Spa to Rita White. Ade closed the Rockcrest Shop a week ago when the boats stopped carrying "trippers". Barbara Rothstein has just about shut down operations in her Monhegan Cottage Industries shop, although she is retaining some things for islanders who may want to Christmas shop there. Sales of my book still go on, 'though at a slow rate. Every so often there is a spurt in mail order sales, which helps. As soon as the Museum finances are straightened out and it becomes financially independent of Monhegan Assoc., Inc. and donations are tax deductible I'll send off a check for over $1300, representing the dollar a book I promised. I won't be quite so generous another time but I'm glad to be able to help during its first year of struggle. If and when another book comes out I'll have to keep pretty good books not to cheat the Museum.

The air is so beautifully warm and clear today my wash dried in just about three hours. Things smell so good when hung outdoors I never want a drier. I hope I can have a clothes line when I build my little house in Damariscotta, which I hope to start planning as soon, or before, the ink is dry on the papers on this place. It would be nice to be able to start the foundation this fall.

Speaking of houses, the Odom cottage is really progressing and they are in hopes of moving in before snow flies. The John Murdock shop with apartment above, where they plan to live until the main house is built, is already closed in so they should be leaving the Jessie Dunbar cottage before winter sets in. Paul Horn has the roof on and the sides finished where he has added a second floor to their cottage. While he was here for several weeks this summer Harry T. Bone raised his shed to give head room to the cellar and was able to complete that work before being called back to his ship, thanks to a heavy storm which kept the Laura B in port and delayed his departure a couple of days two weeks ago.

(Oct. '84)

Doug Boynton finished covering those awful Kelly green shingles on my roof so now the entire place is silver gray. (I need a word processor!) Kevin Nelson shingled the entire east wall of the cottage and finished shingling the south end, so the cottage looks great.

David Sleeper has decided not to separate the energy systems here but to put meters on the 110 volt power lines going to each side, thus measuring the generator use, which sounds like a good plan. There are now fourteen batteries in the cellar and ten panels on the roof, enough for the whole Monhegan House, it would seem. Conversion to 24 vDC from 12 means I have a lot of bulbs and fluorescent lamps to trade in. I just bought 6 bulbs at $2.69 each, but there are several people here still on 12 v who are interested in a trade. According to David the 24 v system will be kinder on lamps and bulbs. Right now I have quite a collection of lamps for my nephew Tommy "The Lamp Doctor" (complete with one of my old stethescopes) to work on.

Chris Roberts at Lincoln County Press looked over those letters dating from 1963 to 1977 and thought they were worth printing, so here I go again rereading and editing as necessary, cutting out repetitive materials and fitting photographs to the text. Let's hope this will be at least half as successful as "Under the Grey Gull's Wing". I'm still fishing for a catchy title and possibly a poem suggesting it.

Rusty Court's house has been sold to an Italian family who were so impressed by our church and the service, during which Mrs. Boylan sang a solo, they wanted a place to live in on Monhegan. Maybe they are friends of my soon-to-be landlord.

It is now October 24th and papers have not come through on the financing of the sale of this property so the deal has been called off. I'm in hopes that something can be worked out with another buyer so I can go ahead with the solar house in Damariscotta to be built on the lot I am to close on in the next couple of weeks. It's beginning to sound like the delays Doug and Harry have been through with their house which really looks as though will be habitable in a few weeks, we trust before Trap Day!

David Sleeper had to leave for a job in Labrador before finishing things up here. So there are no fluorescent bedside lights, lights in the bathrooms or any at the foot of the stairs. He had to rig up a temporary system so I can run this typewriter and the TV without resorting to the generator, but I have to run it to charge the batteries for the system and have no dials to tell me what their charge is. What a mess! I hope he doesn't want to stay in the North once he gets there.

It is now the 30th, so I had better finish this and get on to other things. I'm hoping the new deal on the house will come through before I leave to pass papers on the lot in Damariscotta. The other day I drove by, saw that the river is visible in Winter and met my next door neighbor, a Mr. Shackleford, who seemed pleased to have me as a future neighbor. The heat is on in the Nobleboro house, the electric door to the garage works so I may stay over for a while when I go in next week. I didn't mean to stay until the first of the year but there is no sense in heating two houses. Meanwhile I'm struggling with wondering what to do with what switches on my control panel and hope David comes back soon before I do any irreparable damage.

They are laying wall-to-wall carpet in Doug's house today so he and Harry should be moving in shortly, I hope before I leave. Bill Sweetland has been working nights to get things done. He had to go to Boston because of his mother's illness - inoperable cancer, all very sad.

Many of the young people went on the special and last cruise of the Scotia Prince over the weekend. The weather was horrible and the seas must have been rough. The Laura B danced around a lot when we came over on Saturday and dense fog settled in for two days. The last Sunday trip for the Laura B was none too pleasant for the couple who came over just for the ride!

Victor Fuller is here doing studies on the energy requirement for the telephone system. As it is set up now they need to run the generator six hours a day, not the two or three they thought would be needed. So the cost of the system is more than expected. But more phones are being applied for. Don Cundy has moved back and has ordered a phone and Sally thinks she'll get a private one separate from the Store. Danny Bates has dug the trench for the Sullivan and Winship phones which will be put in by spring. There are only five phone books left but I don't plan to print more until spring and the new phones are in.

No killing frost has hit my garden yet although low lying ones probably were nipped last night. The deer have demolished my syringa bush again and also the chrysanthemum which Lon Gibbs sent me in the hospital and was full of buds, ready to be taken in doors.

It will be a relief to have the election over next week. Signs for Democrat candidates went up on Newt's and Mary Beth's lawn last week but the former ones disappeared. I have the makings of a Republican sign which I may put up down by the Spa over the weekend, unless it rains again.

HAPPY HALLOWE'EN!

Alta

Phoebe was quite sick a couple of weeks ago. I found what appeared to be a Lyme tick on her and gave her some tetracylcine and she recovered after a couple of days. She had me and the Bakers mighty worried for a while, though. Now she is back to her bouncy self.

Hope you can make sense of this. Hope to get a down payment on my condo. Am offering a mortgage at 10% to help with construction costs of new house. We had 4-5 whales playing around all morning.

Nov. 18, 1984

Dear Friends:

Here I sit in the midst of utter, absolute, and complete disorder, if anything can be said to be all that. The other day I began **thinking** of packing up for my winter move inshore some time after Trap Day, with long spells here and there until then. It was a good time to hunt for long lost articles, such as my blue wrapper, a pair of Bean moccasin stockings which are so good to slip into on getting out of bed, and one or two other things I can't now remember. This meant emptying my closets and shelves and even much of the attic, filling all the rooms with boxes overflowing with outgrown and worn out clothing, old sneakers, photographs and what have you. Peter Dirks stopped by to pick up some furniture he had promised to repair and said "Good gracious!" and quietly backed out in panic.

Some of the stuff has been repacked neatly into new cartons ready to go ashore tomorrow. As far as I know now, in this on again, off again deal with the sale of this property I'll be meeting with my lawyers tomorrow afternoon to pass papers for the buyers to take possession immediately. Things all came to a head early this week when I heard that a restriction had been placed on the sale because of a "specific performance" (which I don't understand too well) and that its removal would entail a long court battle which I might lose. Since I have neither time or money to lose I was advised to go ahead with the sale to the first prospect, even though there were imminent prospects of a sale of the cottage as part of a condominium with cash on the barrelhead by the end of the week. What a mess!!! I wonder how pleasant it will be to live close to someone who threatens a court action. I feel out in the cold, alone. People on the island are apprehensive, to say the least, and my prospective buyers are disappointed beyond words. They would have made such good neighbors, too.

This winter Phoebe and I'll be back on the shore of Pemaquid Lake in the same cottage as we were in last year. The dining room table will be covered with plans for the solar house which I had hoped to get started on this fall, but if it goes up in the spring it should be ready for next winter's occupancy.

The island has slowed down. Practically all human robins have left. Faryl Henderson and Fred Wiley are here but last night's cold may send them in tomorrow. Chris Nicholson is still here. The family who bought Rusty's place are here now and plan to spend as much as time as possible here. Their five year old girl is attending our school. He (father) is in NY legislature (or maybe a representative in Congress). They are happy with the house, the island and the ability to stay here a good part of the year.

Doug and Harry are away now but hope to move into their new

home on their return from Thanksgiving in Cleveland with the Schweiers the middle of the following week. Phoebe and I shall be back then, too, in time for a house warming. David Sleeper says he'll be on the same boat. Let's hope so. I have been left with no bathroom or bedroom (fluorescent) lights or any light at the foot of the stairs since the 24V system was put in.

I've just come back from taking a house-warming gift of sundries to the Odom cottage. There is still a lot be be done - all trim around the windows, doors to be stained, plumbing to be finished and Sweetland has gone for until after Thanksgiving.

John and Winnie Murdock have moved into their apartment, Bob Burton is working on the inside of the new P.O. Don Cundy will be back for lobstering and will have a new phone, Bryan is back so Sonny Remick has moved out of the Novelty; he's away now but will probably move back to June's with Courtney. June has been hospitalized. Sonny's boat is still here in the water.

The Laura B is in Stonington for repairs to her shaft and for winterizing. Last week the Desparado made the run, this week it will be the Balmy Days, thank goodness, because so many are going in and I alone have a shipload of luggage.

The kids came around today to collect cans and bottles to raise money to send to Ethiopian relief. Someday I hope the Ethiopians will try to help their own people. All they seem to be able to do is spawn children.

In cleaning out things I discovered I had about 1,000 sheets of this paper so I'm trying to use some of it up with these letters. I haven't figured out, though, how to put it through the copier without repeating the heading. Sorry for the mess at the bottom of the page. I guess I should have covered up the heading on the original. Better luck next time.

Sales of "Under the Grey Gull's Wing" are still going on and will pick up in December. I'm already to send to the Museum $1,400 as soon as I know it is independent of Monhegan Associates, Inc. A collection of my letters from 1963 to 1977 has been taken to the printers and the gallies should be ready for proof-reading next week, at least in part. What shall I support this time? My solar system, perhaps. I can hardly wait to get my ideas put into bricks, and mortar - better, shingles and glass. Purchase of the land went off without a hitch. Now to see something on it. There will be plenty of room for Phoebe. The lot goes back nearly 300 feet to a row of trees. Most of the lot is clear except for a few small weed trees and one which looks like a cherry. I think I can send flash signals to Bob and Margaret Baker whose house is in Newcastle on the other side of the river.

Eleanor Bickford, who owns the Nobleboro house, had a mechanical garage door put in, which is a great joy. No more stepping out into snow, ice, or puddles and straining to open that huge garage door. And no more fumbling for the key in a dark entry. The light stays on for five minutes, time enough to get inside the house. There are ducks on the pond now and Phoebe's three canine friends have come around to play, along with the goat.

Tuesday we'll be off to Concord en route to the Syracuse area to be with the Jacksons for Thanksgiving. Mary Alice and Earl hope to join the family and bring young Phillip, my great grand nephew #3 over from Rochester for the day at least. I think Tommy Ashley, David and Elizabeth Gardner will be going out, too. I don't know how many Jackson offspring will be home other than Amy. It will be a crowd around the table, anyway and should be great fun. I'm the matriarch of the family!

HAPPY THANKSGIVING!! (probably already passed)
Alta Ashley and Phoebe

Nov. '84

Thanksgiving is over, Phoebe and I are back on the island, no longer landowners on Monhegan. Papers were passed on the Tuesday before Thanksgiving, but the money didn't reach the bank until the following Monday, and I had written a sizable check against a practically non-existant account! Fortunately it hadn't been cashed. I now am a life tenant in my side of the Mooring complex I have been calling High Mooring (once dubbed the Pill Box). It seems a bit strange not to own any Monhegan property.

But I do have my 1 1/4 acre in Damariscotta, have house plans, a builder and am ready to go. I saw just the house I want to build in Waldoboro yesterday - an envelope house which has traditional New England farmhouse lines.

Elizabeth, David, Tommy and I drove out to the Jacksons last Wednesday, Libby, Amy and Stephen were back and Susan flew in late Thanksgiving from Boca Raton, having missed her plane because of the floods, etc. Mary Alice, Earl and young Phillip Westerlund came over from Rochester so there were 17 at table and one on the floor with a bottle. What a good baby, he hardly fussed at all even though he was handed around from one admiring aunt, uncle or cousin to another for three days. At less than four months he almost filled out the 18 month sized jacket I had bought him.

On Sunday we went over to Weston to see the Johns, other cousins, where Elizabeth John was on from Milwaukee. Theirs is a lovely home and their three children delightful. It was their wedding I attended in the 70's in New Canaan.

The Balmy is still on the run, a good thing, too, because she was loaded with freight and passengers - but no David Sleeper today. June Day came back with Sonny, Robbie is now living here with Bill Sweetland. Doug and Harry's house is still not done. Harry Bone is back on vacation. Bill Baker is building a greenhouse on the old Cundy place. Don is living with Patsy in the Cundy cottage. Piles of traps with toggles and warp are everywhere. Another five weeks and they'll be under water. Again.

Love,
Alta

SEASON'S GREETINGS 1984

On this December Day a grey gull
flies o'er head to the tall tower on Lighthouse Hill -
Monhegan's link with you throughout the world.
the whirring of her wings joins angels songs
Which bring the seasons' messages of hope and love
To each of you this day and for the coming year.

Dear Friends and folks:

It didn't seem possible that 1984 could ever measure up to 1983 in excitement and fulfillment, but it did, and even outdid it. The year started actually with a picture postcard Christmas spent in Naugatuck, Ct, at the homes of Bob and Scott Sutherlands. A local storm had frosted everything which sparkled as the sun emerged on our arrival. On my return to the island I stopped off in Nobleboro to see the house I had thought of "house-sitting" in for the winter, met Eleanor Bickford, the owner and found it was a delightful rustic cottage on the edge of Pemaquid Lake, two tenths of a mile through the woods from Route One. Ideal! There were three dogs and a goat next door for Phoebe to play with, a cellar full of cut wood, and a big black kitchen wood stove for auxiliary heat **and** a piano. We moved in the day after Trap Day, Jan. 3rd.

In short order I got busy revising the manuscript of Monhegan essays which had made its round of publishers, found that Lincoln County Press would print it. So all spring pages streamed to the press. On May 12th the first 100 volumes were delivered and off we went to deliver them to bookstores from Brunswick to Camden. Every store was interested and it wasn't long before that first box was exhausted.

But that wasn't the only thing exhausted. I felt strangely weary and thought sleeping in my own bed would cure everything so went out to the island only to be brought back by Coast Guard ambulance to Pen Bay with one of my recurrent bouts of pneumonia, this for once not dubbed Tb.

It took much of the summer to regain vitality. But the book sale went like magic. The first 1,000 copies were gone by July and half the second printing has been sold. This is great because I decided to donate a dollar from the sale of each book to the Monhegan Museum and have sent off a check for $1,500 Meanwhile Lincoln County is working on printing a collection of my letters from the island written from 1963 through 1976. The present book is "Under the Grey Gull's Wing" so I formed the Grey Gull Publications under which all my writings from now on will be printed, including my letters, the Monhegan brochure and anything else, it simplifies bookkeeping and IRS records.

Feb. '85
RR 1 Box 52
Nobleboro, ME 04555

Dear Friends:

I'm down with another cold! Just what is going on? Last Tuesday Dr. Williams pronounced me "disgustingly healthy" and here I am five days later barely able to squeak to the McLoons that I could neither sing in choir today nor attend the rehearsal of the Sheepscot Valley chorus tomorrow night. I wanted so to be in church today because Bobbie Ives was the preacher in Ed Berry's absence and the chorus is working on Brahm's Liebeslieder Walzer (sung in German) and Schubert's Mass in C to be sung in Latin.

Added to the cold is the annoyance of acne-like rash on my face - I never had it when a teenager. Well, it could be worse and "this, too, will pass".

It is **spring**! The temperature has climbed above freezing for the first time in weeks making it possible to open doors and windows to air out the house.

Kathy Harris from Lincoln County News has been here this afternoon and has just left with my manuscript and scads of photographs under her arm and a promise to get the presses rolling by the beginning of next week. We have decided on "Time Before the boat" as the title. This time I'll have an ISBN, which may help in distribution of the book to wholesalers. The other day I received a request from Texas for "Under The Grey Gull's Wing". If sales of that book pick up in the summer I may have to go into a third printing.

David Sleeper and I finally got out to the island a couple of weeks ago and found that pipes had parted in the cellar so we had no running water. David managed to hookup a hose to the pump and filled two barrels for our use. I also found that the hot water tank was full so we had a supply of water handy to the kitchen and the john, no shower water, however, so we returned to the mainland on Friday in the ranks of the great unwashed. Oh how good a shower felt!

It took David all five days to finish up work on the photo/voltaic system. We had planned to leave on Wednesday, then on Thursday on a special trip of the Laura B, which didn't pan out anyway, and David was still running from cellar to attic on Friday morning doing last minute adjustments. But now the system really works well and the inverter is large enough to run the vacuum cleaner, and maybe the washing machine but probably not the dishwasher. What a relief. Now all that is needed are gadgets to step down the current from 24 to 12 volts for the TV's and a line to the generator from the cottage so tenants can turn on the generator, if necessary. It has taken a lot of time but it is worth it.

Danny Bates is back into the raising of lambs. He brought out a huge ram with great big horns, according to Rita White, early in January and a month later out came a ewe. She had twins shortly after in one of the coldest nights of the year. Only one lamb survived. Danny is living in the Bates cottage during the winter with his new girl friend whose name escapes me for the moment.

Apparently I took all my extra copies of my last letter back to the island so I can't look back to see what I wrote then. Sorry for any repetitions. Pam Krause didn't have just an ordinary cold at Christmas. She has been hospitalized and is now under chemotherapy. We wish her well.

The day we went to the island was Doug's birthday. Islanders had planned a birthday party for him so we merged my plans with theirs. David and I were at the Odoms' for supper and just at dessert time about twenty people showed up at the door, mainly the younger set, don't ask me why. Maybe it was too cold for the oldies. It was near zero and the snow squeaked under our feet as we made our ways homeward in bright moonlight. Unfortunately, someone left a permanent momento of the occasion in the shape of a cigarette burn in the wall-to-wall carpet in the front room.

Paul Horn and Bob Boody have been working much of the winter on the foundation of the Stineford house. Bob had bought a new back hoe in the late fall, expecting to use it on that job, but the thing had to be sent back once and then languished on the Port Clyde dock until last week, waiting for suitable weather and tides. Paul and Bill Baker have shingled the roof of Three Chimneys (the Nelsons' main house). When Harry T. Bone was home he finished up work on his shed turned apartment. He raised it so there is now head room in the cellar.

More P/V work is in the offing. Lexi and Sarah are putting in a 24 volt system and Alfred and Sherm Stanley are expanding theirs. After all he has said about one's being crazy to work on an island I believe David Sleeper will be doing those jobs.

Jeanne Rollins had a trip to Texas "between cold snaps", with a side trip into Oklahoma where she saw buffalo and prairie dogs. She apparently had a great time.

Lance Burton is back from three weeks in the USSR, which he

enjoyed immensely. He was there with a group from Colby College where he is a student. Mary jokingly said she hoped he would be at the airport when they went to get him, and not having defected!

Work on my new house will start as soon as test holes have been made and a building permit issued. It will be a Cape with a greenhouse on the south side. A guest suite will be on the ground floor, opening directly to the west. I think it will be quite attractive **and** comfortable.

It is now Tuesday and I must rush to Rockland to have this run off so I can beat the new postal rates and the big storm predicted for this evening. We have had little snow but it has been so cold, until yesterday's 40's, that nearly every flake which has fallen is still with us. Phoebe has covered the back yard with tunnelled pathways to her favorite haunts. My next letter will probably be written on the island when I go over for town meeting in the middle of March.

HAPPY VALENTINE'S DAY

Feb. 12, 1985

R.R. 1, Box 52
Nobleboro, ME 04555
March 1985

Dear Friends:

Phoebe just yelped at the door and when I opened it what should stare me in the face but a couple of golden eyes and two pointed horns belonging to the goat next door. Underneath was a trembling golden brown form with a plumy tail belonging to Phoebe. The goat tried to get in the door and hung around for quite a while, no matter what I did. Even Phoebe recovered her bravery after coming indoors and dashed out to challenge the creature, all to no avail. The goat has tired of his game and is down by the shore with his canine palls (sans Phoebe).

My life has taken a sudden turn. A week ago the local pediatrician was killed in a car accident. After hearing how much she was needed I thought it only right to volunteer to cover her office until a replacement could be found. So, after a week of filling out forms, meeting of medical staffs, etc. I am to take up the reins, or better, get back into harness tomorrow (3/5) with a somewhat trembling heart. I hope I can manage a trip to the island for Town Meeting next week so I can report that and run this letter through the copier.

All this has come about just as I am up to my ears in other activities. I have joined the Sheepscot Valley Chorus and am playing Mrs. Chauvanet in "Harvey!" **And** the first proof of my new book-"Time Before the Boat" is due to arrive at my door this afternoon. It's a bit much for someone two months short of eighty years! The other day I called on George Dow, Nobleboro's local historian, and found that this house is on land which once belonged to the Kuenzels (Kinsel), grandparents or grand uncles of my grandmother Emma Hussey Ashley. They came to this country in 1753 and settled on the grant from the King of England to the King of Prussia to settle a debt. The land ran from the Medomak River west. Grandma used to attend the German Church in Bremen, Maine.

Before I forget it, the Monhegan segment of "House for All Seasons" has been preempted by fund-raising, but will be aired on March 30 at 10 a.m. There is also a program at 2 p.m., according to PULSE, so you Mainers and viewers of Channel 10 had better check on both times. After my effort to add hand-written note to about 100 letters! Must wash the dishes and get ready for church.

The high winds we have had for the past two weeks or more have played havoc with the schedule of the Laura B and have nearly covered the grounds and roadway into the cottage with pine cones. I took three shopping bags full to the wreath committee of Pen Bay Medical Auxilliary the other day and have many more stashed in the entryway, and yet haven't made a dent in the number on the ground. Perhaps I should gather some to dip in chemicals for fire-starters as hostess gifts or even Christmas presents. But where to store all I've gathered for the new house as well as those cones?

This coming Thursday we are having a retirement, surprise party for Glenda Truhell, one of the old faithful of the choir. It is to be here at this house right after choir rehearsal, which also will be here. I hope I won't be called out for a home or hospital visit, and that the piano isn't too much out of tune. Also that I'll remember to bring over some chairs from the church; and too, that the predicted snows won't be too thick or too slippery.

It is now March 13th. Monhegan Plantation Meeting took place on the 11th without my presence but after treating children with colds, bronchitis, and red ears I succumbed to their illnesses and have been house-bound since Sunday evening. This is the third time since Christmas and I'm tired of it. There is so much to do rather than lying on the couch wrapped in a blanket.

I've been working on the dummy of the new book. The cover is going to be great - two of my pictures of the harbor taken the afternoon of a trap day, showing all the boats out. It was very calm so the reflections are very clear. There will be about 300 pages, twenty or so pictures. I'm using the picture you took of me, Jim at Thanksgiving in the "about the author" segment. The price will be $8.95, retail.

The party for Glenda came off well and was a complete surprise to her. She thought we were trying to get her back in the choir for Easter! One of the girls wrote a clever poem, which I read to the group. We chipped in a had a G-clef made out of leaded glass by Dorothy Conant to hang in the window and a cake with a staff with hens and roosters sitting on the bars as icing decorations.

Things went quite well in the office, but I was appalled at the amount of antibiotics which are used. Gone are the days when we pondered the relative good of pills vs. rest in bed and forced fluids. There is a pediatrician in Portland that may take over for a while to relieve me at least a day or two a week until a permanent replacement can be found. I've forgotten a lot, but am surprised how much I have recalled. but was I scared that first day! A nurse-assistant works in the office, too, which helps. On Thursday I saw fifteen patients, mostly with earaches. They say this area has the highest incidence of ear problems in the USA!

Jeanne Rollins reported the results of Town meeting. The appropriation is about $13,000 less than last year; assessors are Billie Boynton, Katy Boegel, and Winnie Murdock. Dave Boegel is constable, Shermie Stanley is harbor master and Danny Bates road commissioner. Amy Melenbacker replaced Peter Boehmer on the school board. It seems strange not to be a tax-payer any more.

Things are about set to start on my house. But as long as the money stays in the bank it is earning interest of around 10%, which helps. I'll probably have to use crates for chairs and mattresses on the floor until I can afford new furniture. Much of the stuff on the island will have to stay there. No Round-the-World trips for a while.

My nieces, Elizabeth Gardner and Priscilla Jackson, spent the presidents' weekend here. We attended an art show in Damariscotta and there saw the poster for "Harvey" tryouts. With their encouragement I tried out, got the part of Mrs. Chumly but changed to Mrs. Chauvenet because of rehearsal timings.

Jeanne Rollins is preparing to handle rentals and sales on the island. She is now studying for her exam which will be taken in April. Amy

Melenbacker says she will handle rentals, but she is not planning to obtain a realtor's license.

I now have specs for my P/V system and final house plans are in the mail. Digging will start as sooon as the back-hoe man returns from vacation. One more item: the Rothsteins have bought a house in Rockland and have been living there for the past few weeks. I don't know how much time they intend to spend on the island, but I don't want to open any office for medical care..

Don't forget a "House for All Seasons"
Happy St. Patrick's Day!

Tuesday, June 18, 1985
From the Courier-Gazette
Rockland, Maine
Marilis Hornidge's Book Bag

"Readers of the Courier-Gazette will need no introduction to **Alta Ashley**, correspondent from Monhegan and championess of alternate energy systems and the Monhegan phone company. Her book from last year, **Under the Grey Gull's Wing** which is a somewhat personalized modern history of the island she loves and lives on, has been joined by companion volume **Time Before the Boat**, letters about what it is like to live on an island ... not an isolated island but one which is in summer the home of many glamorous from-away-ers and, in the winter, a close by community which does in the winter what other Mainers do in the summer, namely lobster. It makes for a contrast of almost theatrical size (I have always wondered why someone didn't write at least a one-acter about life on Monhegan) and Dr. Ashley does it proud in this pair. If you've a friend who's always wanted to live on an island, buy these and a ticket on the Laura B and send them as a very welcome gift ... with the warning that all Maine islands aren't like Monhegan (with which Dr. Ashley would concur)."

Box 87
Monhegan, ME 04852
July, 1985

Dear Friends:

Here it is July and I am almost a month late with this letter. The first page of my May letter is missing, so I don't know how much I'll be repeating.

It took me until June 6th to clear out of the Nobleboro place and I left only then because Eleanor Bickford, the owner, came up to use the cottage. In many ways I hated leaving the lake and pine grove, but the progress on my house was such I felt I'd soon be able to stay there overnight, if not for longer during the summer.

I waited for the builder to tell me when to order rugs. He said the other day he wanted to clean up operations before the end of July and now I can't find anyone with time to lay linoleum or wall-to-wall carpeting I want in the bedrooms and on the stairs. Most people can't get the carpenters out fast enough! I shall have to go in again on Monday to see about things.

Monhegan is buzzing. The museum opened on Monday; the special exhibit for the summer is "Monhegan Scenes" a collection of painting of eighteen former artists, from Charles Ebert to Henry Kallem. Henry died here a week before the show was to open, thus his work could be included in the exhibit.

Last night the Plantation Gallery celebrated the tenth year of its existence with a special showing of works of Jacquelaine Hudson, Charlie Martin and Dyan Fitzpatrick. EVERYONE was there. The show will be up until Sunday. Dyan has certainly come a long way since she attended Cooper Union. Jackie had a variety of media, her latest lithograph is of Cathedral Woods. If it disappears from the wall they had better look in my house for it!

When the Kallems came over Henry was pretty feeble and I thought he wouldn't make it to his house, but he lived just about a week and the last day he was out riding around in his golf cart. His ashes are to be buried here later in the season when Anne and Michael return.

The Courts are here. I haven't seen Lee. Ruby is busy with the museum and Monhegan Associates. She took a spill after having breakfast at the Monhegan House last week and thought she had injured a rib, but was only badly bruised and now recovered. Victor is serving meals as "Mr. Vic's" an Eye-Opener from 6-8 a.m., followed by a complete breakfast with varied menu 8-10, and possibly lunch.

Winnie and John Murdock have opened "Sea Hag, Sea Food" and are offering take out food at noon, or one can eat at picnic tables on the premises. Not to be outdone, Sally Murdock is preparing gourmet picnic lunches in a basket complete with wine - a 48 hr. prior order, and North End Pizza is going great guns.

The Laura B is running three trips Fridays, Saturdays and Labor Day, the first leaving Port Clyde at 7:00 a.m, and Monhegan at 8:30. The Hardy Boat, which some dubbed the "Hardly Boat", has come out on occasion. Of course when I wanted to see the captain today he didn't make a trip.

Last week we had so much fog and bad weather reports the Balmy missed a trip, the Laura B didn't make one afternoon trip nor the one the next morning so there were a lot of confused persons wandering around.

Noel Farland is back with hat and staff. He came out in early June or maybe late May, but had to go back in for a spell and I'm not sure where he is living. I hope no one tells him I have a parlor organ. He said earlier that he intended to stay at Teco's but it now belongs to the Carmichaels, and is being extensively remodelled and Lilith Pincus is staying there for a while.

It is now nearly two weeks since I wrote the first page of this epistle, and much has happened in the meantime. I rushed in shore in a futile attempt to get vinal for the bathrooms and laundry, returning the next day because Tommy was to come over. He spent much of the week of the Fourth here and had a great time podging around by himself and painting scenes on rocks and selling them for packs of cigarettes or matches, talking to everyone who went by and sending them up to the Museum and Whitehead - taking over what Ade Mersfelder used to do . He left on Friday and then came Walter and Louise John with Walter, Louise and Stuart, and Carol and Arthur Vash. They were among the many cousins who came to my birthday party and apparently cooked up the trip at the time of the party. The Vashes had spent the first part of the weekend at their place in Denmark, ME and the Johns has joined them there. They stayed at the Inn overnight on Sunday and wined and dined me as I never have been - every meal at the Inn! We covered about half the island while they were here and they had a great time. Doug and Harry Odom and Jennie Noll came over here for cocktails.

The fog was so thick while they were here it wasn't until after supper on Sunday that they caught even a glimpse of Manana Island, a quarter of a mile away across the harbor. It did lift a bit when we went out to White Head, so they saw the cliffs, but not Matinicus or the other islands. Aggie had a great time, found every swamp on the way back through the Red Ribbon trail, so we had to dunk her in the Ice Pond.

On Monday Aggie and I went in with the John-Vash group. We had lunch at the East Wind. It was so late that the chef was out of almost everything we ordered, but what we got was good. Then, we all drove down to the new house, which the cousins admired greatly. I finally was able to get Floor Magic to promise to lay vinal this week and tile and carpet after the carpenters are through. Aggie and I stayed with the Bakers again, so Aggie had a chance to play with their Phoebe. There was so much to do I didn't get back until the late boat on Wednesday, forgetting entirely that I was supposed to be up at the Museum with Jennie Noll that afternoon.

That night we had a squall which capsized a fishing boat out of Portland dumping six people a few miles south of Monhegan. they had only one life jacket! Fortunately a trawler was close by and rescued the people and took them to Boothbay Harbor. Those on the island knew about the accident, learned on the morning news!

Marjorie Mitchener came over on Thursday and just went back today (Monday). We had a very relaxing time. She had a pulled tendon in her right ankle, so we did a minimum of walking. We did get to the Gummeres' and on the way back Aggie found the brook and came out more black than brown so she had another bath in the Ice Pond. We had dinner with Jackie and Julie Hudson one night and got caught in a downpour on the way home. Last night we had dinner (Sunday night buffet at the Inn (her treat). Jackie and Julie were there as was Doris Rich Bowers and her husband and son. That made four Girls' Latin School alumnae there (Julie, Doris, Marge and I). Had Anna Partridge not gone home Thursday there would have been five of us, all from that small school on this tiny island!

The new owner of this establishment started improving the back yard. He took down the deer fence around the vegetable garden and while I was in shore last week the deer came and practically demolished my tomatoes which were just setting fruit. I have salvaged some of the fence and have what is left of the plants surrounding also bits placed over my bush beans and lettuce, so maybe I'll have a bit of a crop. He also built a deck around the generator house so I can lunch there.

It gives a great view of the Meadow and the southern end of the harbor. Thank you, Tony.

My apologies to you, Betsy Halbrooks, for the following, I know you would like to receive at least one letter without its horror story, but here goes.

When the new owners were here for a few days early in the summer the generator kept going on every fifteen minutes or so. They finally found a large leak in their toilet, but not before their entire bathroom was flooded. For eighteen months I had tried to get town water into the cottage, without success. Shortly after the toilet was fixed and the new tenants arrived the pump failed! It didn't take long to hook the place up to the town. Even this part of the house was supplied until the pump motor was replaced. Then the drains stopped up and had to be reamed by a snake and now the toilet downstairs has sprung a leak so that water flows through it at a fast rate, and the pressure tank isn't holding its pressure. The generator went on three times to pump water last night, waking everyone. What next? My idea of selling was to be rid of all these problems!! "Tis better to suffer the slings and arrows----" perhaps is something I should have heeded. Let us hope the Damariscotta house will function a bit better.

The wind has been in the south for so long it looks as though the weather vane is stuck, but not really. Right now the fog is blowing in like smoke from a forest fire, and we have had no sunsets for two weeks. But all three boats bring out their quota of "trippers" each day. All I can think of is army ants as they make their way up the hill from the dock. Zimmie is busier than ever, now the Rockcrest Shop is no more. Winter Works is back in Billy Pynes's part of the Davis house where Rita lives. Richard Farrell has added a room to his studio so he has space for the pictures of other artists. Bryan Hitchcock has remodelled the Black Duck where Joe DeMartini had his studio for so long, and is offering photographic services. Barbara Rothstein has opened her shop Monhegan Cottage Industries again this summer and will be open until Labor Day. Their house is up for sale, however.

Early in June a military funeral was held up at the cemetery for Ray Orne following a service led by Bobby Ives at the church. A week later Philip Orne died; he had been too ill to come out for Ray's service. I don't know when there will be a service for him. Duty Hall came out last week and a service for Ruth Lewis Hall was held at the dock on Wednesday morning with about twenty-five persons attending the Quaker service. Duty went out in Sam Rothstein's skiff and scattered the ashes into the harbor. A stone will be placed in the cemetery later. I'm expecting the Page Marble people to come out any day with that stone, the one for Joe Walsh and the stones they took last year for repairs. They hope to finish the work or restoration in one day. Then it is up to the Plantation to see that the grass is kept cut and the fence repaired. There should be quite a bit of money in the cemetery account now. Monhegan telephone books, sold for the benefit of the fund, are still hot items. We are now well over seventy phones and more than ninety listings, with so many Lucy Stoners and POSSLQ's.

Let us hope I can get these letters mailed before I leave again to pick up furniture, etc. for the new house on Wednesday afternoon-after I sit at the Museum!

I wish I could send you all a whiff of rose perfume and spruce, but Aggie says Arf; she is adapting better and doesn't spend her days growling at all passers-by to and from the Museum.

As ever, Alta

P.S. Work is progressing on the new post office and the original roof on Bob and Mary Burtons' house is being raised. No lack of building on Monhegan. Even Zimmie has a new roof.

Let's hope things will go well at the White House.

P.O. Box 87
Monhegan, ME 04852
Aug. 1985

Dear Friends:

Before August passes completely out of the picture I hope I can finish this letter and get at least a few copies into the mail. Everytime I have tried to get at it there was something which stymied me. While I was back on the mainland the end of July David Sleeper had come out and enlarged the P/V system so that the cottage was put on the inverter, just as I had asked him not to do last winter. By doing that he rearranged power lines to this part of the house and, David-like, didn't leave explanation or diagram. So I put off writing until I could get back to my machine in the new house. When I tried to type, you guessed it, the electricians had been in but had left with the work only half done and no 110 AC power available for the typewriter there, either!

Well, I finally found that in order to power my machine here on the island I have to put on an AC light. Work on the new house should be all done by now so that when I go in on Friday things **should** be in working order. Jay LeGore, the contractor will be there to pull all lose threads together.

The summer has gone so fast my head is in a whirl. With all the typewriter troubles I didn't get two of my columns in and two of the papers didn't print a special article on the Libra-point showing Robert Semple held in Joanne Scott's (once Elsie Rowe's) cottage. That was quite an interesting show, although it was quite dominated by things Robert had done. Joanne had some interesting pieces she had done in stitchery - copies of watercolors she had done previously. One of roses and rocks was exquisite. Helga Houmere's family crest featuring an M, which was formed by interlocking tree trunks on a path in Cathedral Woods was most ingenious.

The island has had its usual water shortage, but it has been particularly bad this year because the big pump has broken down and the severe drop in the water table has put a strain on the lesser pump. Marion Wiley and all those living above me and close to the level of the water tanks have been particularly badly affected. Marion left for her Yarmouth home about the middle of the month. Bill and Frances Vaughn have also left, though possibly for other reasons, and Ruth Lothrop has returned to Reading until around Columbus Day.

Sally Murdock has been in the Ebert cottage - Gull Shores - all summer. She did catering and prepared gourmet picnic baskets all summer. Victor Lord's breakfasts have been very popular. In fact, the Periwinkle hasn't opened until 11 a.m. Sally has bought a house in Martinsville and will be moving off soon. Max Rosenthall is trying to sell his place and the Lewis house is up for sale.

The Carmichaels are delighted with the alterations to Teco's cottage which were done by Billie Boynton, Peter Dirks, Bill Baker and others. Billie Boynton and Jackie Boegel have transformed Ida Proper's old shed into an attractive place and have moved their framing business to it, as well as having Bruce Kornbluth's ceramics and paintings by others there on sale. They have dubbed it The Lupine Gallery.

Zoe Zanidakis was married to Ronnie (?who) a week ago Saturday by Marion Watson. The reception was at the Monhegan House. I had to go ashore to meet with the electricians, rug layers, et al, so missed the festivities. Also missed the gala opening of the Lupine Gallery and the annual meeting of the Library Association. Next year I hope

I can time my mainland visits a bit better.

Fishing has been very poor. There have been so many pogies (menhaden) around the fish completely eschew bait on a hook. The other day, when I was visiting the Bakers, we watched huge schools of pogies being chased by blue fish up the Damariscotta River. Several boats were right in the middle of the schools and didn't pull a single blue all the while we were watching. Thank goodness the pogies didn't suffocate, as they have done in Boothbay Harbor (Southport) and South Bristol. I remember the stench they made when they died in the Kenebec River in Hallowell one year.

The books are going fairly well. I've had to order another printing of Under the Grey Gull's Wing and hope I can make enough profit to **think** about getting a word processor and printer. Then will come the addresso-graph! I'll never be rich at this rate!!

The demand for island telephone books has kept up. I'm now printing them on blue paper which makes them easier to find when covered up on one's desk. The phone system is being powered by the big generator in the Meadow until Bob switches to his smaller one. They haven't yet considered putting in P/V.

The Rothsteins have sold their house and moved to Rockland "so the twins can get a good education". Since I gave most of my supplies to Dan I'm going to have to go into hiding. There are several people who have taken Red Cross courses and can get people onto a boat or helicopter if necessary. Back to the old days before either Danny or I were here.

Ray Orne died early in the summer and a week later Philip died. The service for Ray was most impressive with an honor guard from Brunswick Naval Air Station and two Coasties from Manana in dress uniform. A flag was presented to Elaine. There was a gun salute and taps (by recording). Fred Bolman died in mid-August and Velma Sutter, Ruth Ives' mother, died this week. All but Philip died of cancer, he from heart disease. On a happier note, Sarah Burnham says that Pam Krause is doing better. She has been having chemotherapy.

This morning I decided to try to use the dishwasher, the first time since it spilt all over downstairs last fall. both Paul Horne and Billie Boynton assured me there was no leak. Well, there is. At first I couldn't get it gong and tried both generators without success. Then I found the water had been turned off. Finally the thing began to function on "quick wash", but on the rinse cycle the water went all over the floor downstairs again. Let's hope things are in better condition in shore.

I left the place full of electricians and ruglayers when I came out a week ago.

This summer has been cut up so with trips in shore to check on the house, etc. I've missed many of you on the island. Next year I hope to stay here more. I'll be going in permanently by Oct. 18th - have a speaking engagement on the 19th and other meetings to attend.

As you see from the envelopes I have put these letters into my Grey Gull Publications account. I decided to put all my writings there, including the "Introduction to Monhegan" which was revised extensively this year. So many enterprises wanted to advertise this year a two-page insert had to be made. That means the cost to each one is greatly reduced. Also, more money has been collected from the boxes in which the booklets are displayed. I think for the first time they have met expenses.

Some of you have asked about the Belgian hare. In the report given at the Monhegan Associates meeting it was said that they have been sighted all over the wildlands areas. I haven't seen any but many have been caught in the have-a-heart traps. People concerned feel the problem is under control. More lyme ticks have been found, one on one of the island children before it had taken up residence, fortunately.

If I dont stop now I'll never get these letters run off and in today's (Aug. 28th) mail.

My address after October 19th will be:

HC 61 Box 069
Damariscotta, ME 04543

I am happy to report that Agatha and I are getting along quite well. She likes to go swimming in the Ice Pond each day, wakes me up in the morning with my shoe or some other treasure in her mouth through which she sings a complicated song, the meaning of which I have yet to learn.

Best wishes from Aggie and
Alta

HC Box 069
Damariscotta, ME 04543
Sept. 27, 1985

Dear Friends:

Here I sit in my new work room in Damariscotta wondering if and when Gloria will strike here or Monhegan. We are having heavy rains right now but there is still very little wind.

Right after I wrote the above Jay LeGore, the contractor, arrived to finish things up. He was here about four hours and got everything done except hitching up the fan which drives the heat from the sunroom to the rest of the house - so another day. He left to beat the storm home (Liberty).

The storm is now picking up and much rain has fallen. People are saying that I'll be the only one in the area with electric power. Let's hope the roof stays on! Later on I'll try to get through to Monhegan to see how things are over there. If the eye passes through Bridgton it means the coast and islands will bare the brunt of the storm. At noon Doug said little was happening, not even any heavy seas. He and Harry just returned last night to the island after taking Jenny home to Quincy.

The past two weeks have been a bit difficult for me. I met with David Sleeper and my present landlord on the 13th to see if we could iron out some of the difficulties with the new P/V system. I had written that I couldn't get the small generator to run and the big one to stop! He wrote me three letters, one set with return receipt requested, a plain mailing and one here in Damariscotta, the essence of which was that if I "continued with destructive and unreasonable" behavior he would instigate eviction proceedings. He accused me of playing a "shell game" and of serving only my own part of the house with the expanded P/V system, and so refused to pay his portion of the bill.

Well, the upshot was that **he** had left the battery charger on, which threw the inverter to "overcharge," requiring a trip to the cellar to reset it and also kept enough demand to keep the big generator going. David explained in great detail the workings of the system, absolving me of the accused behavior and putting things on C.'s shoulders. No more threat of eviction and a check was forthcoming.

While I was ashore the last time I dipped into capital and had, as I thought, a sizable check deposited to my account, so wrote six checks two of which were quite large and found the check to me resting comfortably in the Damariscotta P.O. and a notice to me that my account was overdrawn, with a fine of $12 per check!!! I think the bank is going to make a small adjustment, however.

Now for happier thoughts. I came over on Tuesday, hoping to beat Gloria and to be ready for the choir rehearsal which was to be held here Thursday evening. I brought over a load of furniture, too, so now have some bureaus in which to store things, some mirrors, and the marble-topped commode. The rug people brought the two orientals I had cleaned and had them laid shortly after noon. (They look wonderful, all bright and colorful again.) Bob and Margaret Baker were coming to dinner so I baked some pork chops, potatoes and squash, and made applesauce from my apples and set the table for **three**.

About 5:30 Sally Haines arrived, asking if the Bakers were here. I said they were coming and asked if she was coming to dinner, too. When she said "yes" I hastily set another place. Then several cars showed up and I about went into a collapse until I saw all the salads and casseroles. Then came two huge bunches of flowers and a beautiful light mahongany coffee table, of which I was very much in need. The whole thing was a terrific surprise. Elizabeth was in on it but never let on. Had it not been for the hurricane she would have been here, too. But she stayed home to watch an old oak. It survived but two others didn't.

We really did have a choir rehearsal and Marilyn pronounced the piano A-1. Margaret Baker engineered the party, but just what have I done to deserve such. Pat Dubord wrote a poem to me and had it framed! My wounded spirits from the P/V incident were now very well healed.

It is now Saturday. My house was the only one to have lights on last evening. I took Aggie for a flea dip and a bath yesterday and got home just as it began to really pour and blow. Things banged and whistled but the house was as steady as a rock. I thought the deck was bouncing up and down so went out, and nearly blew off, but the deck never moved. No trees went down on my lot but the sound of chain saws has resounded all day in the area.

I just took Aggie to the river for a walk. Nice, clean, fluffy Aggie found a huge mud puddle and lay down in it! She didn't go near the river because there were white caps on it. Oaks and popples took a beating, also some white pines. The road is covered with leaves, pine needles and tree branches. All the apples blew off my trees. An ornamental crab tree down the road is simply laden with bright fruit which miraculously stayed on the tree. Trees were just beginning to put on fall colors, which now cover the streets. But actually, it didn't look like the day after a hurricane when I went out this morning.

I called Doug Odom today to find out how the island fared. Steve Rollins lost his lobster car which he had hauled up on Middle (Swim) Beach, but not far enough. Jamie Wyeth's boat dragged its mooring and his boat rubbed on Vernon Burton's boat, causing some damage.

Alfred Stanley's car dragged its mooring, but no boats were lost and no other damage done. The only damage to Bob and Mary Burton's new addition was some slight damage to the deck, as usual. They are on their fall trip to the Cape and islands so probably got it worse there than on Monhegan. Elizabeth said that David is on the Vineyard and that East Falmouth took a beating, so Nancy's house may have been damaged. She hadn't been able to contact her or David.

Among the people leaving the island a few days early because of the storm were Buddy and Sylvia Alberts, Lulu and Ralph Hichman. Ina and Ray Sturgis will go in with Sherm on Sunday. Charlie and Natalie Johnson were here only a couple of weeks this year. They left last Monday. The two granddaughters, Dana and Leslie Allen, were to come over on Tuesday. If they did I wonder how well they weathered the storm in the Wik Wak. John Brandt down at Pemaquid said the waves were awesome, but Doug said they weren't as huge as expected on Monhegan. The harbor was in the lee of the SE winds. Even the roads stood up well in the downpour. Vi Sturner and Nancy had left the Log Cabin on Tuesday, otherwise they might have had quite a blow up there.

Since I last wrote we have had another wedding and a funeral. Lori Hogan and Bill Sweetland were married on the 14th. Each of their blond daughters dressed in raspberry satin served as flower girls and Bill's son Bobbie as assistant to Alfred Stanley as Best Man. I don't know who the maid of honor was. It was a colorful wedding, performed by Bobbie Ives up by the Museum on a gorgeous day. The reception was held in the McDonough (formerly Meissner) cottage where Bill and Lori have been living. Bill's mother was well enough to attend, which was nice.

HC 61 Box 069
Damariscotta, ME 04543
Nov. 21, 1985

Dear Friends:

Can it be that I haven't sent one of these letters since late August, the date on the last letter I have on my files here on the mainland?!

Here I sit in my new study typing on my new typing desk by the new desk brought over by Rockland Business Machines yesterday.

Maybe in a month or so I'll have all those scraps of paper, now in piles all over this study, in neat files.

We are now a family of three, Agatha (the golden), Christy (a tan tiger kitten) and I. Agatha and Christy are going through spasms of getting acquainted. All was fine until Christy tried to eat Aggie's milk bone a short while ago, now Christy eyes Agatha with much suspicion. But Aggie has taken over the job of bathing Christy, so all should be well soon.

It seemed for a while that I was doing nothing but sashaying from Monhegan to Damariscotta and back. I had to be here to see that last things were done, then back on the island to gather up clothing, records, treasures and what have you I wanted over here. I finally locked up the Monhegan place on October 17th so I could be ready for a talk I gave to the Nobleboro Historical Society about the island on October 19th. Woody and Betty Weinhagens' son Fred came over to the island on the 13th and had two days of rain, but so many parties to attend he and his friend Diane were quite impressed with Monhegan's social life. They were a tremendous help to me in getting all my cartons etc. to the boat and over here to the new house. Most things were too big for my car, but everything did fit into Fred's hatch-back. Doug and Harry have brought over things, too, in their station wagon. (One should never try to move with only a compact sedan!)

I seem to be on the lecture series and interview list. MBBN sent their team here last month for a story on my solar house. The reporter, Mr. Greenman, spent two hours here - no, it was over three hours - and finally said he hoped the network would let him use the material for a full program, instead of the seven minutes scheduled. Unfortunately it is to be (or maybe has been) aired on Channels 12 and 26, neither of which channels are received here or on the island. Mainland people with cable can get it, if and when. It may not be aired until mid-January.

Not to be outdone by TV the Portland Press Herald and Sunday Telegram feature writer, Farris, came out to do a story which will come out in the Telegram a week from this Sunday, December 1st. This week I spoke to a retiree club in Boothbay Harbor about the solar house. I'm sure no one there will be building one because none was under 70. Mr. Clifford, who was the superintendent of schools when Rita White was in school, livened the meeting up with his piano playing, despite his 93 years. He remembered Rita and sent her regards via me. We sang Rosie O'Grady and some other oldies, but not Pony Boy, but we did sing all verses of My Grandfather's Clock with great gusto **and** Let the

Lower Lights Be Burning, I suppose in honor of Monhegan.

The Pemaquid Chapter of the DAR had me talk on my visits to Russia on the 9th. I dashed over to Monhegan for over a boat to pick up my slides and couldn't find them, so came back with the slides from the visit in '68 and my photograph album. On my return Margaret Baker said she had found my '83 slides so all was well. Jackie Hudson came to Pemaquid to hear me. Julie was still in Princeton but is due back anytime now.

The Red House has tenants this winter. Four girls, Marion Chioffe, Colleen Faulkingham, Karen Beckler, and Amy Hirakis are staying there for the winter. Colleen is the new assistant teacher. Mary Beth Dolan is the full time teacher this year. There are nine children in school. That leaves only Ben Murdock as a preschooler. Heidi Boody is at Gould Academy, Becka Hitchcock is at Philllips Andover, Louisa Boehmer is spending the year in graduate study in Japan. She graduated from Bowdoin in June. Kathy Wincapaw is at Colby. Lance Burton will enter U. New Hampshire in February, transferring to their hotel management course from Colby. Tim Boehmer is at Procter.

First we had the problem of deer browsing, and still do; then came the Belgian hare invasion (humanly assisted) and now there is a muskrat population explosion. These animals have pretty well cleaned out the cat tails in the Ice Pond and have moved on to the Meadow and even along the swampy areas all the way from there to Lobster Cove. Everyone suspected has denied bringing in the creatures so their presence is probably a natural and sudden spurt in population from those who have been on the island for many years. Dint and the Stanleys remember hunting muskrats in their youth. The hare population seems to have stabilized. Larry Cooper and the Boehmers lost all their carrots and other things to the muskrats which they can't keep out of their gardens by fences. Larry has been using Ernest Lewis' garden while the Lewises have been off the island.

Lexi Krause gave his usual Hallowe'en party after the children had done their visiting. Martha and Richard entertained the children at a party at the Tribler Cottage at which Sarah McPherson's grandson, Robert Reiss, put on one of his magic shows, much to the delight of all ten children. Ben Murdock, in his clown costume, stole the show as to costumes.

The Athletic Club (volley ball players) held their Fourth Annual dinner and dance at the Monhegan House a couple of weeks ago. Zimmie was quite overcome by the get-ups the girls concocted and named

Jackie Boegel as the Belle of the Ball.

I forgot to say that the girls at the Red House gave a surprise baby shower for Zoe Zanidakis last week. She is now in shore awaiting the birth of her baby. She and her husband have been living on the island, instead of in Portland out of which port he has been fishing.

I stopped to see Diane and Bob Campbell after speaking to the Boothbay Club. They said that they had a boatload of people planning to spend the night on Monhegan a few weeks ago. It was a group of gourmet cooks and friends who loaded the Balmy Days with all sorts of casseroles and things which they had planned to eat on the island, then to spend the night at the Monhegan House and return the following day. But when they got out as far as White Island they encountered winds and waves so high they had to turn back and so they had their feast at Ocean Point, much to everyone's disappointment and relief.

According to the paper, Ricky Boehmer is to appear in a program here in Damariscotta on the 3rd. She will be giving readings interspersed in a musical program. She is also being written up for Yankee Magazine's section on foods. Also National Geographic will be doing an article on Trap Day to come out in Nov. or Dec. '87.

All the human robins have gone. The Bogdanoves are now in their house in Warren, Sally Murdock is settled in Martinsville. Larry Cooper finished painting the church and he and Anna are back in South Harpswell. Doug and Harry Odom stopped by yesterday, will go to Cleveland for the holidays and be back on the island early in December.

The end of last month I had intended going over to the Riddle farm in Center Tuftonboro, but the night before I was to go I received a call from Rue Riddle saying John had had a heart attack and was at Huggins Hospital, so I didn't go over until later. John had had a couple of periods of cardiac arrest on arrival at the hospital and also a severe reaction to medication for which he was transferred to Maine Medical in Portland. But a call from Rue today brought the good news that he is home and doing well. Rue came over here one day while John was in Portland; but they hope both of them can come over before they leave for San Fransisco early in December (not far away) so John can see how the house works. By the way, I finally had to resort to Central Maine Power last night after several days of clouds, rain and almost no sun. I've run my dishwasher, clothes washer and even cooked in my GE toaster oven entirely on solar energy. Last night I planned to run off this letter on the copier and charge up the batteries but a drop-in guest came just when I planned to do it so I used the charger for only half an hour or so.

So far I've used two cylinders of gas for cooking and hot water, as well as what was used by the workmen in setting up the system which they left with a small leak!

Before I got back here for good in October, Margaret Baker called to say that one of my solar room windows was cracked so it was not a great surprise when I found a pane of a triple glazed window had been shattered - really very pretty and looking like a thousand diamonds in the sun. It took quite a while to find out what had happened. It was a small stone hurled by a weed-eater by the people who mowed the lawn. They didn't know how to reach me. The replacement was finally put in two days ago, much to my relief. Anyone interested in a double glazed seven foot window? Probably I can make use of it as a cold frame.

Danny Bates arrived on the island last week with a new puppy which he had dubbed Lowly Slugs! What next?

More wedding bells are to ring on December 8th when Dougie Boynton and Alice Becker are to be married in New York. Danny Bates and Diane Pulsifer were married near Auburn in October. Bill Sweetland and Lauri Hogan were married on the island September 14th up at the Lighthouse with a reception at the McDonough's (the twin Jenney house where the Meissners and then the Sprouls lived). It was a gorgeous day and a lovely wedding. Of course the entire island, except those who couldn't manage the trip to the lighthouse, attended the ceremony conducted by Bobbie Ives. Paul Krause catered the reception. Too bad all weddings can't be held on the island, everyone has such a good time at them.

Billie Boynton and Jackie Boegel have transformed Ida Proper's former studio shed into the Lupine Gallery where they are conducting their frame shop and have art work on sale, including colorful painted wooden birds made in the Dominican Republic and sent up by Bruce Kornbluth where he and Jan are now living. Frances is there now on a grant, along with five other women artists.

Will and Jane Curtis have another book out "Return to These Hills" dealing with the years Calvin Coolidge spent in Vermont, with a foreword by John Coolidge who is married to one of my Holyoke classmates, the former Florence Trumble whose father was Governor of Connecticut.

For 76 years I've waited to see Haley's comet and just can't find it. Can you? I don't remember it back when I was five - too sleepy to keep my eyes open and now I can't wait another 76 years! Maybe later will come success.

And so with that I'll close with best wishes for a Happy Thanksgiving. Elizabeth, David and Tommy will be here.

As ever,
Alta

1985 SEASON'S GREETINGS 1986

Dear Friends:

Anyone who has moved within the year should have the thoughtfulness to send out Christmas greetings early with the new address to help the already over-burdened postal service. My intentions were of the best but here I am on the 19th of December just beginning this letter!

What a year this has been! January was spent working over house plans. In February, while looking over my lot in Damariscotta, I stepped into a test hole and ruptured my Achilles tendon and limped around for weeks until I turned the same ankle, heard a crack, and was able to walk, miraculously, without pain. The day I had the accident was the day Lonna Bunting, a local pediatrician met with a fatal car accident. After hearing the news I volunteered to cover her practice until a substitute could be found. After much paper work and meetings with staff members I took up the reins and worked out of her office for a month. Malpractice insurance was the stickler.

Just about then Phoebe Starling began to fail and died of esophageal paralysis on April 21st, leaving a terribly empty space in my life. I didn't realize how much that little dog meant and how many tears I would shed. Fortunately, I was busy being "Aunt Ethel" in Harvey with the Lincoln County Players. They gave me a premature 80th birthday party a week later at Bob Dunbar's barn in Nobleboro.

Work began on the house in late March, but really didn't get far until I left for Monhegan in June, which meant many trips back and forth to meet with workmen and the contractor. The house, an energy efficient solar affair, with photo-voltaics and hot water system, was ready to move into by the end of July, although I didn't leave the island permanently until mid-October. Since then I have been asked to give talks, interviews and even was taped for a TV show on solar energy for Maine Public Broadcasting, whose channel I can't receive - 12 and

26. I think it aired last week. The one taped on the island is to appear on House for All Seasons on PBS in January. The latest is a four-column spread in today's Lincoln County News. I don't dare show my face in town today!

My second book, which went to the printers in December of '84, finally was distributed to bookstores in early June. Sales have been rather slow, but the first book, "Under the Grey Gull's Wing," has gone into third printing. The second book, "Time Before the Boat," is a collection of my early "Dear Friends" letters. I'm now waiting for it to pay for itself before I work seriously on the next one, which will be another collection of essays on Monhegan.

The joys of summering on the island in my new status as life tenant in my part of the Mooring were somewhat marred by unpleasant encounters with my new land lord. Let us hope all problems have now been resolved. I am enjoying things here in Westview Acres so much I don't miss the island much but do miss the sound and sight of the ocean. However, salt water is less than a mile away, and Pemaquid, with Monhegan on the horizon, only a short drive down Bristol Road. It is good to be on the mainland in winter!

People have come to the house in ones, twos and droves. In late September the choir had its first rehearsal here, which turned out to be a surprise housewarming. Bob Baker had brought over a nice stack of firewood with which he lined one end of the garage. The choir brought a lovely handmade coffee table, as well as casseroles of food - I had thought only the Bakers were coming for dinner, not eighteen! My cousin Fred Weinhagen and his friend, Diane Maugham, came out to the island for a few days, then we all left with eleven cartons of my possessions, plus other furniture, which filled their station wagon to the roof. Fortunately for me they weren't driving a Volkswagen sedan! They christened my guest suite.

Elizabeth Ashley Gardner and Tommy Ashley (niece and nephew) were here for Thanksgiving. David didn't make it because of other commitments. Just this last week Jim and Priscilla Ashley Jackson were here for overnight and our first real snow storm. I accused them of bringing their snow from Syracuse. While they were here all but four of the people who went on the China-Russia trip in '83 made it here for a buffet in the Jackson's honor. I spent all the day before cooking and mixing up hors d'ouvres. The table still groaned with food after the last guest left, even though everyone left well-filled. I'm still working on the egg nog and two dips. Much of the rest went to Bobbie and

Ruth Ives and the people at their Carpenter's Workshop, a boat-building rehabilitation effort of the Ives for young people in need. They build skiffs and dories, many of which are bought by Monheganites.

All the while the house was a building people from Monhegan and the Rockland area, as well as those from around here, made many visits to the site to keep tabs on the progress of the work. I guess that is why I've had so many interviews and write-ups. If it weren't for my interest in energy saving I'd say "no," but I'm convinced that solar energy is the answer to future energy needs. So far my electric bill has been the minimum - $4 - and until the dark days of November I had used no Central Maine Power at all. From here on it should be an upward climb now the shortest day is here.

For 76 years I've looked forward to the revisit of Halley's comet, which I was too sleepy to see in 1910. Last night I think I finally saw it as a faint fuzzy ball to the SW of Jupiter. Let's hope it will be brighter come spring. What a disappointment it has been so far.

A few weeks after we buried Phoebe I acquired a golden retriever named Agatha. I took her on trial and then couldn't convince her former owner I couldn't handle her. Then a month ago I rescued a taffy tiger kitten from the gas chamber and of course dubbed him Christy. Now someone has turned up that wants Agatha so we are parting company next week. Later I hope to find a smaller and short haired dog in a pound somewhere. Aggie has been a very good watch dog, which I feel I need, but I just couldn't get used to her size, long hair, and proclivity to roll in smelly things.

I've saved the best 'til last - my 80th birthday which Elizabeth, Priscilla, David, Tommy and Cindy put on for me on May 5th at Elizabeth's house in Concord. All were asked to bring a silk flower so you can imagine the size of the bouquet the gifts of 64 guests made. My niece Susan Peters Kezer who works for a florist made a beautiful arrangement which is now in the sunroom of the new house after spending the summer at the local health care center while the house was being built. I thought the party was going to be for all of us who were born in the five's or decades, of which I think there are six in the family, and was quite overwhelmed when I found it was just for me. It was wonderful to see so many old and new friends and relatives - thirty-two of each! It is wonderful to be eighty!! But I think someone made a bad mistake in my curriculun vitae!!!

And so it is time for me to hope you had a wonderful Christmas with relatives and friends and that 1986 holds many blessings for you.

I'll be off to Concord to be with the Gardners a few days.

Please put my new address where you can find it.

HC 61 Box 069
Damariscotta, ME 04543

I'll be here at least through May and back in October.

Alta Ashley

Agatha & Christy

Damariscotta — Solar home

Solar cells — Monhegan

High Mooring — 1983